REHABILITATION OF CONCRETE STRUCTURES

Dr. B. VIDIVELLI
Professor, Structural Engineering Department
Annamalai University, Annamalai Nagar-608002

STANDARD PUBLISHERS DISTRIBUTORS
1705-B, NAI SARAK, POST BOX No.: 1066, DELHI-110006
Phones: 23262700, 23285798, Fax: 011-23243180
www.standardpublishers.com www.engineeringbooks.co.in email: stpub@vsnl.

Published by:
A.K. Jain
For Standard Publishers Distributors
1705-B, Nai Sarak, Delhi-110006.

First Edition	:	2007
Reprint	:	2009
Reprint	:	2011
Reprint	:	2013
Reprint	:	2014
Reprint	:	2015
Reprint	:	2016
Reprint	:	2017
Reprint	:	2018
Reprint	:	2019
Second Edition	:	2021
Reprint	:	2024

This book has been published in good faith that the material provided by the author is original. Every effort is made to ensure accuracy of material, but the publisher, printer and author will not be held responsible for any inadvertent error(s). Readers are advised to confirm any doubts with the standard books.

Price: Rs. 350-00

ISBN: 81-8014-110-1

Laser Typesetting by: Bhargave Laser Printers, Delhi.
Printed by: Prabhat Offset Press, Delhi.

PREFACE

The rapid progress and changes in the field of repair and rehabilitation of concrete structures in the recent past have motivated me to write this text. This text would suit the needs of undergraduate and postgraduate courses of various engineering colleges and universities. Most of the institutions have introduced separate courses as the repair and rehabilitation of concrete structures. The main purpose of this text is to provide sufficient information on the state of art relating to repair and rehabilitation of concrete structures. This field has become a major branch of Civil and Structural Engineering and is getting a momentum everyday. Thus to keep with the developments, I have tried to consolidate in a concise form, which would be useful for students preparing for their degree, diploma and competitive examinations.

This book has divided into five parts, and each part consists of three chapters, total of fifteen chapters. The five parts consists of durability and deterioration, damage assessment, repair materials, repair and rehabilitation, maintenance and demolition respectively. At the end of each part objective type questions, short answer questions and theory questions have been provided to enhance the knowledge of the students. Hope this book will be useful to students, faculty members, and practicing engineers.

As most of the information has been derived from many sources including those are specially mentioned in the bibliography and wish to express my indebtedness to these individuals and publications contributing the information, acknowledgements and thanks are recorded to the Bureau of Indian Standards to quote certain regulations, efforts have been made to avoid errors and mistakes. I will be grateful if mistakes or errors, if any, are pointed out and will also welcome suggestions for improvement. I express my sincere thanks to the proprietor of M/s Standard Publishers Distributors to bring out this book.

Special thanks goes to Vice-Chancellor and Registrar, Annamalai University, and Dr. C. Antony Jeyasehar, Head of the Department of Structural Engineering, Annamalai University for their encouragement and motivation.

Finally, I am grateful to my husband and children for their forbearance and patience in putting up with the inconvenience when I was engaged in the preparation of this text.

Mahavir Jayanti

Dr. B. VIDIVELLI

CONTENTS

PART V: MAINTENANCE AND DEMOLITION

PART I
DURABILITY AND DETERIORATION

CHAPTER 1
Physical Causes

1.1 INTRODUCTION

Cement concrete is the most extensively used material for construction of different types of structures/components such as buildings, bridges and shell roofs and also for precast products such as pipes, poles, sleepers, etc. in the past and even at the present times, too much emphasis is placed on concrete compressive strength rather than on environmental factors, which are known to affect concrete durability. This is one of the main reasons for serious deterioration of concrete structures that is prevalent today. Maintenance and repair of constructed facilities/infrastructures is presently the most significant challenge facing the concrete industry in the country, the issue of durability has replaced concerns about strength as the most pressing problem of the day.

1.2 DURABILITY OF CONCRETE

A concrete is said to be durable if it withstands the conditions for which it has been designed, without deterioration, over a period of years. The term durability of concrete is used to characterize in broad terms the resistance of a concrete to a variety of physical or chemical attacks due to the external causes or by the internal causes. The External causes may be due to weathering, occurrence of extreme temperature, abrasion, electrolytic action and attack by natural or industrial liquids and gases. The Internal causes include the alkali aggregate reaction, volume change due to the differences in thermal properties of aggregate and cement paste and above all the permeability of concrete. Environmental penetrations which affect the durability of the concrete are shown in figure 1.1.

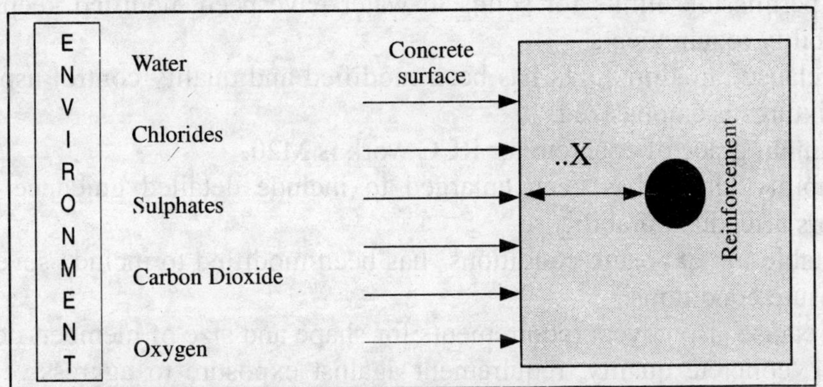

Figure 1.1. Environmental penetrations into concrete

In general, the following factors influence the durability of the concrete:

(a) Water (w/c ratio & water content)

(b) Curing period
(c) Cover to reinforcement
(d) Cement content & its properties
(e) Aggregates
(f) Mix design
(g) Workability
(h) Admixtures
(i) Thermal incompatibility of concrete constituents
(j) Transition zone between aggregate & cement matrix
(k) Environmental interaction

Each of these factors jointly or in isolation controls the susceptibility to deterioration mechanism in concrete. Except cover, all the other factors mentioned above control the pore structure of the concrete, which are directly related to the permeation properties. **Therefore measurement of permeation properties of concrete would result in assessment of durability of concrete. Permeability determines:**

(a) Saturation of concrete with water
(b) Ingress of moisture and air
(c) Corrosion of steel
(d) Cracking and spelling of cover concrete
(e) Water – tightness of the structure
(f) Thermal insulation properties

1.2.1 Changes in the draft IS 456-2000 in terms of durability

Some significant change to improve durability of concrete structures proposed in the draft IS 456-2000 are as follows:

(a) All three grades of OPC, 33 grade, 43 grade, 53 grade and sulphate resisting Portland cement have been included.
(b) The permissible limits for solids in water have been modified keeping in view the durability requirements.
(c) The clause on admixtures has been modified and quality control aspect while using admixtures is emphasized.
(d) Minimum grade of concrete for RCC work is M20.
(e) Durability clause has been enlarged to include detailed guidance concerning the factors affecting durability.
(f) The table of 'Exposure conditions' has been modified to include severe and extreme exposure conditions.
(g) This clause also covers requirements for shape and size of member, depth of concrete cover, concrete quality, requirement against exposure to aggressive chemicals and sulphate attack, minimum cement requirement and maximum w/c ratio, limits of chloride content, alkali silica reaction, and importance of compaction, finishing and curing.
(h) Clause on quality assurance measures has been incorporated.

(i) Cover to reinforcement has been specified based on durability requirement for different exposure conditions.

1.2.2 Protective measures may be established for durable design

(a) The selected structural form
(b) The concrete composition including special additions or admixtures
(c) The reinforcement detailing including concrete cover
(d) A special skin quality concrete including skin reinforcement
(e) Limiting or avoiding crack development and limiting crack widths
(f) Additional protective measures such as coatings
(g) Specified inspection and maintenance procedures including monitoring procedures
(h) Special active protective measures such as cathodic protection or monitoring by way of sensors

1.3 CAUSES OF DISTRESS IN CONCRETE STRUCTURES

Once the evaluation phase has been completed for a structure, the next step is to establish the cause or causes for the damage that has been detected. Since many of the symptoms may be caused by more than one mechanism acting upon the concrete, it is necessary to have an understanding of the basic underlying causes of damage and deterioration. This chapter presents information on the common causes of problems in concrete. Distress/deterioration in concrete structures may arise from a variety of causes as shown in figure 1.2.

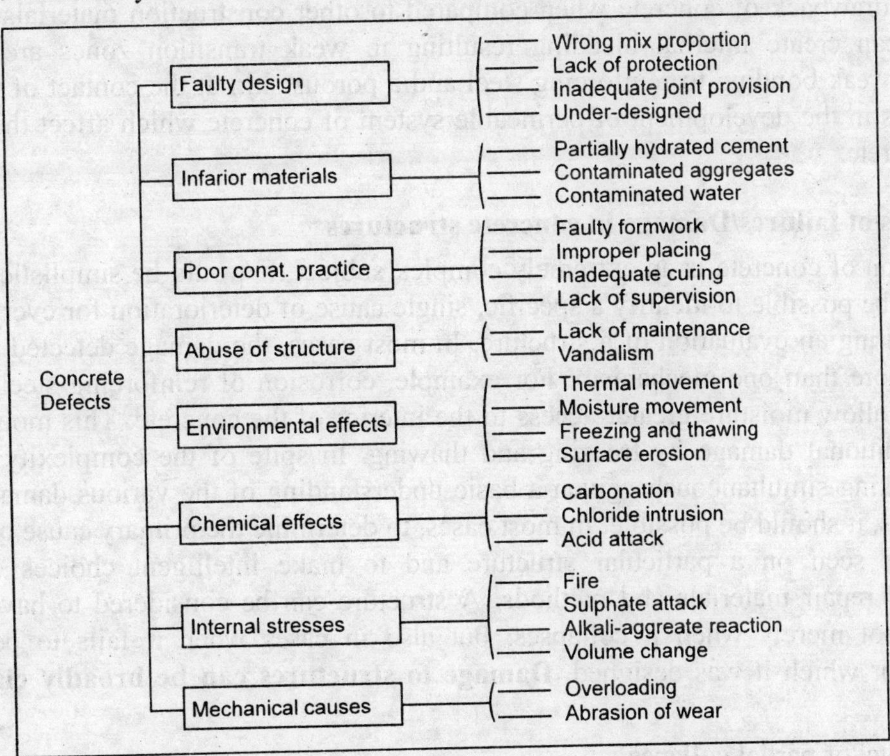

Figure 1.2. Factors causing distress in concrete structures

The following are the major causes of distress in concrete structures:

(a) Structural deficiency arising out of faulty design and detailing as well as wrong assumption in the loading criteria
(b) Structural deficiency due to defects in construction, use of inferior and substandard materials
(c) Damages caused due to fire, floods, earthquakes
(d) Physical deterioration and creep
(e) Chemical deterioration and marine environments
(f) Damages caused due to abrasion, wear and tear dampness
(g) Damages due to impact, vibration, fatigue
(h) Settlement of foundation, thermal expansion

Distress in concrete structures due to faulty design and/or deficiency in detailing and its effect on durability of concrete could be prevented through proper training and understanding of design concepts, detailing and adhering to codes of practice. Factors such as complication in geometric/structural form of the structure leading to difficult execution, congested reinforcement detailing, and difficult access for concrete to flow, increase the risks of inferior in-situ quality.

Deficiencies in construction practices are transportation, placing, finishing and curing of concrete affect the durability of concrete. A good concrete mix from a sound design can have its durability severely impaired by improper placement and curing. This is a particular noticeable drawback of concrete when compared to other construction materials. Excessive vibration can create internal bleeding resulting in weak transition zones around coarse aggregate, weak bonding to reinforcing steel and a porous skin at the contact of form work. This results in the development of permeable system of concrete which affect the durability of the concrete.

1.3.1 Types of failures/Damage in concrete structures

Deterioration of concrete is an extremely complex subject. It would be simplistic to suggest that it will be possible to identify a specific, single cause of deterioration for every symptom detected during an evaluation of a structure. In most cases, the damage detected will be the result of more than one mechanism. For example, corrosion of reinforcing steel may open cracks that allow moisture greater access to the interior of the concrete. This moisture could lead to additional damage by freezing and thawing. In spite of the complexity of several causes working simultaneously, given a basic understanding of the various damage causing mechanisms, it should be possible, in most cases, to determine the primary cause or causes of the damage seen on a particular structure and to make intelligent choices concerning selection of repair materials and methods. A structure can be considered to have failed or damaged, not merely when it collapses, but also in cases when it fails to perform the functions for which it was designed. **Damage to structures can be broadly classified as under:**

(a) A total or partial collapse
(b) Discolouration of materials

(c) Cracking (includes pattern cracks)
(d) Spelling of materials resulting in reduction in size of members
(e) Deformations such as deflection, buckling, twisting, and distortion

1.3.2 Common causes and deterioration of concrete

(a) Accidental Loadings
(b) Chemical Reactions
- Acid attack
- Aggressive-water attack
- Alkali-carbonate rock reaction
- Alkali-silica reaction
- Miscellaneous chemical attack
- Sulphate attack

(c) Construction Errors
(d) Corrosion of Embedded Metals
(e) Design Errors
- Inadequate structural design
- Poor design details

(f) Erosion
- Abrasion
- Cavitations

(g) Freezing and Thawing
(h) Settlement and Movement
(i) Shrinkage
- Plastic
- Drying

(j) Temperature Changes
- Internally generated
- Externally generated
- Fire

(k) Weathering

1.4 SHRINKAGE IN CONCRETE

Shrinkage is caused by the loss of moisture from concrete. It may be divided into two general categories: that which occurs before setting (plastic shrinkage) and that which occurs after setting (drying shrinkage). Shrinkage in concrete means moisture movement in concrete. Shrinkage may be defined as the volume changes in concrete due to loss of water or moisture due to evaporation or by hydration of cement or by carbonation. In practice, the shrinkage is simply measured as linear strain. Its units are thus mm/mm, expressed in 10^{-6} **Shrinkage can be further classified in to the following categories:**

(a) Plastic shrinkage (b) Drying shrinkage
(c) Autogenous shrinkage (d) Carbonation shrinkage

1.4.1 Plastic shrinkage

Mechanism:

During the period between placing and setting, most concrete will exhibit bleeding to some degree. Bleeding is the appearance of moisture on the surface of the concrete; it is caused by the settling of the heavier components of the mixture. Usually, the bleed water evaporates slowly from the concrete surface. If environmental conditions are such that evaporation is occurring faster than water is being supplied to the surface by bleeding, high tensile stresses can develop. These stresses can lead to the development of cracks on the concrete surface. Typically, the cracks are isolated rather than patterned. These cracks are generally wide and shallow. A typical plastic shrinkage cracks as shown in figure 1.3.

(a) The primary cause of plastic shrinkage cracks is the rapid evaporation of water from the surface of concrete.
(b) Plastic shrinkage cracks occur within few hours after placing concrete while it is still plastic and before it has attained any significant strength.
(c) There occur almost entirely on horizontal surfaces exposed to the atmosphere. It is seen that the cracks are parallel to one another and are spaced 0.3 m to 1.0 m apart.

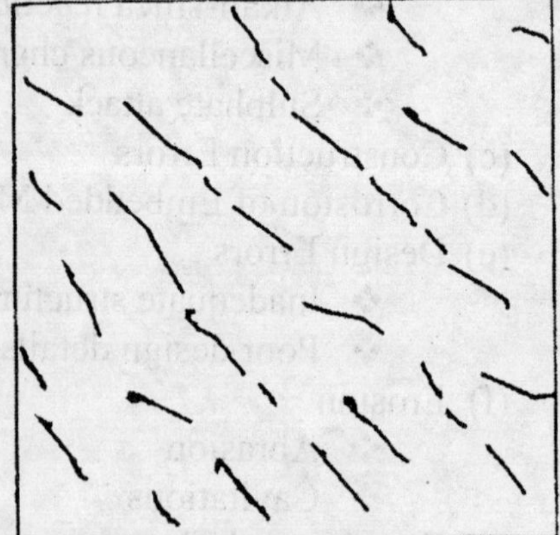

Figure 1.3. Plastic shrinkage cracks

(d) They can be deep and the width varying from 0.1mm to 3 mm.
(e) The magnitude of plastic shrinkage and plastic shrinkage cracks are depending upon ambient temperature, relative humidity and wind velocity.
(f) In other words, it depends up on the rate of evaporation of water (bleeding) from the surface of concrete. (1 kg / m^2 / hr)

The following measures could be taken to reduce or eliminate plastic shrinkage cracks:

(a) Moisten the sub-grade and formworks
(b) Erect temporary wind-breakers to reduce the wind velocity over concrete
(c) Erect temporary roof to protect green concrete from hot sun
(d) Reduce the time between placing and finishing. If there is delay cover the concrete with polyethylene sheets
(e) Minimise evaporation by covering concrete with burlap, fog spray and curing compounds.

Additionally, it will be beneficial to minimize the loss of moisture from the concrete surface between placing and finishing. Finally, curing should be started as soon as is practical. If cracking caused by plastic shrinkage does occur and if it is detected early enough, revibration and refinishing of the cracked area will resolve the immediate problem of the cracks.

Plastic settlement cracks and plastic cracks:

If there is any obstruction to uniform settlement by way of reinforcement or larger piece of aggregate, then it creates some voids or cracks. This is called **plastic settlement cracks.** A typical plastic settlement crack over reinforcing bar is as shown in figure 1.4.

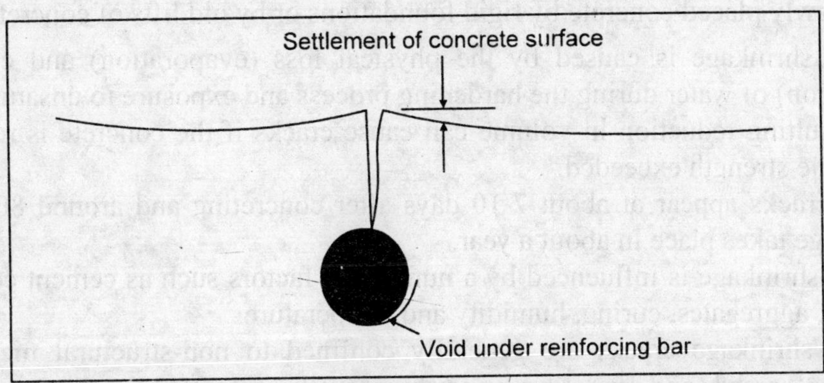

Figure 1.4. Typical plastic settlement cracks over reinforcing bar

(a) This is generally happens in a deep beam.

(b) These settlement cracks and voids are so severe it needs grouting operators to seal them off.

(c) Such cracks can be prevented by use of air–entraining admixtures or water reducing admixtures.

(d) Revibration, if possible is an effective step.

(e) **Plastic cracks** develop before the concrete has hardened (*i.e.*) between 1 to 8 hours of placement of concrete.

(f) As the evaporation of water takes place from the surface of concrete, it contracts resulting inducement of tensile stresses in concrete causing cracks.

(g) These cracks can be prevented by restricting the rate of evaporation of water from the surface to less than 0.5 kg/hr/m^2

1.4.2 Drying shrinkage

Mechanism:

Drying shrinkage is the long-term change in volume of concrete caused by the loss of moisture. If this shrinkage could take place without any restraint, there would be no damage to the concrete. However, the concrete in a structure is always subject to some degree of restraint by either the foundation, by another part of the structure, or by the difference in shrinkage between the concrete at the surface and that in the interior of a member. This restraint may also be attributed to purely physical conditions such as the placement of a footing on a rough foundation or to chemical bonding of new concrete to earlier placements or to both. The combination of shrinkage and restraints cause tensile stresses that can ultimately lead to cracking.

Symptoms:

Visual examination will typically show cracks that are characterized by their fineness and absence of any indication of movement. They are usually shallow, a few inches in depth. The crack pattern is typically orthogonal or blocky. This type of surface cracking should not be confused with thermally induced deep cracking which occurs when dimensional change is restrained in newly placed concrete by rigid foundations or by old lifts of concrete.

(a) Drying shrinkage is caused by the physical loss (evaporation) and chemical loss (hydration) of water during the hardening process and exposure to unsaturated air.

(b) The resulting reduction in volume can cause cracks if the concrete is restrained and its tensile strength exceeded.

(c) These cracks appear at about 7-10 days after concreting and around 80% of drying shrinkage takes place in about a year.

(d) Drying shrinkage is influenced by a number of factors such as cement content, water content, aggregates, curing, humidity and temperature.

(e) Drying shrinkage cracks are generally confined to non-structural members, floor toppings, screeds and rendering, and parapet walls.

(f) The total drying shrinkage is made up of irreversible shrinkage and reversible shrinkage. On initial drying out, an appreciable amount of the total shrinkage in irreversible, but after several cycles of wetting and drying, the shrinkage becomes entirely reversible.

Some of the preventive measures include:

(a) Use of minimum water content,

(b) Use of highest possible aggregate content,

(c) Providing adequate and early curing,

(d) Eliminating the external restraints as much as possible,

(e) Providing sufficient close spaced reinforcement.

Additionally, placing the concrete at as low a temperature as practical; dampening the subgrade and the forms; dampening aggregates if they are dry and absorptive; and providing an adequate amount of reinforcement to distribute and reduce the size of cracks that do occur. Restraint can be reduced by providing adequate contraction joints.

1.4.3 Autogenous shrinkage

(a) If no movement of water to or from the set paste of concrete is allowed, then the shrinkage developed is known as **autogenous shrinkage.**

(b) This shrinkage is caused by the loss of water consumed or used up in the hydration of cement.

(c) The magnitude of this shrinkage is very small and is not much of significance.

1.4.4 Carbonation shrinkage

(a) Carbonation is the reaction of carbon dioxide present in the atmosphere, with the hydrated cement minerals in the presence of moisture.

(b) The simultaneous reaction of CO_2 with hydrated cement minerals in concrete induces contraction of concrete, which is known as **carbonation shrinkage.**

1.5 FREEZE AND THAW ON CONCRETE

Freeze-thaw disintegration or deterioration takes place when the following conditions are present:

(a) Freezing and thawing temperature cycles within the concrete
(b) Porous concrete that absorbs water (water-filled pores and capillaries)

1.5.1 Mechanism of deterioration

As the temperature of a critically saturated concrete is lowered during cold weather, the freezable water held in the capillary pores of the cement paste and aggregates expands upon freezing. If subsequent thawing is followed by refreezing the concrete is further expanded, so that repeated cycles of freezing and thawing have a cumulative effect. By their very nature, concrete hydraulic structures are particularly vulnerable to freezing and thawing simply because there is ample opportunity for portions of these structures to become critically saturated. Concrete is especially vulnerable in areas of fluctuating water levels or under spraying conditions. Exposure in such areas as the tops of walls, piers, parapets, and slabs enhances the vulnerability of concrete to the harmful effects of repeated cycles of freezing and thawing. The use of deicing chemicals on concrete surfaces may also accelerate damage caused by freezing and thawing and may lead to pitting and scaling. It involves the development of osmotic and hydraulic pressures during freezing, principally in the paste, similar to ordinary frost action. The freezing and thawing mechanism on concrete is shown in figure 1.5.

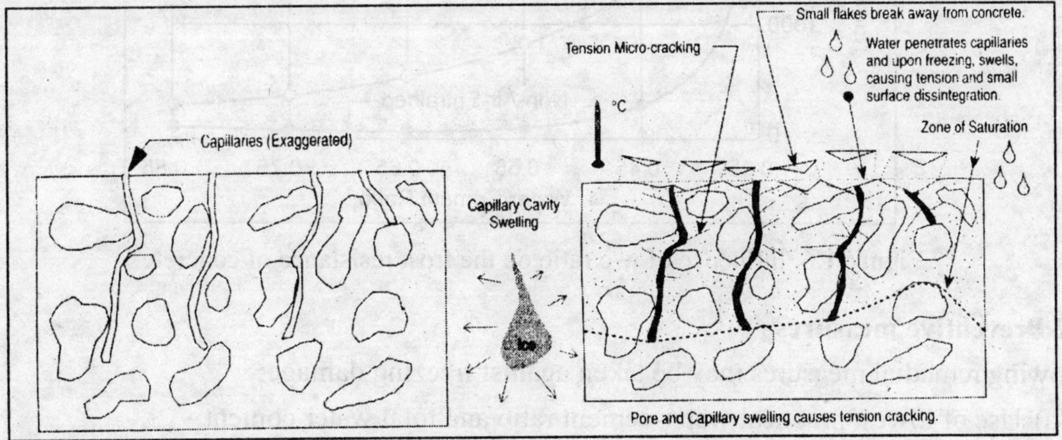

Figure 1.5. Mechanism of freeze and thawing action

(a) Freeze and thaw deterioration generally occurs on horizontal surfaces that are exposed to water, or on vertical surfaces that are at the water line in submerged portions of structures.
(b) The freezing water contained in the pore structure expands as it is converted into ice.

The expansion causes localized tension forces that fracture the surrounding concrete matrix.

(c) The first stage is the development of fine closely spaced cracks parallel to the edge of the exposed concrete.

(d) The cracks soon become filled with a dark deposit of calcium carbonate and are commonly called **D-cracks.**

(e) As the deterioration continues small pieces of concrete between the cracks separate from the body of concrete.

(f) The deterioration is reduced as the water cement ratio is reduced but the only positive way to prevent the problem is to protect concrete by the adequate air-void system.

(g) The superiority of air entrained concrete with respect to freezing and thawing action is shown in figure 1.6.

(h) Visual examination of concrete damaged by freezing and thawing may reveal symptoms ranging from surface scaling to extensive disintegration.

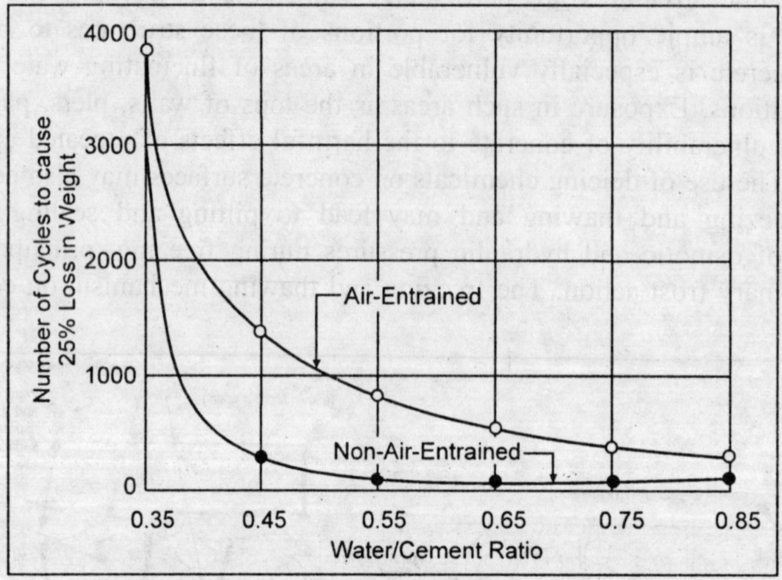

Figure 1.6. Influence of w/c ratio on the frost resistance of concrete

1.5.2 Preventive measures

Following remedial measures may be taken against freezing damage:

(a) Use of lowest practical water-cement ratio and total water content

(b) Adequate air entrainment has been found effective to control the freezing damage

(c) Use of durable aggregate also is useful to check the freezing effect

(d) Adequate curing of concrete prior to exposure to freezing conditions is also important.

(e) Designing the structure to minimize the exposure to moisture. For example, providing positive drainage rather than flat surfaces whenever possible.

1.5.3 Rate of freeze-thaw deterioration

(a) Increased porosity (increases rate)
(b) Increased moisture saturation (increases rate)
(c) Increased number of freeze - thaw cycles (increases rate)
(d) Air entrainment (reduces rate)
(e) Horizontal surfaces that trap standing water (increases rate)
(f) Aggregate with small capillary structure and high absorption (increases rate)

One type of cracking of concrete road and airfield surfaces called **D-cracking** is particularly linked to aggregate. In consists of development of fine cracks near free edges of slabs, but the initial cracking starts lower in the slab where moisture accumulates and the coarse aggregate becomes saturated to the critical level. With cyclic freezing and thawing, the mortar surrounding the aggregate fails leading to the D-cracking. Aggregates of sedimentary, calcareous or siliceous origin (gravel, crushed rock) lead to D-cracking due to frost action.

1.6 WEATHERING ON CONCRETE

Weathering is frequently referred to as a cause of concrete deterioration. It defines as Changes in color, texture, strength, chemical composition, or other properties of a natural or artificial material due to the action of the weather. The environmental factors that can cause cracking include: **Freezing and thawing, wetting and drying, & cooling and heating.**

(a) The damage from freezing and thawing is the most common weather related physical deterioration.
(b) Other weathering processes that may cause cracking in concrete are alternate wetting and drying, and heating and cooling.
(c) If the volume changes due to these processes are excessive, cracks may develop and give the impression that the concrete is on the verge of disintegration.
(d) The damage due to these factors may appear in the form of general flacking and spalling of concrete from the surface of concrete.
(e) Concrete generally loses strength with increase in temperature about 300'c, damage being greater with aggregate having high coefficient of thermal expansion.

1.7. CRAZING ON CONCRETE

(a) Crazing is the development of a network of fine random cracks on the surface of concrete or mortar caused by shrinkage and in usually related to finishing and curing procedures.
(b) Excessive floating and traveling bring water, cement and dust from the aggregates to the surface to produce a surface skin, which has a higher drying shrinkage than the underlying concrete.
(c) Spreading dry cement on a surface that in too wet to trowel, and sprinkling water on concrete that in too dry to trowel, both produce a skin likely to suffer from crazing.
(d) Use of highly absorptive aggregates batched in the dry state may contribute to crazing
(e) The use of over-wet concrete contributes to all the problems associated with drying shrinkage since shrinkage is almost directly proportional to the amount of mixing

water in concrete.

(f) Improper curing can cause crazing even if the concrete is properly batched, mixed, placed and finished.

(g) Use of unvented heater (Salamanders) during the curing period is also a cause of crazing.

1.7.1 Preventive measures

Some of the preventive measures include:

(a) Damping the subgrade before placing concrete,
(b) Avoiding over-finishing of the surface,
(c) Delaying trowelling until the surface moisture has disappeared,
(d) Avoiding sprinkling of dry cement or water on the surface during finishing operation,
(e) Starting curing as soon as possible,
(f) Avoiding the use of unvented heaters,
(g) Batching absorptive aggregates in a moist condition,
(h) Avoiding higher temperature differentials between the concrete and curing water.

1.8 HONEYCOMBING ON CONCRETE

Honeycomb consists of exposed pockets of coarse aggregates not covered by a surface layer of mortar. This may be caused by inadequate consolidation, presence of excess water in concrete or by leaky forms, which allow the mortar to escape.

1.8.1 Preventive measures

Some of the preventive measure includes:

(a) To follow good concrete construction practice
(b) To use workable concrete
(c) To provide good forms (tight)
(d) To give proper vibration (compaction)
(e) To be planned sequence of placing.

1.9 SWELLING ON CONCRETE

(a) Concrete used continuously in water from the time of casting exhibits a net increase in volume and its weight.
(b) This increase in volume due to continued hydration of cement is known as swelling.
(c) The swelling is due to the absorption of water by the cement gel.

1.10 POPOUTS ON CONCRETE

(a) A popout is a conical-shaped hole in the surface with a portion of coarse aggregate particle exposed.
(b) These occur outdoors on the horizontal and vertical surfaces.
(c) They are caused by freezing of water in aggregate particles that have an internal pore structure which causes them to expand unduly upon freezing
(d) Popouts do not appear during construction but they start to appear during the first

winter following construction and may continue to form for several years.

(e) These do not harm the concrete but they are unsightly.

(f) Rocks that have produced popouts include: Chert, Shale, Clay stone, Mudstone, Argillaceous limestone, and Siltstone.

(g) Popouts can be prevented only by avoiding aggregates which cause them.

1.11 CREEP ON CONCRETE

Concrete, brickwork and timber when subjected to sustained loads not only undergoes instaneous elastic deformation but also exhibit **a time-dependent deformation known as creep.** In concrete creep results in a sustained increase in elastic deformation, which sometimes leads to formation of cracks in brick masonry of framed and load bearing structures. Deferring the removal of centering and application of external load can reduce it. When the sustained load is removed, the strain decreases immediately by an amount equal to the elastic strain at the given age (Instaneous recovery). This instantaneous recovery is then followed by a gradual decrease in strain, called creep recovery, which is a part of total creep strain suffered by the concrete. If a loaded concrete specimen is viewed as being subjected to a constant strain, the creep decreases the stress progressively with time. This is called relaxation. While 80 to 85 percent shrinkage strains occur in 6 months, only about 75 percent of creep strains occur in 12 months.

1.11.1 Influencing factors

All the factors which influence shrinkage are influence creep includes:

(a) Types of aggregate,

(b) Type of cement and cement content,

(c) Admixture used,

(d) Entrained air,

(e) Mix proportions,

(f) Mixing time,

(g) Age of concrete,

(h) Level of sustained stress,

(i) Ambient humidity,

(j) Temperature and

(k) Size of the specimen

1.12 ABRASION, EROSION AND CAVITATIONS ON CONCRETE

(a) Abrasion refers to wearing away of the surface by friction.

(b) Erosion refers to wearing away of the surface by fluids.

(c) The cavitations refer to the damage due to non-linear flow of water at velocities more than 12 m/sec.

(d) The concrete used in the roads, floors, the pavements and the concrete used in the hydraulic structures should exhibit resistance against abrasion, erosion and cavitations.

(e) The more the compressive strength the higher is the resistance to abrasion, erosion and cavitations.

(f) Due to abrasion, dusting problem arises, which may be harmful in many situations, e.g., industrial floors having material handling system or robots.

(g) If the aggregate in concrete is wear resistant, it is the properties of the cement matrix that controls the abrasion resistance.

(h) Use of steel fibres in concrete matrix improves abrasion resistance of concrete and also polymer based systems when applied to concrete to improves the abrasion resistance.

(i) The shape and surface texture of aggregate plays an important part in the abrasion resistance of the concrete

(j) Epoxy screeding and polymer application to the surface is said to be effective against cavitations.

1.12.1 Abrasion-erosion

Abrasion damage caused by waterborne debris and the techniques used to repair the damage on several Corps' structures.

Mechanism:

Abrasion-erosion damage is caused by the action of debris rolling and grinding against a concrete surface. In hydraulic structures, the areas most likely to be damaged are spillway aprons, stilling basin slabs, and lock culverts and laterals. The sources of the debris include construction trash left in a structure, riprap brought back into a basin by eddy currents because of poor hydraulic design or asymmetrical discharge, and riprap r other debris thrown into a basin by the public. Also barges and towboats impacting or scraping on lock wells and guide wells can cause abrasions erosion damage.

Symptoms:

Concrete surfaces abraded by waterborne debris are generally smooth and may contain localized depressions. Most of the debris remaining in the structure will be spherical and smooth. Mechanical abrasion is usually characterized by long shallow grooves in the concrete surface and spelling along monolith joints. Armor plates is often torn away or bent.

Prevention:

The following measures should be followed to prevent or minimize abrasion-erosion damage to concrete hydraulic structures.

Design - It appears that given appropriate flow conditions in the presence of debris, all of the construction materials currently being used in hydraulic structures are to some degree susceptible to erosion. While improvements in materials should reduce the rate of concrete damage caused by erosion, this improvement alone will not solve the problem. Until the adverse hydraulic conditions that can cause abrasion-erosion damage are minimized or eliminated, it will be extremely difficult for any of the construction materials currently being used to avoid damage by erosion. Prior to construction or repair of major structures, hydraulic model studies of the structure may be required to identify potential causes of erosion damage and to evaluate the effectiveness of various modifications in eliminating those undesirable hydraulic conditions. Many older structures have spillways designed with a vertical end-sill. This design is usually efficient in trapping the erosion-causing debris within the spillway. In some structures, a 45-deg fillet installed on the upstream side of the end sill has resulted in a self cleaning stilling basin. Recessing monolith joints in lock walls and guide walls will minimize stilling basin spalling caused by barge impact and abrasion.

Operation - In existing structures, balanced flows should be maintained into basins by using all gates to avoid discharge conditions where eddy action is prevalent. Substantial discharges that can provide a good hydraulic jump without creating eddy action should be released periodically in an attempt to flush debris from the stilling basin. Guidance as to discharge and tail water relations required for flushing should be developed through model and prototype tests. Periodic inspections should be required to determine the presence of debris in the stilling basin and the extent of erosion. If the debris cannot be removed by flushing operations, the basin should be cleaned by other means.

1.12.2 Cavitations-erosion

Cavitations-erosion is the result of relatively complex flow characteristics of water over concrete surfaces.

Mechanism:

There is little evidence to show that water flowing over concrete surfaces at velocities less than 12.2 m/sec causes any cavitations damage to the concrete. However, when the flow is fast enough and where there is surface irregularity in the concrete, cavitations damage may occur. Whenever there is surface irregularity, the flowing water will separate from the concrete surface. In the area of separation from the concrete, vapor bubbles will develop because of the lowered vapor pressure in the region. As these bubbles are carried downstream, they will soon reach areas of normal pressure. These bubbles will collapse with an almost instantaneous reduction in volume. This collapse, or implosion, creates a shock wave which, upon reaching a concrete surface, induces very high stresses over a small area. The repeated collapse of vapor bubbles on or near the concrete surface will cause pitting. Concrete spillways and outlet works of many high dams have been severely damaged by cavitations.

Symptoms:

Concrete that has been damaged will be severely pitted and extremely rough. As the damage progresses, the roughness of the damaged area may induce additional cavitations.

Prevention:

The following measures should be followed to prevent or minimize cavitations -erosion damage to concrete hydraulic structures.

Hydraulic design - Even the strongest materials cannot withstand the forces of cavitations indefinitely. Therefore, proper hydraulic design and the use of aeration to reduce or eliminate the parameters that trigger cavitations are extremely important.

Construction practices - Construction practices are of paramount importance when concrete surfaces are exposed to high-velocity flow, particularly if aeration devices are not incorporated in the design. Such surfaces must be as smooth as can be obtained under practical conditions. More restrictive tolerances on surfaces should be avoided since they become highly expensive to construct and often impractical to achieve, despite the use of modern equipment and good construction practices. Where possible, transverse joints in concrete conduits or chutes should be minimized. These joints are generally in a location where the greatest problem exists in maintaining a continuously smooth hydraulic surface. One construction technique, which has proven satisfactory in placement of reasonably smooth hydraulic surfaces, is the traveling slip form screed. This technique can be applied to tunnel inverts and to spillway chute slabs. Since surface hardness improves cavitations resistance, proper curing of these surfaces is essential.

1.13 TEMPERATURE CHANGES

Changes in temperature cause a corresponding change in the volume of concrete. As was true for moisture-induced volume change, temperature-induced volume changes must be combined with restraint before damage can occur. Basically, there are three temperature change phenomena that may cause damage to concrete. First, there are the temperature changes that are generated internally by the heat of hydration of cement in large placements. Second, there are the temperature changes generated by variations in climatic conditions. Finally, there is a special case of externally generated temperature change-fire damage. Internally and externally generated temperature changes are discussed in subsequent paragraphs.

1.13.1 Internally generated temperature differences

Mechanism:

The hydration of Portland cement is an exothermic chemical reaction. In large volume placements, significant amounts of heat may be generated and the temperature of the concrete may be raised by more than 38 °C over the concrete temperature at placement. Usually, this temperature rise is not uniform throughout the mass of the concrete, and steep temperature gradients may develop. These temperature gradients give rise to a situation known as internal restraint-the outer portions of the concrete may be losing heat while the inner portions are gaining. If the differential is great, cracking may occur. Simultaneously with the development of this internal restraint condition, as the concrete mass begins to cool, a reduction in volume takes place. If the reduction in volume is prevented by external conditions (such as by chemical bonding, by mechanical interlock, or by piles or dowels extending into the concrete), the concrete is externally restrained. If the strains induced by the external restraint are great enough, cracking may occur. There is increasing evidence, particularly for rehabilitation work, that relatively minor temperature differences in thin, highly restrained overlays can lead to cracking. Such cracking has been seen repeatedly in lock wall resurfacing and in stilling basin overlays. Measured temperature differentials have typically been much below those normally associated with thermally induced cracking.

Symptoms:
Cracking resulting from internal restraint will be relatively shallow and isolated. Cracking resulting from external restraint will usually extend through the full section. Thermally induced cracking may be expected to be regularly spaced and perpendicular to the larger dimensions of the concrete.

Prevention:
In general, the following may be beneficial: using as low a cement content as possible; using a low-heat cement or combination of cement and pozzolans; placing the concrete at the minimum practical temperature; selecting aggregates with low moduli of elasticity and low coefficients of thermal expansion; cooling internally or insulating the placement as appropriate to minimizing temperature differentials; and minimizing the effects of stress concentrators that may instigate cracking.

1.13.2 Externally generated temperature differences

Mechanism:
The basic failure mechanism in this case is the same as that for internally generated temperature differences-the tensile strength of the concrete is exceeded. In this case the temperature change leading to the concrete volume change is caused by external factors, usually changing climatic conditions. This cause of deterioration is best described by the following examples:

First, a pavement slab cast in the summer. As the air and ground temperatures drop in the fall and winter, the slab may undergo a temperature drop of 27 °C, or more. Typical parameters for such a temperature drop (coefficient of thermal expansion of 10.8 × 10-6/°C) indicate a 30m slab would experience a shortening of more than 13 mm. If the slab were restrained, such movement would certainly lead to cracking.

Second, a foundation or retaining wall that is cast in the summer, as the weather cools, the concrete may cool at different rates-exposed concrete will cool faster than that insulated by soil or other backfill. The restraint provided by this differential cooling may lead to cracking if adequate contraction joints have not been provided.

Third, concrete that experience significant expansion during the warmer portions of the year. Spalling may occur if there are no adequate expansion joints. In severe cases, pavement slabs may be lifted out of alignment, resulting in so-called blowups.

Fourth, concretes that have been repaired or overlayed with materials that do not have the same coefficient of thermal expansion as the underlying material. Annual heating and cooling may lead to cracking or de-bonding of the two materials.

Symptoms:
Visual examination will show regularly spaced cracking in the case of restrained contraction. Similarly, spalling at expansion joints will be seen in the case of restrained expansion. Problems resulting from expansion-contraction caused by thermal differences will be seen as pattern cracking, individual cracking, or spalling.

Prevention:
The best prevention is obviously to make provision for the use of contraction and expansion joints. Providing reinforcing steel (temperature steel) will help to distribute cracks and minimize the size of those that do occur. Careful review of the properties of all repair materials will help to eliminate problems caused by temperature changes.

1.14 FIRE ON CONCRETE

A fire in a concrete structure causes damage. The extent of which depends upon the intensity and duration of the fire. **The principal types of damages are**:

(a) Reduction in strength of concrete
(b) Cracking and spalling of concrete
(c) Deflection and deformation of members
(d) Discolouration
(e) Other miscellaneous functional failures

Concrete is not a refractory material, but it has good fire resistant properties. **Fire resistance of concrete structure is determined by three main factors:**

(a) The capacity of concrete itself to withstand heat
(b) The conductivity of the concrete to heat
(c) The coefficient of thermal expansion of concrete

In the case of RC, the fire resistance is not only dependent upon the type of concrete but also the thickness of cover to reinforcement. The fire introduces high temperature gradients and as a result of it, the surface layers tend to separate and spall off from the cooler interior. The heating of reinforcement aggregates the expansion both laterally and longitudinally of the reinforcement bars resulting in loss of bond and loss of strength of reinforcement. The effect of increase in temperature on the strength of concrete is not much up to a temperature of about 250°C but above 300°C, definite loss of strength takes place.

In mortar and concrete, the aggregates undergo a progressive expansion on heating while the hydrated products of the set cement, beyond the point of maximum expansion shrinks. These two opposing actions progressively weaken ad crack the concrete. Quartz, the principal mineral in sand, granites and gravels expands steadily up to about 573°C. At this temperature it undergoes a sudden expansion of 0.85 %. This expansion has a disruptive action on the stability of concrete. By the use of fire resistance aggregates, to improve the fire resistance capacity of concrete. The best fire resistant aggregates, amongst the igneous rocks are the basalts and dolerites.

Limestone expands steadily until temperature of about 900°C and then begins to decomposition; hence it has been found that dense limestone is considered as a good fire resistant aggregate. Blast furnace slag aggregate and broken bricks aggregates are good resistant to fire. The large series of tests indicated that even the best fire resistant concrete have been found to fail if concrete is exposed for a considerable period to a temperature exceeding 900°C, while serious reduction in strength occurs at a temperature of about 600°C.

1.15 THERMAL MOVEMENT IN CONCRETE

Thermal movement due to considerable amount of heat: due to heat of hydration, atmospheric temperature and external fire. Due to thermal movement, changes in shape and volume of concrete causes cracks on the concrete structure. The extent of temperature rise depends on the properties of cement used and the shape and size of the component. The heat of hydration may not be significant but in mass concrete works it is an important factor to be contended with. The control of heat and avoidance of cracks to maintain the integrity of concrete structures is a subject by itself.

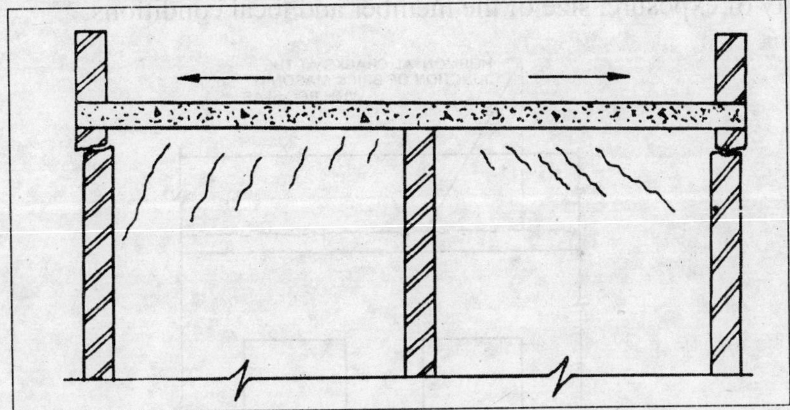

Figure 1.7. Cracking in top most storey of a load bearing structure

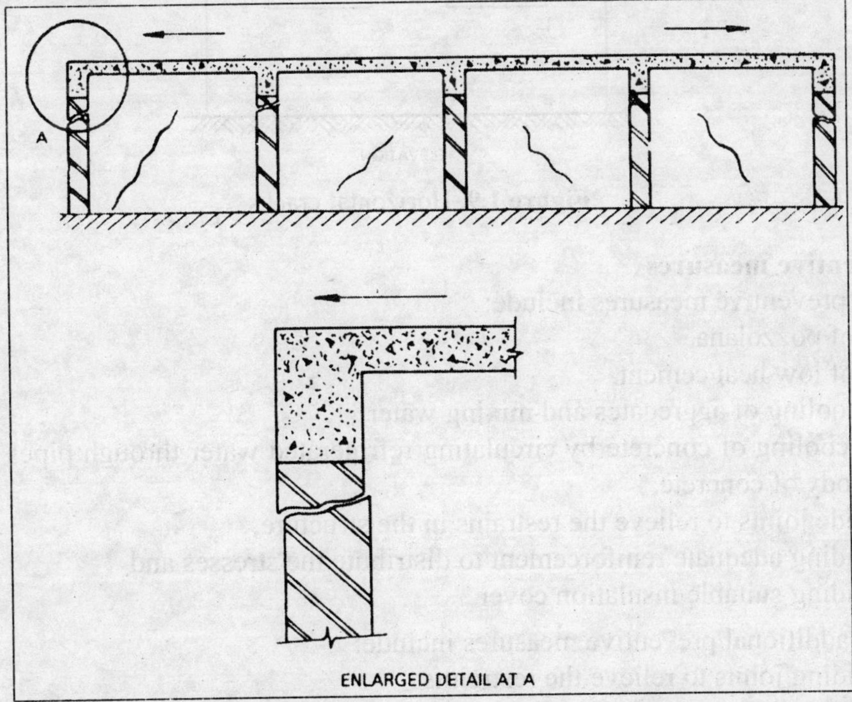

Figure 1.8. Cracking in cladding and cross walls of a framed structure

In load bearing structures, horizontal cracks may appear in the cross walls as shown in figure 1.7. In framed structures, roof slab along with beams and columns moves jointly causing diagonal cracks in walls which are located parallel to movement and horizontal cracks below beams in walls which are right angles to movement as shown in figure 1.8. Horizontal cracks also appear at the support of a brick parapet wall or brick-cum-iron railing over RCC cantilever slabs as shown in figure 1.9 due to differential thermal movement. Members which are directly exposed to sun such as sunshades, canopies, balconies, facia, fins, railings etc. should be provided with enhanced (50-100%) temperature reinforcement depending upon the severity of exposure, size of the member and local conditions.

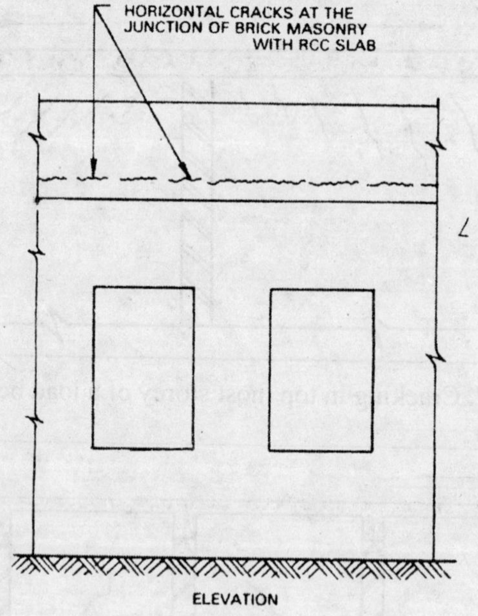

Figure 1.9. Horizontal cracks

1.15.1 Preventive measures

Some of the preventive measures include:

(a) Use of pozzolana,
(b) Use of low heat cement,
(c) Pre-cooling of aggregates and mixing water,
(d) Post cooling of concrete by circulating refrigerated water through pipes embedded in the body of concrete,
(e) Provide joints to relieve the restrains in the structure,
(f) Providing adequate reinforcement to distribute the stresses and
(g) Providing suitable insulation cover

Some of the additional preventive measures include:

(a) Providing joints to relieve the restraints
(b) Providing adequate reinforcement to distribute the stresses
(c) Providing suitable insulation cover

1.16 SUB-GRADE MOVEMENT

If there are local soft pockets in the sub-grade on which the concrete in placed or if there are any air pockets or hollows under the building paper, there will be localized settlement of concrete due to the weight of the plastic mass. If this settlement occurs after finishing of the concrete surface, cracks will ensure. The occurrence can be prevented by proper attention to compacting and draining the sub-grade.

1.17 FORMWORK MOVEMENT

Movement of formwork during the period while the concrete is going from a fluid to a rigid material may induce cracking and separation within the concrete. A crack open to the surface will allow access of water to the interior of the concrete. An internal void may give rise to freezing or corrosion problems if the void becomes saturated. Any movement of formwork, which occurs between the time when concrete begins to loose its fluidity and the time when it has fully set will cause internal, cracks to appear in the structure. These cracks are potentially dangerous in that they form a water pocket in the concrete mass, which upon freezing, will spall the concrete surface. Reber corrosion can also result from such water pockets. Such cracks can be prevented by a proper design of the forms with respect to the details and deflections.

1.18 SETTLEMENT AND MOVEMENT

There is two types of mechanism in settlement and movement in concrete structural members: differential movement and subsidence.

1.18.1 Mechanism

Differential movement:

Situations in which the various elements of a structure are moving with respect to one another are caused by differential movements. Since concrete structures are typically very rigid, they can tolerate very little differential movement. As the differential movement increases, concrete members can be expected to be subjected to an overstressed condition. Ultimately, the members will crack or spall.

Mechanism of subsidence:

Situations in which an entire structure is moving or a single element of a structure, such as a monolith, is moving with respect to the remainder of the structure are caused by subsidence. In these cases, the concerns are not over cracking or spalling but rather stability against overturning or sliding. Whether portions of a single structural element are moving with respect to one another or whether entire elements are moving, the underlying cause is more than likely to be a failure of the foundation material. This failure may be attributed to long-term consolidations, new loading conditions, or to a wide variety of other mechanisms. In situations in which structural movement is diagnosed as a cause of concrete deterioration, a thorough geotechnical investigation should be conducted.

1.18.2 Symptoms

Visual examination of structures undergoing settlement or movement will usually reveal cracking or spalling or faulty alignment of structural members. Very often, movement will be apparent in nonstructural members such as block or brick masonry walls. Another good indication of structural movement is an increase in the amount of water leaking into the structure. Since differential settlement of the foundation of a structure is usually a long-term phenomenon, review of instrumentation data will be helpful in determining whether apparent movement is real. Review by structural and geotechnical engineering specialists will be required.

1.18.3 Prevention

Prevention of settlements and movements or corrective measures is beyond the scope of this manual. Appropriate structural and geotechnical engineering manuals should be consulted for guidance.

1.19 FOUNDATION SETTLEMENT

Shear cracks occur when there is differential settlement of foundation. Shear failure is predominant in made-up ground.

Some of the preventive measures:

(a) Providing an impermeable apron all around the building,
(b) Prevention of water stagnation around the building,
(c) Providing adequate drainage system,
(d) Avoiding the plantation of fast growing trees in the immediate vicinity of the building.

1.20 CONSTRUCTION ERRORS

Failure to follow specified procedures and good practice or outright carelessness may lead to a number of conditions that may be grouped together as construction errors. Typically, most of these errors do not lead directly to failure or deterioration of concrete. Instead, they enhance the adverse impacts of other mechanisms. Each error will be briefly described below along with preventative methods. In general, the best preventive measure is a thorough knowledge of hat these construction errors are plus an aggressive inspection program.

1.20.1 Adding water to concrete

Water is usually added to concrete in one or both of the following circumstances: First, water is added to the concrete in a delivery truck to increase slump and decrease emplacement effort. This practice will generally lead to concrete with lowered strength and reduced durability. As the w/c of the concrete increases, the strength and durability will decrease. In the second case, water is commonly added during finishing of flatwork. This practice leads to scaling, crazing, and dusting of the concrete in service.

1.20.2 Improper alignment of formwork

Improper alignment of the formwork will lead to discontinuities on the surface of the concrete. While these discontinuities are unsightly in all circumstances, their occurrence may be more critical in areas that are subjected to high velocity flow of water, where cavitations-erosion may be induced, or in lock chambers where the "rubbing" surfaces must be straight.

1.20.3 Improper consolidation

Improper consolidation of concrete may result in a variety of defects, the most common being bugholes, honeycombing, and cold joints. "Bugholes" are formed when small pockets of air or water are trapped against the forms. A change in the mixture to make it less "sticky" or the use of small vibrators worked near the form has been used to help eliminate bugholes. Honeycombing can be reduced by inserting the vibrator more frequently, inserting the vibrator as close as possible to the form face without touching the form, and slower withdrawal of the vibrator. Obviously, any or all of these defects make it much easier for any damage-causing mechanism to initiate deterioration of the concrete. Frequently, a fear of "overconsolidation" is used to justify a lack of effort in consolidating concrete. Overconsolidation is usually defined as a situation in which the consolidation effort causes all of the coarse aggregate to settle to the bottom while the paste rises to the surface. If this situation occurs, it is reasonable to conclude that there is a problem of a poorly proportioned concrete rather than too much consolidation.

1.20.4 Improper curing

Curing is probably the most abused aspect of the concrete construction process. Unless concrete is given adequate time to cure at a proper humidity and temperature, it will not develop the characteristics that are expected and that are necessary to provide durability. Symptoms of improperly cured concrete can include various types of cracking and surface disintegration. In extreme cases where poor curing leads to failure to achieve anticipated concrete strengths, structural cracking may occur.

1.20.5 Improper location of reinforcing steel

This section refers to reinforcing steel that is improperly located or is not adequately secured in the proper location. Either of these faults may lead to two general types of problems. First, the steel may not function structurally as intended, resulting in structural cracking or failure. A particularly prevalent example is the placement of welded wire mesh in floor slabs. In many cases, the mesh ends up on the bottom of the slab which will subsequently crack because the steel is not in the proper location. The second type of problem stemming from improperly located or tied reinforcing steel is one of durability. The tendency seems to be for the steel to end up near the surface of the concrete. As the concrete cover over the steel is reduced, it is much easier for corrosion to begin.

1.20.6 Premature removal of shores or reshores

If shores or reshores are removed too soon, the concrete affected may become overstressed and cracked. In extreme cases there may be major failures. Lack of support for forms/ premature removal of forms or in adequate compaction can result in the settlement cracking of concrete.

1.20.7 Settling of the concrete

During the period between placing and initial setting of the concrete, the heavier components of the concrete will settle under the influence of gravity. This situation may be aggravated by the use of highly fluid concretes. If any restraint tends to prevent this settling, cracking or separations may result. These cracks or separations may also develop problems of corrosion or freezing if saturated.

1.20.8 Settling of the subgrade

If there is any settling of the subgrade during the period after the concrete begins to become rigid but before it gains enough strength to support its own weight, cracking may also occur.

1.20.9 Vibration of freshly placed concrete

Most construction sites are subjected to vibration from various sources, such as blasting, pile driving, and from the operation of construction equipment. Freshly placed concrete is vulnerable to weakening of its properties if subjected to forces which disrupt the concrete matrix during setting. The vibration limits for concrete, expressed in terms of peak particle velocity.

1.20.10 Improper finishing of flat work

The most common improper finishing procedures, which are detrimental to the durability of flat work, are discussed below.

Adding water to the surface - Evidence that water is being added to the surface is the presence of a large paint brush, along with other finishing tools. The brush is dipped in water and water is "slung" onto the surface being finished.

Timing of finishing- Final finishing operations must be done after the concrete has taken its initial set and bleeding has stopped. The waiting period depends on the amounts of water, cement, and admixtures in the mixture but primarily on the temperature of the concrete surface. On a partially shaded slab, the part in the sun will usually be ready to finish before the part in the shade.

Adding cement to the surface - This practice is often done to dry up bleed water to allow finishing to proceed and will result in a thin cement-rich coating which will craze or flake off easily.

Use of tamper - A tamper or "jitterbug" is unnecessarily used on many jobs. This tool forces the coarse aggregate away from the surface and can make finishing easier. This practice, however, creates a cement-rich mortar surface layer which can scale or craze. A jitterbug should not be allowed with a well-designed mixture. If a harsh mixture must be finished, the judicious use of a jitterbug could be useful.

Jointing - The most frequent cause of cracking in flatwork is the incorrect spacing and location of joints.

1.21 CRACKING DUE TO CONSTRUCTION OVERLOADS

The loads induced during construction can be far more severe than those experienced in service. Unfortunately, these conditions may occur at the early ages when the concrete is most susceptible to damage and often result in permanent cracks. A common error occurs when the precast members are not properly supported during transportation and erection. The use of arbitrary or convenient lifting points may cause severe damage. A big element lowered too fast, and stopped suddenly carries significant momentum, which is translated into an impact load that may be several times the dead weight of the element.

Storage of materials and equipment can easily result in loading conditions during construction far more severe than any load for which the structure is designed. Damage from unintentional construction overloads can be prevented only if the designers provide information on load limitations for the structure and if the construction personnel heed to these limitations.

(a) Inadequate provision of main steel reinforcement, or inadequate provision of temperature reinforcement or wrong spacing of bars, or absence of corner reinforcement may cause unacceptable cracks in concrete.

(b) One of the most common occurrences is the displacement of top bars in cantilever thin chajjas, by the movement of concreting gang, causes cracks at the junction point of cantilever chajja.

(c) Innumerable examples may be cited such as congestion of reinforcement and difficulties in proper compacting concrete, particularly at the column and beam junctions, deep beams, the –ve reinforcement over T & L beams, should be taken care. In the absence of such care concrete is sure to crack.

1.22 CRACKS DUE TO EXTERNALLY APPLIED LOADS

Load-induced tensile stresses may result in cracks in concrete elements. A design procedure specifying the use of reinforcing steel, not only to carry tensile forces, but also be obtaining both an adequate crack distribution and a reasonable limit on crack width is recommended. Flexural and tensile crack widths can be expected to increase with time for members subjected to either sustained or repetitive loading. A well-distributed reinforcing arrangement offers the best protection against undesirable cracking.

1.23 ACCIDENTAL LOADINGS

Mechanism:

Accidental loadings may be characterized as short-duration, one-time events such as the impact of a barge against a lock wall or an earthquake. These loadings can generate stresses higher than the strength of the concrete, resulting in localized or general failure. Determination of whether accidental loading caused damage to the concrete will require knowledge of the events preceding discovery of the damage. Usually, damage caused by accidental loading will be easy to diagnose.

Symptoms:

Visual examination will usually show spalling or cracking of concrete, which has been subjected to accidental loadings.

Prevention: Accidental loadings by their very nature cannot be prevented. Minimizing the effects of some occurrences by following proper design procedures (an example is the design for earthquakes) or by proper attention to detailing (wall armor in areas of likely impact) will reduce the impacts of accidental loadings.

1.24 DESIGN ERRORS

Design errors may be divided into two general types: those resulting from inadequate structural design and those resulting from lack of attention to relatively minor design details. Each of the two types of design errors is discussed below.

1.24.1 Inadequate structural design

Mechanism:

The failure mechanism is simple. The concrete is exposed to greater stress than it is capable of carrying or it sustains greater strain than its strain capacity.

Symptoms:

Visual examinations of failures resulting from inadequate structural design will usually show one of two symptoms. First, errors in design resulting in excessively high compressive stresses will result in spalling. Similarly, high torsion or shear stresses may also result in spalling or cracking. Second, high tensile stresses will result in cracking. To identify inadequate design as a cause of damage, the locations of the damage should be compared to the types of stresses that should be present in the concrete. For example, if spalls are present on the underside of a simple-supported beam, high compressive stresses are not present and inadequate design may be eliminated as a cause. However, if the type and location of the damage and the probable stress are in agreement, a detailed stress analysis will be required to determine whether inadequate design is the cause.

Prevention:

Inadequate design is best prevented by thorough and careful review of all design calculations. Any rehabilitation method that makes use of existing concrete structural members must be carefully reviewed.

1.24.2 Poor design details

While a structure may be adequately designed to meet loadings and other overall requirements, poor detailing may result in localized concentrations of high stresses in otherwise satisfactory concrete. These high stresses may result in cracking that allows water or chemicals access to the concrete. In other cases, poor design detailing may simply allow water to pond on a structure, resulting in saturated concrete. In general, poor detailing does not lead directly to concrete failure; rather, it contributes to the action of one of the other causes of concrete deterioration. Several specific types of poor detailing and their possible effects on a structure are described in the following paragraphs. In general, all of these problems can be prevented by a thorough and careful review of plans and specifications for the project. In the case of existing structures, problems resulting from poor detailing should be handled by correcting the detailing and not by simply responding to the symptoms.

Abrupt changes in section:
Abrupt changes in section may cause stress concentrations that may result in cracking. Typical examples would include the use of relatively thin sections such as bridge decks rigidly tied into massive abutments or patches and replacement concrete that are not uniform in plan dimensions.

Insufficient reinforcement at reentrant corners and openings:
Reentrant corners and openings also tend to cause stress concentrations that may cause cracking. In this case, the best prevention is to provide additional reinforcement in areas where stress concentrations are expected to occur.

Inadequate provision for deflection:
Deflections in excess of those anticipated may result in loading of members or sections beyond the capacities for which they were designed. Typically, these loadings will be induced in walls or partitions, resulting in cracking.

Inadequate provision for drainage:
Poor attention to the details of draining a structure may result in the ponding of water. This ponding may result in leakage or saturation of concrete. Leakage may result in damage to the interior of the structure or in staining and encrustations on the structure. Saturation may result in severely damaged concrete if the structure is in an area that is subjected to freezing and thawing.

Insufficient travel in expansion joints:
Inadequately designed expansion joints may result in spalling of concrete adjacent to the joints. The full range of possible temperature differentials that a concrete may be expected to experience should be taken into account in the specification for expansion joints. There is no single expansion joint that will work for all cases of temperature differential.

Incompatibility of materials:
The use of materials with different properties (modulus of elasticity or coefficient of thermal expansion) adjacent to one another may result in cracking or spalling as the structure is loaded or as it is subjected to daily or annual temperature variations.

Neglect of creep effect:
Neglect of creep may have similar effects as noted earlier for inadequate provision for deflections. Additionally, neglect of creep in prestressed concrete members may lead to excessive prestress loss that in turn results in cracking as loads are applied.

Rigid joints between precast units:
Designs utilizing precast elements must provide for movement between adjacent precast

elements or between the precast elements and the supporting frame. Failure to provide for this movement can result in cracking or spalling.

Unanticipated shear stresses in piers, columns, or abutments:
If, through lack of maintenance, expansion bearing assembles are allowed to become frozen, horizontal loading may be transferred to the concrete elements supporting the bearings. The result will be cracking in the concrete, usually compounded by other problems, which will be caused by the entry of water into the concrete.

Inadequate joint spacing in slabs:
This is one of the most frequent causes of cracking of slabs-on-grade. Guidance on joint spacing and depth of contraction joints may be found.

The design and detailing errors that may cause an acceptable cracking as follows:

(a) Improper selection and / or detailing of reinforcement.
(b) Use of poor detailed re-entrant corners in walls, precast members and slabs.
(c) Restraint of members subjected to volume changes due to variations in temperature and moisture.
(d) Lack of adequate contraction joints.
(e) Improper design of foundations results in differential settlement with in the structure.
(f) Re-entrant corners provide a location for stress concentration and these are main locations for initial cracks, as in the case of window and door openings in concrete walls and beams.

The structures, in which cracking may cause major problem of serviceability need special care in the design and detailing. These structures also need continuous inspection during all phases of construction.

CHAPTER 2

Chemical Causes

2.1 CHEMICAL ATTACK ON THE CONCRETE

This category includes several specific causes of deterioration that exhibit a wide variety of symptoms. In general, deleterious chemical reactions may be classified as those that occur as the result of external chemicals attacking the concrete (acid attack, aggressive water attack, miscellaneous chemical attack, and sulphate attack) or those that occur as a result of internal chemical reactions between the constituents of the concrete (alkali-silica and alkali-carbonate rock reactions). Each of these chemical reactions is described below.

Chemical attack on concrete can be classified as follows:

(a) Acid attack
(b) Alkali attack
(c) Carbonation
(d) Chloride attack
(e) Leaching
(f) Salt attack
(g) Sulphate attack.

Ingress of dissolved substances from the external environment may cause various forms of chemically induced deterioration by reaction with the cement paste or aggregate constituents.

Although the resistance of concrete to chemical attack is directly influenced by:

(a) Its porosity,
(b) The cement composition used in the concrete,
(c) Conditions under which the cement paste hardened,
(d) All determine properties of concrete and
(e) Ability to resist various effects of fluids in the environment.

Most of the problems due to chemical attack arise from the aggressive fluid penetrating into the interior pore space in the concrete. Therefore, damage is in many cases dependent on the permeability of the surface layers and not on the body of the concrete.

The penetration may be influenced by the effects of:

(a) Temperature gradient,
(b) Freezing,
(c) Loading,
(d) Electric current, and
(e) Other factors

In general, the main type of aggressive fluid to which concrete will be exposed as one or more of the following:

(a) Mineral acids,
(b) Some organic acids,
(c) Solutions of sulphates, chlorides, sugars, nitrates, phenols, and ammonium compounds

2.1.1 Types of chemical reactions

The most aggressive solutions are those where hot solutions of the above fluids are split on the surface of concrete and are absorbed forming a reservoir of chemicals with in the concrete.

Essentially five main types of chemical reactions can be recognized:

(a) **Type 1:** Simple leaching of free calcium hydroxide (hydrated lime).
(b) **Type 2:** Reaction between attacking solutions and cement compounds resulting in the formation of secondary compounds, which are either leached from the concrete, or remain in a non bound form, resulting in gradual strength loss.
(c) **Type 3:** Reaction similar to type 2, but resulting in the crystallisation of the reaction products giving rises to expansive forces, which disrupt the concrete.
(d) **Type 4:** Crystallisation of salts directly from the attacking solution causing disruption of the concrete
(e) **Type 5:** Corrosion of the embedded steel reinforcement resulting from breakdown of the passivation zone by aggressive solutions.

2.2 CARBONATION ATTACK ON THE CONCRETE

Carbonation can be defined as the reaction of carbon dioxide (CO_2) with the hydrated

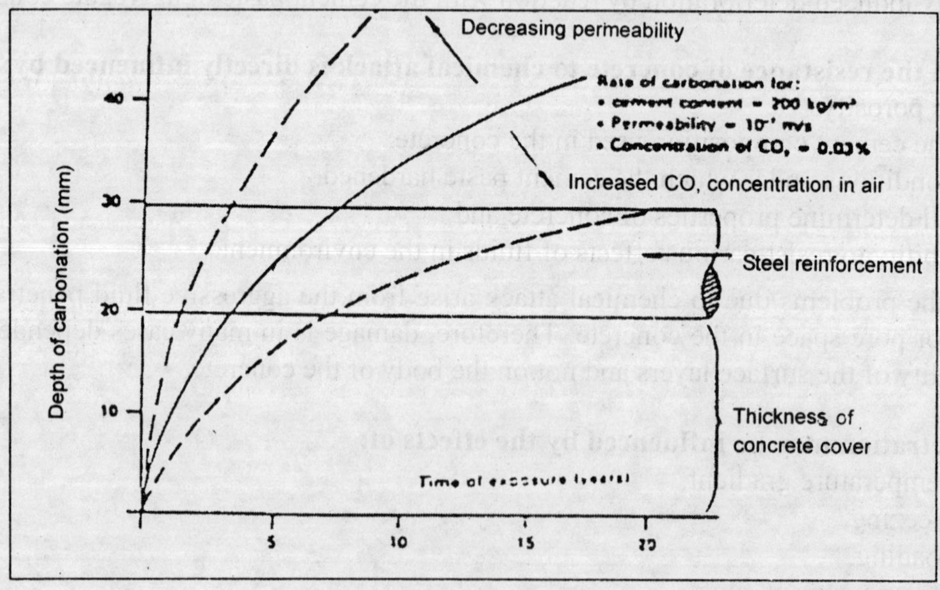

Figure 2.1. Carbonation in concrete with respect to exposure of time

cement. Carbonation is the effect of carbon dioxide in the air on cement products, mainly the hydroxides in the presence of moisture. The calcium hydroxide is converted to calcium carbonate by absorption of Carbon dioxide; Calcium carbonate in slightly soluble in water and therefore, when it is formed, it tends to seal the surface pores of the concrete and destroying the impermeability system of the concrete. As a result of the reaction, the layer of concrete, close to the surface of the concrete becomes carbonated and this carbonated layer is not sufficiently alkaline (pH below 9) to protect reinforcing steel.

Apart from the reaction between CO_2 and $Ca(OH)_2$, the former also dissolves any moisture present to form carbonic acid. This is a weak electrolyte, which allows the steel to be oxidised by the atmospheric oxygen, and then reacts with water to form hydrated ferric oxide. The corrosion product occupies a much greater volume than the original metal from which it was formed and thus sets up bursting forces in the surrounding concrete. The average depth of carbonation due to factors is illustrated in figure 2.1.

2.2.1 Influencing factors

The rate at which carbonation reaches the reinforcement is dependent on the following factors:

(a) **Time :** rate decreases with increasing time of exposure to air
(b) **Cover to reinforcement :** the greater the cover, the better the protection afforded to the steel
(c) **Concentration of carbon-dioxide in the atmosphere :** rate increases with increasing carbon dioxide in the air
(d) **Permeability of concrete :** this depends on concrete quality; carbonation depth increases by four times when the w/c ratio is increased from 0.4 to 0.8
(e) **Alkali content in the concrete :** this depends upon cement content and type of cement
(f) **Condition of concrete cover:** any imperfection in the cover such as segregation, poor compaction or cracking enables carbonation to progress more rapidly.

Concrete of good quality usually carbonates very slowly: even after a period of 50 years, carbonation is to penetrate to a greater depth then 5-10 mm. A permeable concrete may carbonate to depth of 25 mm in less than 10 years. **Carbonation reduced due to reduced moisture content and reduces the accumulation/concentration of carbon dioxide in atmosphere.**

2.3. SULPHATE ATTACK ON THE CONCRETE

Mechanism–Sulphates are found in most of the soils as calcium, potassium, sodium and magnesium sulphates. Solid salts do not attack concrete, but when present in solution they can react with hardened cement paste. Sulphates are present in seawater, industrial effluents and some ground water. Sulphate attack occurs when pore system in concrete is penetrated by solution of sulphates.

Sulphate reaction is dependent on the following parameters:

(a) Concentration of sulphate ions

(b) Cations present in the sulphate solution
(c) C_3A content of cement
(d) Density, Permeability of the concrete

2.3.1 Chemical mechanism

The effect of sulphate on concrete can be mainly, chemical and physical and they are closely related. The sulphate attack or reaction is indicated by the characteristic whitish appearance on the surface. As a result of the chemical reactions between sulphate and hydration products, changes in the microstructure and pore size distribution of the cement paste takes place. Sulphate converts calcium hydroxide in to large volume of calcium sulphate (gypsum).

$$Na_2SO_4.10H_2O + Ca(OH)_2 \rightarrow CaSO_4.2H_2O + 2NaOH + 8H_2O$$

The second hydration product, tricalcium aluminates hydrate reacts with sulphate solution to form sulpho aluminates hydrate (ettringite), which has a greater volume than that of the original compound.

$$2(3CaO.Al_2O_3.12H_2O) + 3(Na_2SO_4\ 10H_2O) \rightarrow 3CaO.Al_2O_3.3CaSo_4.31H_2O + 2Al(OH)_3 + 6NaOH + 17H_2O$$

Due to this, the resultant internal expansive stress may be great enough to cause deformation, cracking and eventually loss of cohesion. When concrete cracks, its permeability increases and the aggressive water penetrates more easily in to the interior, thus accelerating the process of deterioration. The following figure 2.2 illustrates the influence of permeability on the sulphate attack mechanism and the increase of permeability due to crack formation.

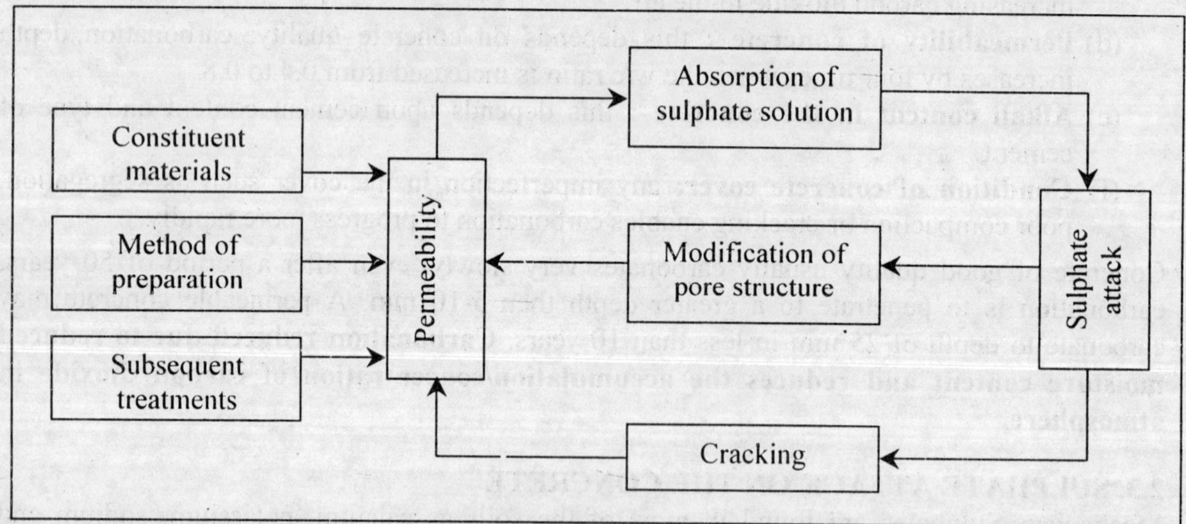

Figure 2.2. Sulphate attack – Permeability interaction

2.3.2 Physical mechanism

In addition to the two chemical reactions, there may also be a purely physical phenomenon in which the growth of crystals of sulphate salts disrupts the concrete. The damage usually starts at edges and corners and is followed by progressive cracking and spalling which reduce the concrete to a friable or even soft state. The rate of sulphate attack increases with an increase in the strength of the solution (concentration of sulphate). The rate of sulphate attack can be reduced by the use of cement low in tricalcium aluminates and by the addition of pozzolanic materials. A typical sulphate attack damage is as shown in figure 2.3.

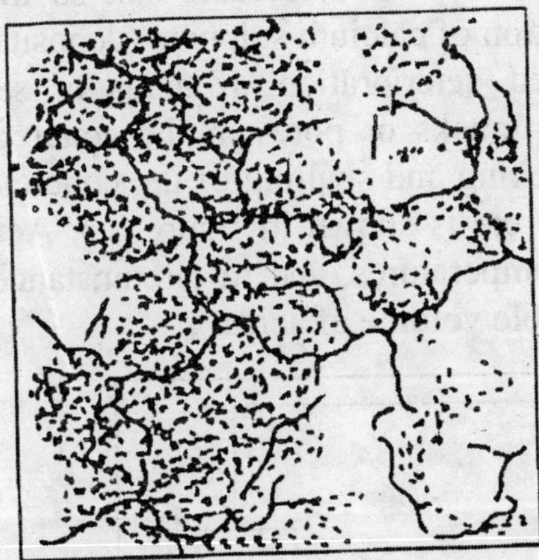

Visual examination will show map and pattern cracking as well as a general disintegration of the concrete.

Figure 2.3. A typical sulphate attack damage

2.3.3 Preventive measures

Some of the preventive measures include:

(a) Using dense concrete, high-quality concrete with a low water-cement ratio
(b) Using sulphate resisting Portland cement / super sulphated cement,
(c) Strict adherence to the minimum cement content and max water cement ratio specified for concrete and mortar in the code of practice and
(d) Provision for expansion through joints.

2.4 ACID ATTACK ON THE CONCRETE

Portland cement concrete is a highly alkaline material and is not very resistant to attack by acids. The deterioration of concrete by acids is primarily the result of a reaction between the acid and the products of the hydration of cement. Calcium silicate hydrate may be attacked if highly concentrated acid exists in the environment of the concrete structures. In most cases, the chemical reaction results in the formation of water-soluble calcium compounds that are then leached away. In the case of sulphuric acid attack, additional or accelerated deterioration results because the calcium sulphate formed may affect the concrete by the sulphate attack mechanism. If the acid is able to reach the reinforcing steel through cracks or pores in the concrete, corrosion of the reinforcing steel will result and will cause further deterioration of the concrete.

2.4.1 Mechanism

Acid conditions are harmful to the alkaline concrete and the severity of the attack depends on the type of acid. The action of acid on concrete is to dissolve away the cement, and in the case of limestone aggregate, the aggregate also gets dissolved. Due to chemical attack, in most of the cases, the chemical reaction results in formation of water soluble calcium compounds which are then leached away by the aqueous solutions. This results in an increase of the porosity and the permeability of the system.

Oxalic and phosphoric acids are exceptions, because the resulting calcium salts are insoluble in water and the surface of the concrete becomes coated with the insoluble salt so that reaction ceases. In the case of sulphuric acid attack, formation of calcium sulphate (deposited as gypsum) will affect concrete resulting in accelerated deterioration. If acids or salt solutions are able to reach the reinforcing steel through cracks or pores in the concrete, corrosion of steel can occur, which will in turn cause cracking and spalling of the concrete. The influence of acid attack on the permeability of concrete is shown in figure 2.4. Acid attack is a problem in sewers, especially at fairly high temperatures, and in circumstances where the concrete is exposed to rapid flows and considerable volume of acids.

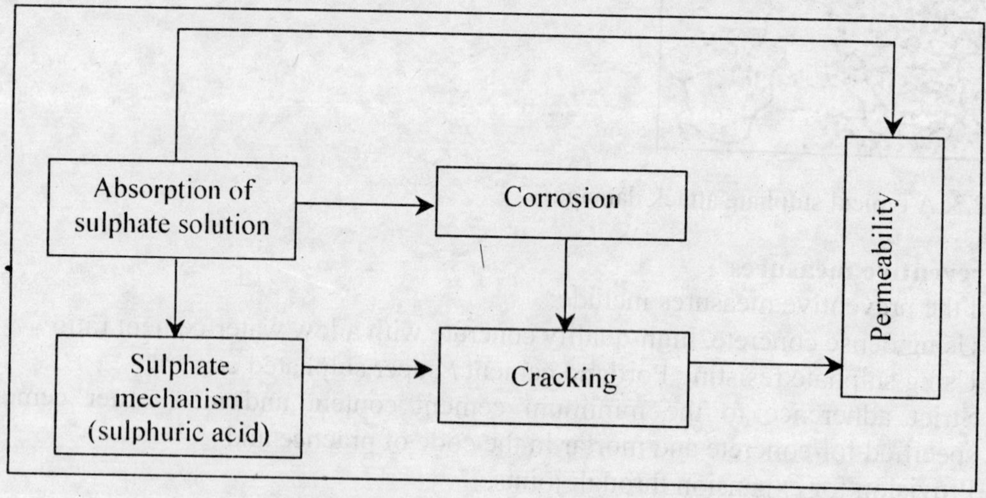

Figure 2.4. Influence of Acid attack on permeability

2.4.2 Symptoms

Visual examination will show disintegration of the concrete evidenced by loss of cement paste and aggregate from the matrix. If reinforcing steel has been reached by the acid, rust staining, cracking, and spalling may be present. All cements except High Alumina Cement (HAC) are equally susceptible to acid attack and have comparatively more resistance to acid attack than other types of cement. It should be noted that density and permeability of the concrete influence its durability to such a degree that they override the influence of the type of cement used.

2.4.3 Preventive measures

The acid attack can be reduced by the following measures:

(a) By increasing cement content and reducing w/c ratio

(b) By improving quality of cover concrete

(c) By treating the surface with sodium silicate known as water glass

(d) By surface treatment with coal, tar, bituminous paints, epoxy resins etc.

2.5 ALKALI REACTION ON THE CONCRETE

The reaction of some forms of silica and carbonates in aggregates with the alkalis in cement produces a gel, which causes expansion and cracks. Portland cement concrete made with non-alkali reactive aggregates is highly resistant to strong solutions of most bases (sodium or potassium hydroxides). However, if sodium hydroxide penetrates concrete and becomes concentrated at an evaporating face, physical damage would result from crystallisation of sodium carbonate formed by the reaction between NaOH and carbon dioxide from the air.

2.5.1 Mechanism of Alkali-aggregate reaction (AAR)

This is also called alkali – carbonate reaction. Certain carbonate rock aggregates have been reactive in concrete. The results of these reactions have been characterized as ranging from beneficial to destructive. The destructive category is apparently limited to reactions with impure dolomitic aggregates and are a result of either dedolomitization or rim-silicification reactions. Visual examination of those reactions that are serious enough to disrupt the concrete in a structure will generally show map or pattern cracking and a general appearance, which indicates that the concrete is swelling. A distinguishing feature which differentiates alkali-carbonate rock reaction from alkali-silica reaction is the lack of silica gel exudations at cracks. Petrographic examination in accordance with ASTM C 295 may be used to confirm the presence of alkali-carbonate rock reaction. Typical alkali aggregate reaction damage is as shown in figure 2.5.

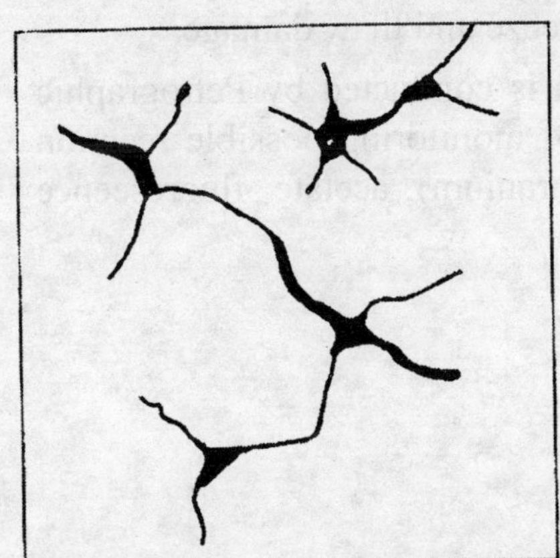

Figure 2.5. Alkali aggregate reaction damage

Influencing factors:

The reactivity of aggregate depends upon the following factors:

(a) Size of the aggregate particles
(b) Porosity of the aggregate particles
(c) Alkali content in cement
(d) Fineness of cement particles
(e) Availability of non-evaporable water in the paste
(f) Alternate wetting and drying of concrete
(g) Temperature in the range of 10 to 40°C accelerates the reaction

Prevention:
In general, the best prevention is to avoid using aggregates that are or suspected of being reactive.

2.5.2 Mechanism of Alkali – silica reaction (ASR)

ASR in concrete results from the sodium and potassium alkalis released during the hydration of Portland cement and certain siliceous constituents in the aggregate. The alkali - silica gel, which is formed, imbibes pore fluid causing its expansion. Expansion of gel induces internal stresses and there stresses may be occasionally base enough to cause cracking and expansion of concrete. Visual examination of those concrete structures that are affected will generally show map or pattern cracking and a general appearance that indicates that the concrete is swelling. Petrographic examination may be used to confirm the presence of alkali-silica reaction.

(a) Apart from cracking, the major effect of ASR is a reduction in compressive strength, tensile strength and modulus of elasticity.
(b) The resulting damage due to ASR depends on the amount of moisture available for the ASR gel to expand [Moisture content of the concrete must be approximately 80 percent; relative humidity at 21°C to 24°C].
(c) Unconfined concrete undergoing AAR exhibits tell-tale signs of surface map cracking on exposed surfaces.
(d) Once the cracking has formed, more moisture penetrates the concrete; accelerating the alkali-aggregate reaction and allowing additional freeze and thaw damage.

Usually, testing for the presence of alkali-aggregate reaction is conducted by Petrographic examination of concrete. Recently, a new method capable of monitoring possible reaction has been developed. This method utilizes the uranyl (uranium) acetate fluorescence technique and is rapid and economical.

Some of the preventive measures:
(a) Use of low alkali cement,
(b) Use of slag cement,
(c) Use of non-reactive aggregates,
(d) Use of silica in concrete mix
(e) Prevention of dampness manifesting around concrete for any significant periods

2.6 CHLORIDE ATTACK ON THE CONCRETE

Chlorides can be introduced into concrete by coming into contact with environments containing chlorides, such as seawater or deicing salts. Penetration of the chlorides starts at the surface and then moves inward. **Chlorides may enter into the concrete from the following sources:**

(a) Cement of the concrete
(b) Water mixed in concrete
(c) Aggregates of the concrete

(d) Admixtures added to the concrete
(e) Chlorides may also enter by diffusion from atmosphere

Penetration takes time, depending upon:

(a) The amount of chlorides coming into contact with the concrete
(b) The permeability of the concrete
(c) The amount of moisture present

2.6.1 Mechanism

Eventually the concentration of chlorides in contact with the reinforcing steel will cause corrosion when moisture and oxygen are present. As the rust layer builds, tensile forces generated by the expansion of the oxide cause the concrete to crack and delaminate. When cracking and delamination progress, accelerated corrosion takes place because of easy access of corrosive salts, oxygen and moisture. The concentration of chlorides necessary to promote corrosion, among other factors, is greatly affected by the concrete's pH. It was demonstrated that a threshold level of 8000 ppm of chloride ions was required to initiate corrosion when the pH was 13.2. As the pH was lowered to 11.6, corrosion was initiated with only 71 ppm of chloride ions. The IS 456–2000 has specified the chloride content as chlorine (Cl), in concrete at the time of placing as shown in the Table 2.1.

Table 2.1. Limits of chloride content of concrete as per IS 456 – 2000

S.No	*Use of Concrete*	*Max. chloride content in kg/m³*
1	Concrete containing metal and steam cured at high temperature and prestressed concrete	0.4
2	R.C.C or plain concrete containing metal (steel)	0.6
3	Concrete not containing embedded metal or any material requiring protection from chlorides	3.0

2.7 HYDROLYSIS AND LEACHING ON THE CONCRETE

Leaching of lime compounds leads to the formation of salt deposits on the surface of the concrete, known as **efflorescence.** When water percolates through poorly compacted concrete or through cracks or along badly made joints and when evaporation takes place at the surface of the concrete, leaching occurs. Calcium carbonate formed by the reaction of $Ca(OH)_2$ with CO_2 is left behind in the form of a white deposit.

Efflorescence is more likely to occur in concrete, which is porous near the surface. Thus the type of formwork plays a vital role in addition to the compaction and w/c ratio. The occurrence of efflorescence is greater when cool wet weather is followed by a dry and hot spell. In this sequence, there is little initial carbonation, the surface moisture dissolves lime and $Ca(OH)_2$ is finally drawn to the surface where it reacts with CO_2 to form calcium carbonate. Efflorescence can also be caused by the use of unwashed seashore aggregate. The salt coating on the surface of the aggregate particles in the course of lime leads to a white deposit on the surface of the concrete.

Early efflorescence can be removed with a brush and water. Heavy deposits may require acid treatment of the surface of the concrete. The acid used is diluted HCl (1:5 to 1:10). Typically, the quantity of a 1:10 solution of acid used would be 200g/m^2, (applied by a sponge) the depth of concrete removed by 0.01mm. There is no danger of continued action of the acid because it is used up by the reaction with lime, but the concrete should be washed in order to remove the salts, which would be formed. Since the acid removes lime, the surface of the concrete becomes darker. Hence acid must be applied uniformly in terms of concentration and duration of action. Thus, the influence of leaching is to increase the permeation properties of concrete due to the removal of calcium hydroxide from the pore solution.

2.8 SALT ATTACK ON THE CONCRETE/SALT WEATHERING

Solid salts do not attack concrete, but when present in solution they can react with hardened concrete. It is a more general problem in masonry structures. Efflorescence is a whitish crystalline deposit on the surface. Efflorescence is the formation of calcium carbonate precipitate on the concrete surface owing to carbonation.

2.8.1 Preventive measures

Some of the preventive measures include:

(a) Using sound materials free from salts,
(b) Proper concrete proportioning,
(c) Consolidation and Curing,
(d) Preventing the access of moisture to the structure.

2.9 AGGRESSIVE-WATER ATTACK/SOFT WATER ATTACK

Some waters have been reported to have extremely low concentrations of dissolved minerals. These soft or aggressive waters will leach calcium from cement paste or aggregates. This phenomenon has been infrequently reported in the United States. From the few cases that have been reported, there are indications that this attack takes place very slowly. For an aggressive water attack to have a serious effect on hydraulic structures, the attack must occur in flowing water. This keeps a constant supply of aggressive water in contact with the concrete and washes away aggregate particles that become loosened as a result of leaching of the paste.

Visual examination will show concrete surfaces that are very rough in areas where the paste has been leached. Sand grains may be present on the surface of the concrete, making it resemble coarse sandpaper. If the aggregate is susceptible to leaching, holes where the coarse aggregate has been dissolved will be evident. Water samples from structures where aggressive-water attack is suspected may be analyzed to calculate the Langlier Index, which is a measure of the aggressiveness of the water.

2.9.1 Prevention

The aggressive nature of water at the site of a structure can be determined before construction or during a major rehabilitation. Additionally, the water-quality evaluation at many structures can be expanded to monitor the aggressiveness of water at the structure. If there are indications that the water is aggressive or is becoming an aggressive, area susceptible to high flows may be coated with a non Portland-cement-based coating.

2.10 CRYSTALLISATION OF SALTS IN PORES

Most of the problems due to salts arise from the aggressive fluid penetrating in to the interior pore space in the concrete. Damage is, therefore, in many causes depending on the permeability of the surface layers and not on the body of the concrete. **In the most aggressive situations, crystallisation of salts in pores, directly from the attacking solution, causes disruption of the concrete.** Due to chemical attack, in most of the cases, the chemical reaction results in formation of water-soluble calcium compounds (salt compounds), which are then leached away by the aqueous solutions. **This results in an increase of the porosity and permeability of the system.**

In the case of sulphuric acid attack, formation of calcium sulphate will affect concrete resulting in accelerated deterioration; calcium sulphate formed by the initial reaction can proceed to react with calcium aluminates to form **calcium sulpho aluminates which on crystallization** can cause expansive disruption of the concrete. If acids or salt solutions are able to reach the reinforcing steel through cracks or pores in the concrete, corrosion of steel can occur, which will in turn cause cracking and spalling of the concrete. **Concrete with increased cement content and a reduced w/c ratio is required to prevent the salt attack.** Further a substantial sacrificial thickness of the same quality of concrete at the surface (cover concrete) has to be provided or protective surface coating has to be given.

Portland cement concrete made with non–alkali reactive aggregates is highly resistant to strong solutions of most bases (sodium or potassium hydroxide solution). How ever, if sodium hydroxide penetrates concrete and becomes concentrated at an evaporating face, physical damage would result from **crystallization of sodium carbonate** formed by the reaction between NaOH and CO_2 from the air.

2.11 SEAWATER ATTACK ON THE CONCRETE

The marine environment is characterized by wave action, which imposes shock loads and causes the erosion of concrete surfaces by abrasion and cavitations. In addition, the concrete is exposed to the aggressive constituents of sea water and subjected to repeated freeze-thaw and wet-dry cycles. Thus the deterioration of concrete structures in such an environment is both chemical and physical in nature and the type of the attack may be demarcated into three zones depending on the tidal lines as shown in figure 2.6.

The upper part above the high tide line is not directly exposed to seawater. However, it is opposed to atmospheric air, frost action, and wind blown sea salts. Consequently, cracking due to reinforcement corrosion and/or freeze-thaw cycles is the main deterioration phenomena in this zone. In the tidal zone, the structure is subjected to alternate wet and dry cycles, freeze ad thaw cycles, impact of waves, and floating ice, abrasion by sand and gravel, and reinforcement corrosion. The lower zone, submerged in water, is a relatively stable environment, where freeze-thaw action and reinforcement corrosion does not occur. Here the predominant deterioration action is chemical attack, sulphate attack, which is, causes strength retrogression.

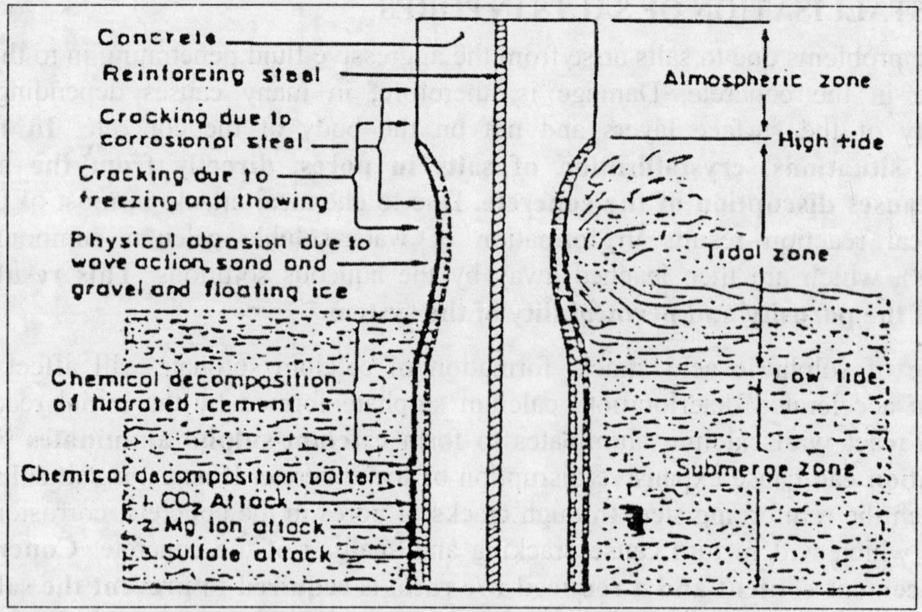

Figure 2.6. Deterioration of concrete in seawater

In-place **concrete density, cement type, and cement content** play a pivotal role in the resistance of concrete to seawater. Concrete made with calcium aluminate, super sulphated cements, and also those containing supplementary cementing materials, resists seawater fairly well. Such improved resistance as compared to OPC stems from reduced free lime content in such concrete. In case of reinforced concrete, the absorption of salt establishes anodic and cathodic areas; the resulting electrolytic action leads to an accumulation of corrosion products in the steel with a consequent rupture of the surrounding concrete. Hence it is essential **to provide sufficient cover to reinforcement.** Adopting **low w/c ratio and high cement content** are essential. A **well compacted concrete, good workmanship, especially the construction joints, and type of cements** is of vital importance.

2.12 BIOLOGICAL ATTACK ON CONCRETE

Existence of vegetation, such as fast growing trees in the vicinity of compound walls can sometimes cause cracks in walls due to expansive action of roots growing under the foundation. Roots of a tree generally spread horizontally on all sides to the extent of height of the tree above the ground and when trees are located close to a wall; these should always be viewed with suspicion. Sometimes plants take root and begin to grow in fissures of walls, because of seeds contained in bird droppings. If these plants are not removed well in time, these may in course of time develop and cause severe cracking of the wall in question.

When soil under the foundation of a building happens to be shrinkable clay, cracking in walls and floors of the building can occur either due to dehydrating action of growing roots on the soil which may shrink and cause foundation settlement, or due to upward thrust on a portion of the building, when old trees are cut off and the soil that had been dehydrated earlier by roots, swells up on getting moisture from some source, such as rain. Even small plants such as lichen, algae and mass growing on concrete surface attract moisture and encourage physical and chemical process to deteriorate the concrete. Besides, humic acid produced by micro–growth reacts with cement. Marine borers and marine plants also contribute to the deterioration of concrete. Sometimes in tropical conditions algae, fungi and bacteria use atmospheric nitrogen to form nitric acid, which attack concrete.

2.12.1 Preventive measures

Following are some general measures for avoidance of cracks due to vegetation:

(a) Do not let trees grow too close to buildings, compound walls, garden walls, etc., taking extra care if soil under the foundation happens to be shrinkable soil/clay. If any saplings of trees start growing in fissures of walls, remove them at the earliest opportunity.

(b) If some large trees exist close to a building and these are not causing any problem, as far as possible, do not disturb these trees if soil under the foundation happens to be shrinkable clay.

(c) If from any site intended for new construction, vegetation including trees is removed and the soil is shrinkable clay, do not commence construction activity on the soil until it has undergone expansion after absorbing moisture and has stabilized.

2.13 MECHANISM OF MISCELLANEOUS CHEMICAL ATTACK

Concrete will resist chemical attack to varying degrees, depending upon the exact nature of the chemical includes an extensive listing of the resistance of concrete to various chemicals. To produce significant attack on concrete, most chemicals must be in solution that is above some minimum concentration. Concrete is seldom attacked by solid dry chemicals. Also, for maximum effect, the chemical solution needs to be circulated in contact with the concrete. Concrete subjected to aggressive solutions under positive differential pressure is particularly vulnerable. The pressure gradients tend to force the aggressive solutions into the matrix. If the low-pressure face of the concrete is exposed to evaporation, a concentration of salts tends to accumulate at that face, resulting in increased attack. In addition to the specific nature of the chemical involved, the degree to which concrete resists attack depends upon the temperature of the aggressive solution, the w/c of the concrete, the type of cement used (in some circumstances), the degree of consolidation of the concrete, the permeability of the concrete, the degree of wetting and drying of the chemical on the concrete, and the extent of chemically induced corrosion of the reinforcing steel.

2.13.1 Symptoms

Visual examination of concrete, which has been subjected to chemical attack, will usually show surface disintegration and spalling and the opening of joints and cracks. There may also be swelling and general disruption of the concrete mass. Coarse aggregate particles are generally more inert than the cement paste matrix; therefore, aggregate particles may be seen as protruding from the matrix.

2.13.2 Prevention

Typically, dense concretes with low w/c (maximum w/c = 0.40) provide the greatest resistance. The best-known method of providing long-term resistance is to provide a suitable coating.

CHAPTER 3
Corrosion

3.1 BASIC PRINCIPLE OF CORROSION

Concrete is a high alkalinity material. The pH of newly produced concrete is usually between 12 and 13. In this range of alkalinity, embedded steel is protected from corrosion by a passivating film bonded to the reinforcing bar surface. However, when the passivating film is disrupted, corrosion may take place. The damage caused by **rebar corrosion** in concrete structure has been considered as one of the major durability problem affecting the **service life of concrete structures.** There is a distinct difference between the corrosion of bare steel exposed to atmosphere and the steel embedded in concrete. Reinforcement corrosion can get initiated through one or more effects from the environment. The nature of reinforcement corrosion mechanism can be attributed to three predominant process, namely **chemical, electro chemical and physical.**

(a) It is **chemical in the sense** that the alkalinity of concrete can get reduced to a pH level of less than 10.0 by the ingress of carbon dioxide or the passivity of steel can be destroyed by the ingress of chloride thereby initiating corrosion in both cases.

(b) It is **electrochemical in the sense** that the galvanic cells get established by forming locally or generally cathodic and anodic sites resulting in a flow of current with moist concrete serving as the electrolyte. In this process, the rate of corrosion is influenced by the oxygen supply.

(c) It is **physical in the sense** that as the corrosion process progresses; the corrosion product (rust) experiences a volume growth as high as six to seven times the original corroding metal. This volume growth exerts physical expansive forces to the concrete surrounding the steel. These forces exceed the tensile strength of concrete, cracking of concrete occurs and as further corrosion takes place, spalling of concrete occurs resulting in exposure of reinforcement.

3.1.1 Influencing factors

The parameters, which influence the corrosion process, are:

(a) The cover thickness
(b) The quality of concrete in the cover region, especially in terms of permeability and diffusivity
(c) Environmental conditions
(d) pH value in concrete
(e) Chloride level in concrete and
(f) Presence of cracks etc

The problem of corrosion of reinforcement in concrete structures needs serious consideration by the designers and constructors. It is emphasised that a good construction practice can minimise the corrosion problem to a large extent.

3.2 CORROSION MECHANISM OF EMBEDDED METALS

Steel reinforcement is deliberately and almost invariably placed within a few inches of a concrete surface. Under most circumstances, Portland cement concrete provides good protection to the embedded reinforcing steel. This protection is generally attributed to the high alkalinity of the concrete adjacent to the steel and to the relatively high electrical resistance of the concrete. Still, corrosion of the reinforcing steel is among the most frequent causes of damage to concrete.

3.2.1 High alkalinity and electrical resistivity of the concrete

The high alkalinity of the concrete pore solution can be reduced over a long period of time by carbonation. The electrical resistivity can be decreased by the presence of chemicals in the concrete. The chemical most commonly applied to concrete is chloride salt in the form of deicers. As the chloride ions penetrate the concrete, the capability of the concrete to carry an electrical current is increased significantly. If there are differences within the concrete such as moisture content, chloride content, oxygen content, or if dissimilar metals are in contact, electrical potential differences will occur and a corrosion cell may be established. The anodes will experience corrosion while the cathodes will be undamaged. On an individual reinforcing bar there may be many anodes and cathodes, some adjacent, and some widely spaced.

3.2.2 Enhanced reduction in load-carrying capacity of concrete

As the corrosion progresses, two things occur: First, the cross-sectional area of the reinforcement is reduced, which in turn reduces the load carrying capacity of the steel. Second, the products of the corrosion, iron oxide (rust), expand since they occupy about eight times the volume of the original material. This increase in volume leads to cracking and ultimately spalling of the concrete. For mild steel reinforcing, the damage to the concrete will become evident long before the capacity of the steel is reduced enough to affect its load-carrying capacity. However, for prestressing steel, slight reductions in section can lead to catastrophic failure.

3.2.3 Other mechanisms of corrosion

In addition to the development of an electrolytic cell, corrosion may be developed under several other situations. The first of these is corrosion produced by the presence of a stray electrical current. In this case, the current necessary for the corrosion reaction is provided from an outside source. A second additional source of corrosion is that produced by chemicals that may be able to act directly on the reinforcing steel.

3.2.4 Symptoms of metal corrosion

Visual examination will typically reveal rust staining of the concrete. This staining will be followed by cracking. Cracks produced by corrosion generally run in straight, parallel lines at uniform intervals corresponding to the spacing of the reinforcement. As deterioration continues, spalling of the concrete over the reinforcing steel will occur with the reinforcing bars becoming visible.

3.2.5 Prevention of metal corrosion

(a) Use of concrete with low permeability
(b) Use of properly proportioned concrete having a low w/c
(c) Use of as low a concrete slump as practical
(d) Use of good workmanship in placing the concrete
(e) Curing the concrete properly
(f) Providing adequate concrete cover over the reinforcing steel
(g) Providing good drainage to prevent water from standing on the concrete
(h) Limiting chlorides in the concrete mixture; and
(i) Paying careful attention to protruding items such as bolts or other anchors

3.3 CORROSION PROCESS

The corrosion process of reinforcement embedded in concrete has two distinct periods namely, **initiation period and propagation period.** Figure 3.1 shows these periods schematically as service life of a structure. The initiation period, during which the metal, have been embedded in concrete remains passive whilst, with the concrete, environmental changes are taking place that may ultimately terminate passively. The propagation period, during which begins at the moment of depassivation and involves the propagation of corrosion at a significant rate, until a final state is reached, when the structure is no longer considered acceptable on grounds of structural integrity, serviceability or appearance.

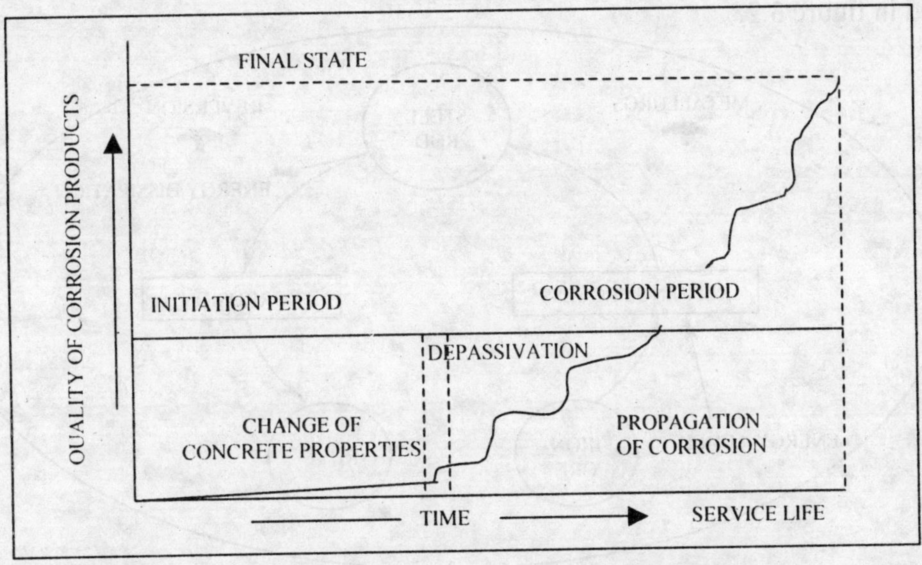

Figure 3.1. Service life model for corrosion affected structures

3.3.1 Initiation period

That is the time taken to initiate corrosion, which can be caused either by ingress of carbon dioxide or chloride ions. The reinforcing steel embedded in concrete is well protected by providing adequate cover thickness and good quality concrete with certain minimum cement content.

Influencing factors:

The factors, which influence the corrosion initiation, are:

(a) The environment to which the structure is exposed
(b) The cover thickness
(c) Quality of cover concrete in terms of its alkalinity, permeability and diffusion characteristics
(d) The type of steel
(e) Critical chloride in concrete and
(f) Presence of cracks

3.3.2 Propagation period

After the initiation of corrosion, the propagation begins and this period has two distinct processes. One is that the corrosion follows **an electrochemical process** and the other is the **physical process** due to which damage to concrete occurs. During the propagation period, the corrosion progresses at a rate depending on the availability of oxygen and moisture.

Electrochemical process:

The **electrochemical process corrosion** can be considered as the metallurgy in reverse. Steel is produced from the basic iron ore which is oxide in nature. Energy is added to make the ore into steel and during the electrochemical process by corrosion, electrons get liberated dissipating the energy added and thereby the steel goes back to its oxidised form. This is illustrated in figure 3.2.

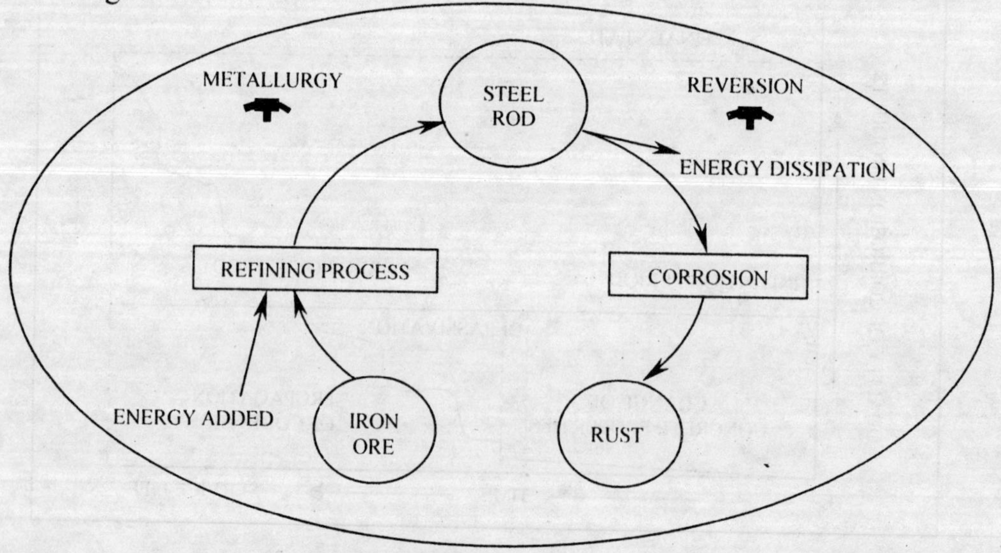

Figure 3.2. Metallurgy in reverse is corrosion

In respect of reinforcing steel, **the electrochemical process can occurs in two situations.** Immediately after production in the factory, the rods come out in light blue colour. During transportation and storage, a thin oxide film gets formed and this acts as a passive layer. However, during handling, it is likely that the passive layer may get mechanically destroyed creating locally depassivated spots. Such spots in the presence of water and oxygen create galvanic cells, forming anodic and cathodic sites and highly localised corrosion takes place. Such corrosion is known as localised pitting corrosion. The pitting corrosion in embedded steel is schematically illustrated in figure 3.3.

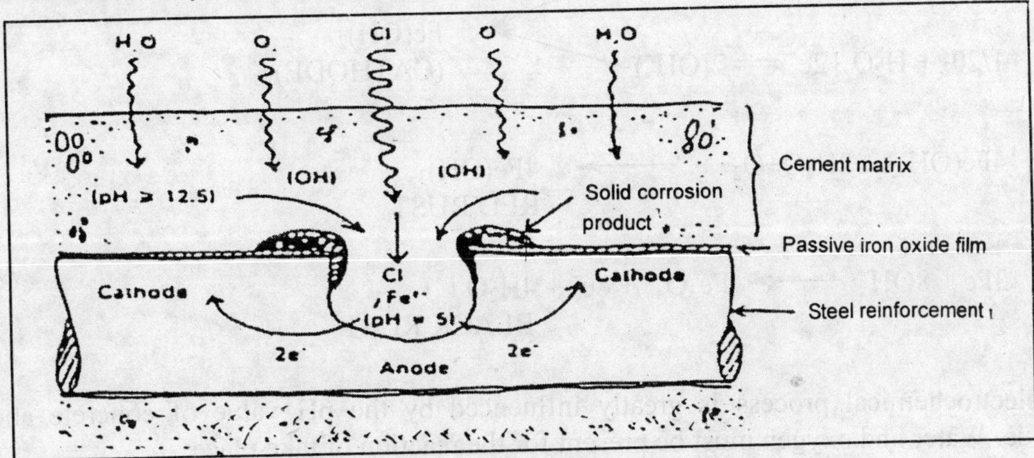

Figure 3.3. Occurrence of pitting corrosion

The process follows an electrochemical phenomenon creating a potential gradient and current flow between the anodic and cathodic location. It is necessary that the reinforcing rods are well protected during storage and this can be achieved by keeping the rods under covered sheds, placing them on wooden supports and providing a cement slurry coating. **Another situation of electrochemical process is when the rod is embedded in concrete.** Corrosion process of rebar in concrete is illustrated in figure 3.4. In this situation, the electrochemical process progress by forming the anodic and cathodic sites, involving chemical reactions.

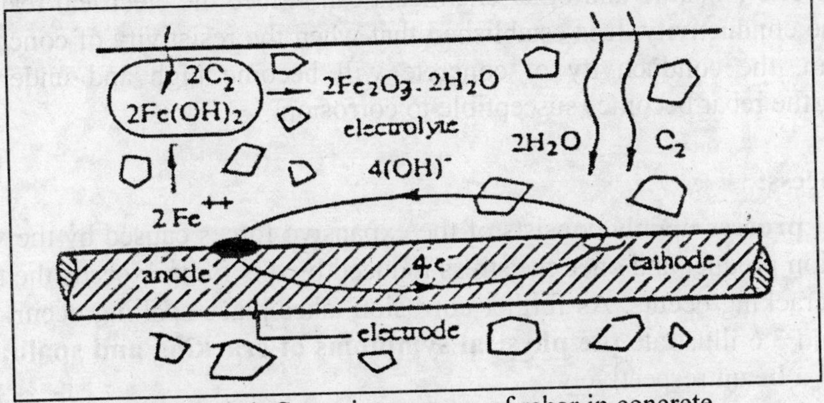

Figure 3.4. Corrosion process of rebar in concrete

A moist concrete matrix forms an acceptable electrolyte and the steel reinforcement provides the anode and cathode. Electrical current flows between the cathode and anode, and the reaction results in an increase in metal volume as the Fe(Iron) is oxidized into $Fe(OH)_2$ and $Fe(OH)_3$ and precipitates as Fe OH (rust).

The electrochemical equations as given below:

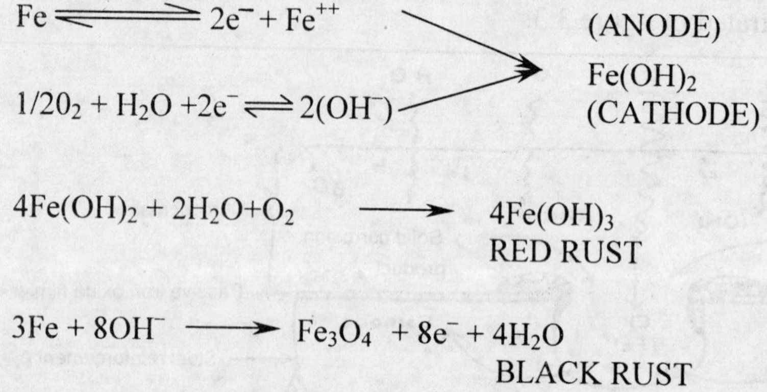

$$Fe \rightleftharpoons 2e^- + Fe^{++} \quad \text{(ANODE)}$$

$$1/2O_2 + H_2O + 2e^- \rightleftharpoons 2(OH^-) \quad \text{(CATHODE)}$$

$$\rightarrow Fe(OH)_2$$

$$4Fe(OH)_2 + 2H_2O + O_2 \longrightarrow 4Fe(OH)_3 \quad \text{RED RUST}$$

$$3Fe + 8OH^- \longrightarrow Fe_3O_4 + 8e^- + 4H_2O \quad \text{BLACK RUST}$$

The electrochemical process is greatly influenced by the pH value of concrete and the chloride. Water and oxygen must be present for the reaction of take place.

Influencing factors:

The factors, which influence the electrochemical process, can be summarised as follows:

(a) pH values
(b) Chloride content
(c) Moisture with in the concrete and
(d) Oxygen supply

In addition to the above factors, electrical resistivity of concrete also influences the electrochemical process. Very dry concrete can have a high resistivity of more than 100 kilo.ohm.cm. The moisture and other chemicals can reduce the electrical resistivity, thereby increasing the conductivity. It is established that when the resistivity of concrete falls below 5000 ohm.cm, the conductivity of concrete will become high and under such internal environment, the rebar becomes susceptible to corrosion.

Physical process:

The physical process mainly consists of the expansive forces caused by the volume growth of the corrosion product and once the stress induced by this force exceeds the tensile strength of concrete, cracking occurs. As further corrosion takes place spalling occur. The following figures 3.5 and 3.6 illustrate **the physical symptoms of cracking and spalling mechanism in a slab and a beam** respectively.

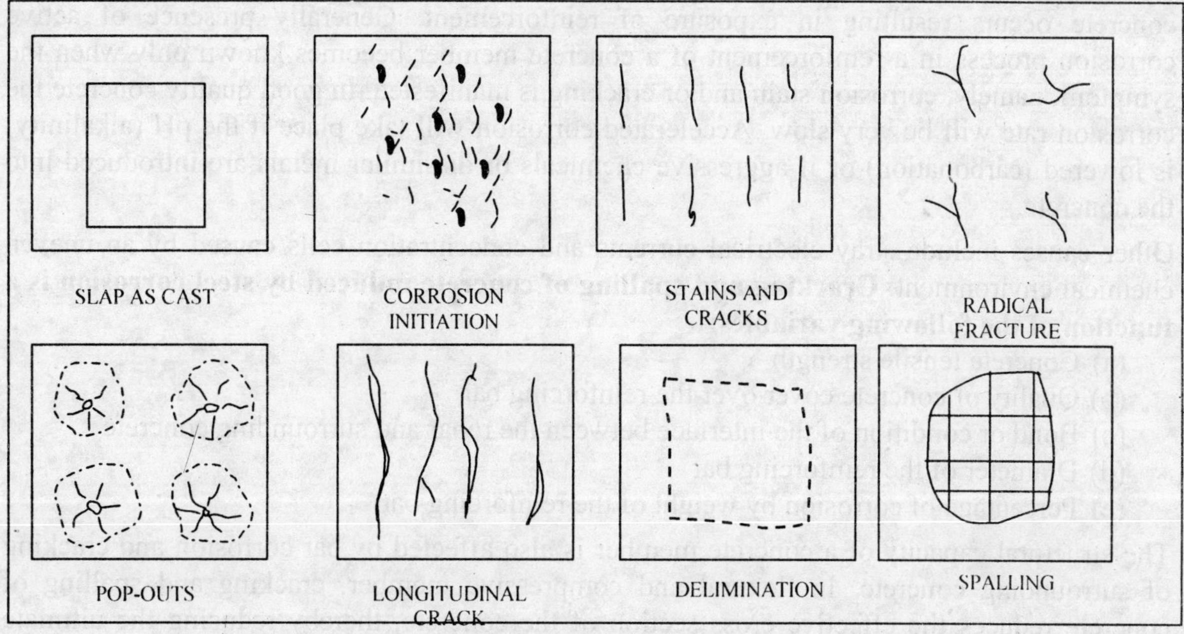

Figure 3.5. Typical symptoms of corrosion in RC slab

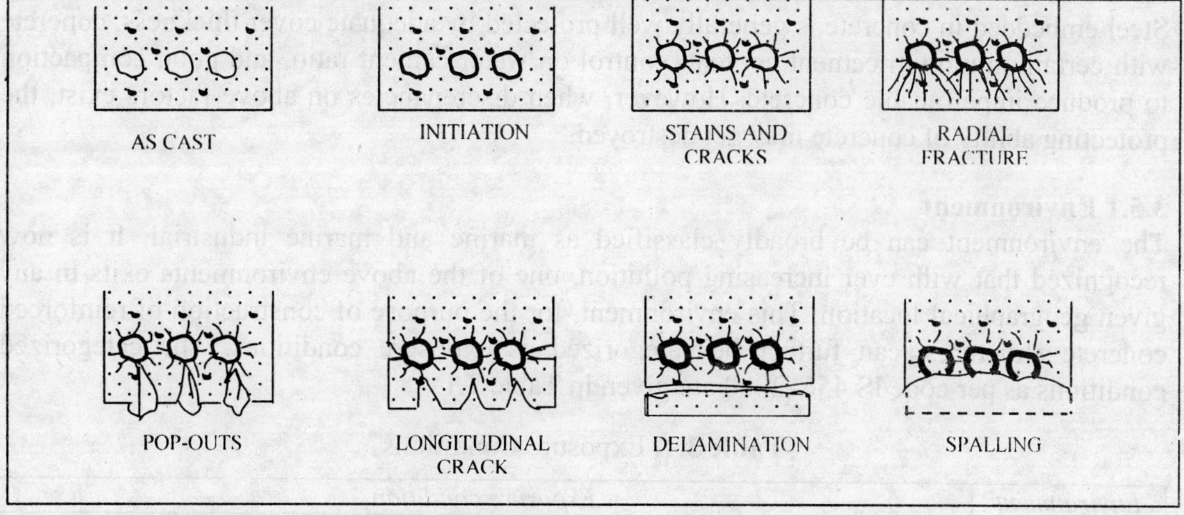

Figure 3.6. Typical symptoms of corrosion in RC beam

3.4 DAMAGES DUE TO CORROSION

In RC structures, the corrosion of reinforcement is unique in the sense that the corrosion process causes extensive damage to the concrete. The corrosion product has a volume growth as high as the original. This volume growth exerts physical expansive forces to the concrete surrounding the steel. Once the stresses induced by these forces exceed the tensile strength of concrete, cracking of concrete occurs, and as further corrosion takes place, spalling of concrete occurs resulting in exposure of reinforcement. Generally presence of active corrosion process in a reinforcement of a concrete member becomes known only when the symptom, namely, corrosion stain and or cracking is manifested. In good quality concrete the corrosion rate will be very slow. Accelerated corrosion will take place if the pH (alkalinity) is lowered (carbonation) or if aggressive chemicals or dissimilar metals are introduced into the concrete.

Other causes include stray electrical currents and concentration cells caused by an uneven chemical environment. **Cracking and spalling of concrete induced by steel corrosion is a function of the following variables:**

(a) Concrete tensile strength
(b) Quality of concrete cover over the reinforcing bar
(c) Bond or condition of the interface between the rebar and surrounding concrete
(d) Diameter of the reinforcing bar
(e) Percentage of corrosion by weight of the reinforcing bar.

The structural capacity of a concrete member is also affected by bar corrosion and cracking of surrounding concrete. In flexural and compressive member, cracking and spalling of concrete reduces the effective cross section of the concrete, thereby reducing the ultimate load carrying capacity.

3.5 CODAL PROVISIONS FOR DIFFERENT EXPOSURE CONDITIONS

Steel embedded in concrete is generally well protected by adequate cover thickness, concrete with certain minimum cement content, control on water cement ratio, and good compaction to produce impermeable concrete. However, when discrepancies on above factors exist, the protecting ability of concrete may get destroyed.

3.5.1 Environment

The environment can be broadly classified as marine and marine industrial. It is now recognized that with ever increasing pollution, one of the above environments exits in any given geographical location. This environment, for the purpose of construction of reinforced concrete structures, can further be categorized as exposure conditions. The categorized conditions as per code IS 456–2000 are given in Table 3.1.

Table 3.1. Exposure Conditions

Environment	*Exposure condition*
Mild	Concrete surfaces protected against weather or aggressive conditions except those situated in coastal area
Moderate	Concrete surfaces sheltered from severe rain or freezing Concrete exposed to condensation and rain Concrete continuously under water Concrete completely immersed in sea water
Severe	Concrete surfaces exposed to severe rain, alternate wetting and drying of occasional freezing or severe condensation Concrete completely immersed in sea water

Environment	*Exposure condition*
Very severe	Concrete surfaces exposed to seawater spray corrosive fumes or severe freezing Concrete exposed to aggressive subsoil ground water or coastal environment
Extreme	Surface of members in tidal zone. Members in direct contact with liquid/solid aggressive chemicals

3.5.2 Cover thickness

The cover thickness is the primary physical protection to steel reinforcement. The importance of providing the stipulated cover thickness should be given due consideration during construction. Larger cover thickness obviously will delay the initiation since the time taken for the aggressive chemical such as CO_2 and chloride to penetrate will be longer. Optimum cover thickness is to be chosen taking into account the structural design aspects. However, certain minimum nominal cover thicknesses are to be provided depending on the exposure conditions (environment) and values stipulated in the recent draft code of IS 456 - 2000 are given in the following Table 3.2.

Table 3.2. Nominal cover to meet durability requirement

S.No.	*Exposure conditions*	*Nominal concrete cover in mm not less than*
1.	Mild	20
2.	Moderate	30
3.	Severe	45
4.	Very severe	50
5.	Extreme	75

3.5.3 Quality of concrete

Corrosion initiation is largely influenced by the quality of concrete especially in the cover region. Both chemical and physical parameters of concrete need to be considered. The primary chemical parameter is its alkalinity expressed in terms of its pH value. The alkalinity is provided by the internal pore solution of concrete which is basically calcium hydroxides. This will give the pH value of more than 12.0 and this can be best achieved by using proper cement and certain minimum quantity. Realizing the importance, the minimum requirement of cement content for concrete for different exposure conditions is stipulated in the draft IS 456–2000 and these are given in Table 3.3.

Table 3.3. minimum cement content for different exposure conditions

S.No	*Exposure*	*Plain concrete in kg/m³*	*Reinforcement concrete in kg/m³*
1	Mild	220	300
2	Moderate	250	300
3	Severe	260	350
4	Very severe	280	375
5	Extreme	300	375

3.5.4. Water–cement ratio

The water–cement ratio plays a predominant role in the permeability characteristics of concrete and therefore, the draft IS 456–2000 stipulates water–cement ratio requirement for different conditions of exposure as given in Table 3.4.

Table 3.4. Maximum water-cement ratio for different exposure conditions

S.No	*Exposure*	*Plain concrete maximum free w/c*	*Reinforcement concrete minimum free w/c*
1	Mild	0.60	0.55
2	Moderate	0.60	0.50
3	Severe	0.50	0.45
4	Very severe	0.45	0.45
5	Extreme	0.40	0.40

3.6 CORROSION PROTECTION TECHNIQUES

The detailed analysis of the factors that influence corrosion mechanism and process indicates that corrosion protection requires a multiple approach. There are many possible approaches as listed below:

(a) Coating to reinforcement
(b) Galvanized reinforcement
(c) Improving metallurgically by addition of certain elements
(d) Using stainless steel
(e) Using non-ferrous reinforcement
(f) Using corrosion inhibitors
(g) Coating to concrete
(h) Cathodic protection, either by means of impressed unit or by sacrificial anodes.
(i) Electrochemical chloride removal
(j) Improving the cover concrete.

3.6.1 Coating to reinforcement

The objective of coating to steel rebar is to provide a sufficiently durable barrier to aggressive materials such as chlorides. Initially, the bar in shot blasted to remove mill-scale. This ensures an adequate bond between the epoxy & the steel. The bar is then heated to a controlled temperature before passing through a spray booth. Hence, electro-statically charged epoxy powder particles are deposited evenly on the surface of the bar. Shortly after spraying, the epoxy starts to use and hardens sufficiently for the bars to be handled. The coating thickness typically varies from 130 micron to 300 micron.

3.6.2 Galvanized reinforcement

Galvanized reinforcement consists of standard black bar, hot dipped in molten zinc. This process forms a coating which is metallurigcally bonded to the surface of the percent metal. The surface of the zinc reacts with calcium hydroxide in the concrete to form a passive layer, preventing corrosion. However, it is a more difficult process to control as uneven coatings can form and it is more expensive.

3.6.3 Improving metallurgically by adding certain elements

The bars can be improved for its corrosion resistance by adding certain elements such as chromium and copper during the formation/process itself.

3.6.4 Stainless steel reinforcement

Stainless steel is the name given to a family of corrosion resistant steels containing a minimum of 12 % chromium. On contact with air, the chromium forms a thin oxide layer on the surface of steel. This is passive and resists corrosion. The austenic group of steels has superior corrosion resistance and is the most suitable for use in reinforced concrete.

3.6.5 Non-ferrous reinforcement

Currently a number of manufacturers who are developing non-ferrous reinforcement as an alternative to the conventional steel in traditional structures. A range of man -made fibers is used, the most common being glass, carbon and aramid. The fibers are used either in the form of ropes or combined with suitable resins to form rods. It may be seen that aramid, glass and carbon fibers have ultimate tensile stresses well in excess of that of reinforcing steel and well resistance to corrosion.

3.6.6 Corrosion inhibitors

Certain admixtures can be made to inhibit corrosion of the reinforcement in the presence of chlorides. Example:

(a) The addition of calcium nitrate extends the time to corrosion initiation.
(b) Total corrosion of samples with calcium nitrate is substantially less
(c) The corrosion rate, once corrosion is initiated is less with calcium nitrate.

3.6.7 Concrete coatings

A concrete surface coating of silane-siloxane type was evaluated for its corrosion performance. By several test conducted and the results show the chloride penetration depth is minimized in coated specimen compared with uncoated specimen.

3.6.8 Cathodic protection

Cathodic protection is a technique by which the electrical potential of the steel is increased to a level at which corrosion cannot take place. It is widely used for both steel and concrete offshore structures, while on land it has been used for the protection of pipelines and similar structures. Two different methods are employed, an impressed current and the use of sacrificial anodes. In the first the structure is connected to the negative terminal of a DC power source, ideally using an anode, which does not corrode. In the second the reinforcement is connected to anodes with a more negative corrosion potential than steel, such as Zinc or aluminum. In both cases electrical continuity of the reinforcement is required. The principal of cathodic protection have been used to remove chlorides from contaminated concrete.

3.6.9 Electrochemical chloride removal

This is another emerging area with lot of potentialities. This technique needs only a temporary installation lasting few days. 20 to 50 % of the chloride present in concrete can be removed. An electrical current in the range of 1 to 5 A/m^2 is needed.

3.6.10 Improving the concrete

Codes and standards aim to achieve good durability of reinforced and prestressed structures in aggressive environments by specifying:

(a) High cement content
(b) Low water/cement ratio
(c) Suitable minimum thickness of cover to the reinforcement
(d) Careful curing.

3.7 RELATING SYMPTOMS TO CAUSES OF DISTRESS AND DETERIORATION

Given a detailed report of the condition of the concrete in a structure and a basic understanding of the various mechanisms that can cause concrete deterioration, the problem becomes one of relating the observations or symptoms to the underlying causes. When many of the different causes of deterioration produce the same symptoms, the task of relating symptoms to causes is more difficult than it first appears. Although there will usually be a combination of causes responsible for the damage detected on a structure, this procedure should provide a starting point for an analysis.

3.7.1 Evaluate structure design to determine adequacy

First consider what types of stress could have caused the observed symptoms. For example, tension will cause cracking, while compression will cause spalling. Torsion or shear will usually result in both cracking and spalling. If the basic symptom is disintegration, then overstress may be eliminated as a cause. Second, attempt to relate the probable types of stress causing the damage noted to the locations of the damage. For example, if cracking resulting from excessive tensile stress is suspected, it would not be consistent to find that type of damage in an area that is under compression. Next, if the damage seems appropriate for the location, attempt to relate the specific orientation of the damage to the stress pattern. Tension cracks should be roughly perpendicular to the line of externally induced stress. Shear usually causes failure by diagonal tension, in which the cracks will run diagonally in the concrete section. Visualizing the basic stress patterns in the structure will aid in this phase of the evaluation. If no inconsistency is encountered during this evaluation, then overstress may be the cause of the observed damage. A thorough stress analysis is warranted to confirm this finding. If an inconsistency has been detected, such as cracking in a compression zone, the next step in the procedure should be followed.

3.7.2 Relate the symptoms to potential causes

Depending upon the symptom, it may be possible to eliminate several possible causes. For example, if the symptom is disintegration or erosion, several potential causes may be eliminated by this procedure.

3.7.3 Eliminate the readily identifiable causes

From the list of possible causes remaining after symptoms have been related to potential causes, it may be possible to eliminate two causes very quickly since they are relatively easy to identify. The first of these is corrosion of embedded metals. It will be easy to verify whether the cracking and spalling noted are a result of corrosion. The second cause that is readily identified is accidental loading, since personnel at the structure should be able to relate the observed symptoms to a specific incident.

3.7.4 Analyze the available clues

If no solution has been reached at this stage, all of the evidences generated by field and laboratory investigations should be carefully reviewed. Attention should be paid to the following points:

(a) If the basic symptom is that of disintegration of the concrete surface, then essentially three possible causes remain: chemical attack, erosion, and freezing and thawing. Attempts should be made to relate the nature and type of the damage to the location in the structure and to the environment of the concrete in determining which of the three possibilities is the most likely to be the cause of the damage.

(b) If there is evidence of swelling of the concrete, then there are two possibilities: chemical reactions and temperature changes. Destructive chemical reactions such as alkali-silica or alkali-carbonate attack that cause swelling will have been identified during the laboratory investigation. Temperature-induced swelling should be ruled out unless there is additional evidence such as spalling at joints.

(c) If the evidence is spalling and corrosion and accidental loadings have been eliminated earlier, the major causes of spalling remaining are construction errors, poor detailing, freezing and thawing, and externally generated temperature changes. Examination of the structure should have provided evidence as to the location and general nature of the spalling that will allow identification of the exact cause.

(d) If the evidence is cracking, then construction errors, shrinkage, temperature changes, settlement and movement, chemical reactions, and poor design details remain as possible causes of distress and deterioration of concrete. Each of these possibilities will have to be reviewed in light of the available laboratory and field observations to establish which is responsible.

(e) If the evidence is seepage and it has not been related to a detrimental internal chemical reaction by this time, then it is probably the result of design errors or construction errors, such as improper location or installation of a water stop.

3.7.5 Determine why the deterioration has occurred

Once the basic cause or causes of the damage have been established, there remains one final requirement: to understand how the causal agent acted upon the concrete. For example, if the symptoms were cracking and spalling and the cause was corrosion of the reinforcing steel, what facilitated the corrosion? Was there chloride in the concrete? Was there inadequate cover over the reinforcing steel? Another example to consider is concrete damage caused by freezing and thawing. Did the damage occur because the concrete did not contain an adequate air-void system, or did the damage occur because the concrete used was not expected to be saturated but, for whatever reason, was saturated? Only when the cause and its mode of action are completely understood should the next step of selecting a repair material be attempted.

OBJECTIVE TYPE QUESTIONS

1. The shrinkage in concrete develops due to
   (a) Loss of water by evaporation
   (b) Hydration of cement
   (c) Withdrawal of water stored in unsaturated air voids in concrete
   (d) All the above.

2. Shrinkage in concrete increases with
   (a) Decrease in humidity of the surroundings
   (b) Increase in cement content
   (c) Increase in w/c ratio
   (d) All the above

3. Identify the false statement of the following
   (a) Plastic shrinkage is not affected by the loss of water from the surface of concrete
   (b) Plastic shrinkage of concrete is caused by the suction of dry concrete below it
   (c) Drying shrinkage is developed due to withdrawal of water from hardened concrete stored in unsaturated air voids
   (d) Initial drying shrinkage is not recoverable even after a prolonged storage in water.

4. Shrinkage of concrete can be reduced by
   (a) Using non-absorbent formwork
   (b) Using low water/cement ratio
   (c) Using of presaturated aggregate
   (d) All the above

5. The rate of carbonation shrinkage depends
   (a) Moisture content concrete
   (b) Permeability of concrete
   (c) Accumulation of CO_2
   (d) All the above

6. Quantify the correct statement
   (a) Size and shape of concrete element have no effect on the shrinkage of concrete.
   (b) The ratio of volume to its exposed surface has no effect on the shrinkage of concrete
   (c) The magnitude of shrinkage varies with the shape and size of concrete element
   (d) Relative humidity of the surroundings has no effect on the shrinkage of concrete

7. Identify the false statement
   (a) Carbonation of concrete increases its weight.
   (b) Hydrates present in the gel contribute substantially to the shrinkage.
   (c) Carbonation increase the strength of concrete
   (d) Carbonation of concrete neutralizes the alkaline nature of hydrated cement paste.

8. Identify the correct statement
   (a) Concrete made with sand stone has the maximum shrinkage
   (b) Concrete made with limestone has no effect on shrinkage
   (c) All aggregates have equal shrinkage
   (d) Quality of cement paste has no influence in concrete shrinkage

9. The permeability of concrete reduces with the
   (a) Carbonation of concrete
   (b) The strength of cement paste
   (c) Decrease in the porosity of concrete
   (d) All the above

10. The durability of concrete is due to resistance to
   (a) Chemical attack
   (b) Deterioration from environmental conditions
   (c) Internal disruptive forces
   (d) All the above

11. High frost-resistance concrete should be
   (a) Free from cracks
   (b) Dense
   (c) Air entrained
   (d) All the above

12. Concrete to be used in contact with alkaline soil or alkaline water should have
   (a) High w/c ratio
   (b) Rich mix
   (c) High alumina content
   (d) Low w/c ratio

13. The porosity of concrete mainly depends upon
   (a) Grading of aggregate
   (b) Cement content
   (c) Quantity of mixing water
   (d) All the above

14. Identify the incorrect statement of the following
   (a) The durability of concrete is independent of its compressive strength
   (b) The compressive strength of concrete is a measure of durability of a concrete
   (c) Environmental conditions also have a great influence on the durability of concrete
   (d) In recent time more importance is given to durability aspect of concrete

15. Identify the correct statement of the followings
   (a) There is no difference in the microstructure of concrete made with low and high w/c ratio.

(b) The chloride ions diffuse at much lower rate in low w/c ratio concrete than that of high w/c ratio concrete
(c) The permeability of concrete increases if made with low w/c ratio
(d) Use of low w/c ratio has not been found effective in controlling the corrosion of reinforcement of concrete.

16. Identify the false statement of the following
(a) Higher w/c ratio cause higher permeability and higher volume changes in concrete
(b) To produce dense and impermeable concrete lowest possible w/c ratio should be used
(c) Water /cement ratio has no effect on the durability of concrete
(d) Higher w/c ratio develops cracks in concrete finally disintegrate it.

17. Identify the correct statement of the followings
(a) The permeability of gel pores is about 1/100 of cement paste as a whole
(b) Gel pores have no contribution in the permeability of cement paste
(c) W/c ratio is the main contributing factor to the permeability of concrete
(d) All are correct

18. Identify the false statement of the followings
(a) Magnesium sulphate is more reactive and causes maximum damage
(b) Dry sulphates react most violently and cause greater damage
(c) Ammonia sulphate is present in agricultural soils and water, due to the use of fertilizers and sewage and industrial effluent cause damage.
(d) The extent of sulphate attack depends upon the concentration of sulphate solution and permeability of concrete.

19. The sulphate attack can be controlled by
(a) Use of air entrainment
(b) Use of high alumina cement
(c) Use of pozzolana
(d) All the above

20. The reaction between the active silica constituents of aggregate and alkalis of cement is called
(a) Alkali aggregate reaction
(b) Soda equivalent
(c) Flocculated reaction
(d) Carbonation reaction

21. The alkali-aggregate reaction is affected by
(a) Porosity of the aggregate particles
(b) Alkali cement content
(c) Fineness of cement particles
(d) All the above

22. Alkali-aggregate reaction can be prevented by
    (a) Adding finely powdered silica to the concrete mix
    (b) Cutting the contact of concrete from external source of moisture
    (c) Reducing the alkali content of cement to less than 0.6 %
    (d) The entire above are correct.

23. The reaction of carbon dioxide with hydrated cement is known as
    (a) Corrosion
    (b) Silica-alkali reaction
    (c) Carbonic reaction
    (d) Unsoundness

24. Carbonation is the reaction between
    (a) Hydrated cement and hydrogen peroxide
    (b) Hydrated cement and carbon dioxide
    (c) Hydrated cement and sulphur dioxide
    (d) Hydrated cement and calcium hydroxide

25. The rate of carbonation depends upon
    (a) Relative humidity of ambient medium
    (b) Permeability of concrete
    (c) Grade of concrete
    (d) All the above

26. Identify the false statement of the following
    (a) The depth of carbonation of protective concrete remains constant
    (b) The depth of carbonation will be more of ill cured and high w/c ratio concrete
    (c) Depth of carbonation of un protective concrete is independent of time
    (d) The depth of carbonation of concrete is dependent on the depth of cover

27. Identify the correct statement of the followings
    (a) Chloride attack on concrete is more serious than any other cause
    (b) The volume of corroded reinforcement increases up to 6 times of its original volume
    (c) The chloride content in cement should not be more than 0.1%
    (d) All the above are correct

28. Identify the false statement of the following
    (a) Deterioration of concrete in sea water is caused due to expansion of concrete
    (b) Deterioration of concrete in sea water is caused due to erosion of concrete mass
    (c) Deterioration of concrete in sea water is caused due to leaching of some lime content
    (d) The most severe attack on concrete in sea water takes place just above the high water level

29. Sea water effects on concrete may be minimised
   (a) By the use of cement with low C_3 A content
   (b) By the use of pozzolanic materials
   (c) By providing adequate cover to reinforcement
   (d) All the above

30. To protect reinforcement from corrosion in moderate exposure condition the main cover should be
   (a) 80 mm (b) 60 mm
   (c) 30 mm (d) 25 mm

31. Identify the incorrect statement
   (a) Durability of concrete is its resistance to the deteriorating influences of both internal and external agencies
   (b) Durability of concrete is not influenced by the permeability of concrete
   (c) The gradual deterioration of concrete by chemically aggressive agents is called corrosion of concrete
   (d) Corrosion of steel reinforcement causes internal stresses which promote destruction of concrete.

32. Identify the false statement
   (a) The principal form of destruction of concrete is decomposition due to leaching action
   (b) The principal form of destruction of concrete is the chemical reaction between hardened concrete and chemicals of solution
   (c) Grinding action of land storm
   (d) Increase in the volume of solid phase due to crystallization with in the pore structure

33. The deterioration of concrete can be prevented by
   (a) Limiting w/c ratio
   (b) Making a dense concrete
   (c) Impregnating the pores of concrete with a suitable polymer
   (d) All the above

34. Identify the false statement
   (a) The corrosion of reinforcement can be prevented by the use of corrosion reducers or coating steel or concrete
   (b) The corrosion of reinforcement can be prevented by the proper design of concrete for the intended environmental exposure
   (c) Increase in thickness of concrete cover to reinforcement
   (d) By the use of stainless steel

35. The corrosion of embedded steel reinforcement is more rapid when the concrete member is immersed in

(a) Alkaline solution
(b) Acidic solution
(c) Water with dissolved oxygen
(d) Sea water

36. To reduce the corrosion of reinforcement, the chloride ions should be limited to its critical or threshold value (% by mass of cement)
(a) 0.5 (b) 0.1
(c) 0.15 (d) 0.2

37. Identify the incorrect statement of the following
(a) Structures immersed in water permanently show little corrosion of the reinforcement
(b) In regions of relative humidity of 50 % or less the corrosive action on reinforcement is negligible
(c) The distress due to corrosive action may be in the form of deep pitting and severe loss of cross section of reinforcement
(d) For prestressed concrete the total amount of chloride ions should be limited to 0.1% by mass of cement

38. Which of the following compounds may be used as iron corrosion reducer
(a) Fly ash
(b) Calcium sulphate
(c) Potassium dichromate, zinc and lead chromates
(d) Calcium and sodium chloride

39. Identify the correct statement
(a) Corrosion to steel not take place at relative humidity less than 40 %
(b) The product of corrosion occupy volume less than 4 times the original volume of steel
(c) Rust contains large quantity of chloride
(d) Ammonium salts can be used for deicing with advantage

40. Identify the correct statement of the following
(a) Gel pores contribute much to the permeability of concrete
(b) The solubility of calcium sulphate is maximum
(c) Magnesium sulphate is least reactive of all sulphates
(d) The permeability of cement paste is governed by w/c ratio and degree of hydration

41. Measurement of permeation properties of concrete would result in
(a) Assessment of durability of concrete
(b) Assessment of strength of concrete
(c) Assessment of fire resistance of concrete
(d) Assessment of cracks of concrete

42. Chemical damage on concrete dependent on the
(a) Body of the concrete
(b) Permeability of the surface layer of concrete
(c) Evaporation of water on the concrete
(d) Leakages in form work

43. Crystallization of acids in concrete can cause
(a) Wetting shrinkage
(b) Drying shrinkage
(c) Expansive disruption
(d) None of the above

44. Chemical reaction in concrete results in
(a) Decrease of the porosity
(b) An increase of the porosity
(c) An increase of the porosity and permeability
(d) All the above

45. The major effect of ASR is a
(a) Reduction in compressive strength
(b) Improvement in tensile strength
(c) Reduction in compressive and tensile strength
(d) None of the above

46. Sulphate reaction is dependent on the
(a) Concentration of sulphate ions
(b) Cations present in the sulphate solution
(c) C_3A content of cement
(d) All the above

47. Carbonation is the effect of
(a) Oxygen in the air on cement products
(b) Carbon dioxide in the air on cement products
(c) Hydrogen in the air on cement products
(d) None of the above

48. The depth of carbonation in approximately proportional to
(a) Square root of time
(b) Cube root of time
(c) Multiple of time
(d) Reciprocal of time

49. The rate of carbonation dependent on the
(a) Cover to reinforcement
(b) Concentration of CO_2 in the atmosphere
(c) Alkali content in the concrete

(d) All the above

50. Leaching of lime compounds lead to the formation of salt deposits on the surface of the concrete known as
(a) Depassivation
(b) Efflorescence
(c) Crazing
(d) None of the above

51. Early efflorescence can be removed
(a) With acid treatment
(b) With grinding
(c) With a brush and water
(d) All the above

52. The initiation period is the time taken to initiate corrosion which can be caused
(a) Either by ingress of CO_2 or chloride ions
(b) Either by ingress of O_2 or hydrogen ions
(c) Either by ingress of H_2O or sulphate ions
(d) None of the above

53. During the propagation period of corrosion depending on the
(a) Availability of oxygen and moisture
(b) Availability of oxygen and hydrogen
(c) Availability of CO_2 and moisture
(d) All the above

54. The factors which influence the corrosion initiation are
(a) The cover thickness
(b) The type of steel and critical chloride
(c) Presence of cracks
(d) All the above

55. The factors which influence the electrochemical corrosion process
(a) pH value
(b) Sulphate content
(c) CO_2 supply
(d) None of the above

56. Plastic settlement cracks can be minimised by
(a) Use of air-entraining admixtures
(b) Water reducing admixtures
(c) Re-vibration
(d) All the above

57. Formwork movement cracks can be prevented by
(a) A proper design of the forms
(b) A proper design of mix
(c) Over compaction
(d) All the above

58. The development of a network of fine random cracks on the surface of concrete is called
(a) Popouts
(b) Crazing
(c) D-Cracking
(d) None of the above

59. Honeycomb may be caused by
(a) In adequate consolidation
(b) Presence of excess water in concrete
(c) Leaky forms
(d) All the above

60. A conical shaped hole in the surface of concrete with a portion of coarse aggregate particle exposed is called
(a) Popouts
(b) Honey comb
(c) Crazing
(d) D-Cracking

61. Popouts can be prevented by avoiding
(a) Less cement content
(b) More w/c ratio
(c) Aggregation which came them
(d) All the above

62. The freeze and thaw cracks filled with a dark deposit of calcium carbonate are called
(a) D-cracking
(b) Joint cracking
(c) Crazing
(d) All the above

63. Differential settlement of foundation causes
(a) Compressive cracks
(b) Tensile cracks
(c) Shear cracks
(d) None of the above

64. Concrete exhibit a time dependent deformation known as
(a) Shrinkage

(b) Creep
(c) Elastic modulus
(d) All the above

65. Creep of concrete increases with increase in
(a) Water and cement content
(b) W/c ratio
(c) Temperature
(d) All the above

66. Creep of concrete decreases with increase in
(a) Humidity of the atmosphere
(b) Age of material
(c) Time of loads
(d) All the above

67. In concrete creep results in a substantial increase in
(a) Elastic deformation
(b) Plastic deformation
(c) Ultimate deformation
(d) None of the above

68. Creep leads to formation of cracks in
(a) Brick masonry of framed structures
(b) Brick masonry of load bearing structures
(c) Concrete framed and load bearing structures
(d) All the above

69. The common distress in concrete structures
(a) Cracking,
(b) Spalling
(c) Disintegration
(d) All the above

70. Freeze and thaw means
(a) Wetting and drying
(b) Drying and wetting
(c) Heating and cooling
(d) All the above

71. Popouts start to appear during the first
(a) Summer following construction
(b) Winter following construction
(c) Spring following construction
(d) None of the above

72. The predominant process of corrosion mechanism is
(a) Chemical, (b) Electrochemical
(c) Physical process (d) All the above

73. The necessity of IS 456–2000 code revision in terms of
(a) Durable aspects (b) Strength aspects
(c) Permeable aspects (d) None of the above

74. The structure exposed to atmospheric condition is
(a) Mild (b) Moderate
(c) Severe (c) All the above

75. The presence of cracks in concrete drastically
(a) Reduces the initiation time of corrosion
(b) Increases the initiation time of corrosion
(c) Reduces the propagation time of corrosion
(d) None of the above

ANSWERS

1. d	2. d	3. a	4. d	5. d	6. c	7. b	8. a	9. d	10. d
11. d	12. c	13. d	14. a	15. b	16. c	17. d	18. b	19. d	20. a
21. d	22. d	23. c	24. b	25. d	26. c	27. d	28. a	29. d	30. c
31. b	32. c	33. d	34.d	35. b	36. c	37. d	38. c	39. a	40. a
41. a	42. b	43. c	44. c	45. c	46. d	47. b	48. a	49. d	50. b
51. c	52. a	53. a	54. d	55. a	56. d	57. a	58. b	59. d	60. a
61. c	62. a	63. c	64. b	65. d	66. d	67. a	68. d	69. d	70. a
71. b	72. d	73. a	74. d	75. a					

SHORT ANSWER QUESTIONS

1. What is the use of cement concrete?
2. What is meant by durable structure?
3. Define the term durability.
4. What are the factors influence the durability of the concrete?
5. Define permeable and impermeable concrete
6. Relate permeability and durability.
7. Is permeability a single function?
8. What are the determinations of permeability?
9. What are the changes in IS 456 – 2000 in the view of durability?
10. What is the measure of durability?
11. What is the effect of external causes on durability of concrete?
12. What is the effect of internal causes on durability of concrete?
13. Name the reasons for deterioration of concrete.
14. What are the distresses in concrete structures?
15. Name the physical causes on concrete
16. Name the chemical causes on concrete
17. Define shrinkage in concrete.
18. What are the classifications of shrinkage?
19. What are the factors influencing the drying shrinkage?
20. Define physical loss and chemical loss in concrete structures?
21. What is differential shrinkage?
22. Explain the term Autogenous shrinkage.
23. Define the term carbonation shrinkage.
24. What are the preventive measures of drying shrinkage?
25. Name the defects normally observed in concrete structures.
26. What is the rate of plastic shrinkage?
27. What is meant by plastic settlement cracks?
28. Define plastic cracks.
29. What is the result of poor construction practices?
30. What are the preventive measures used in thermal movement of concrete?
31. What is the effect of fire on concrete?
32. How to minimize frost damage?
33. D – cracking define.
34. What is the role of weathering on concrete?
35. What is the effect of creep on concrete structures?
36. Classify the chemical attack on concrete.
37. What is the result of crystallization of salts in pores?
38. How to prevent the salt attack?
39. AAR and ASR–Expand

40. What is the result of water permeates by the aggregates?
41. What is meant by swelling?
42. How to classify the damages in structure?
43. What is known as abrasion and erosion in concrete?
44. Define the defect popouts.
45. How to prevent popouts?
46. What is meant by carbonation and what are the factors influenced?
47. What is the rate of carbonation?
48. What are cavitations in concrete?
49. What is meant by crazing?
50. What are the preventive measures used to reduce crazing?
51. Define 'Honey Combing'
52. How to prevent honey combing?
53. What is the effect of hydrolysis & leaching?
54. How to remove early and heavy efflorescence?
55. How sulphate attack occurs and what are the parameters considered?
56. How to reduce the sulphate attack on concrete?
57. What is the effect of chloride attack and How to prevent it?
58. How can be reduced the acid attack in concrete?
59. What is the effect of corrosion?
60. What are the factors influence the corrosion process?
61. Which is ultimate result of rebar corrosion?
62. What are the variables considered in steel corrosion?
63. Chemical process of corrosion – explain
64. Electrochemical process of corrosion – explain
65. Physical process of corrosion – explain
66. Name the periods of corrosion process
67. What are the factors influence the corrosion initiation?
68. In what factors changes in IS 456 – 2000 on durability of concrete?
69. Name the process in propagation period
70. How the bare steel prevent from corrosion?
71. What is meant by pitting corrosion?
72. How the embedded steel get corrosion?
73. Name the physical symptoms of corrosion on beams and slabs.
74. Name the corrosion protection techniques
75. What are the protective measures consider by the durable design?

THEORY QUESTIONS

1. Discuss the factors affecting permeability of concrete structures.
2. Define defect. Explain the defects in concrete.
3. What are the defects normally observed in concrete structures? Discuss briefly their remedies.
4. What are the causes for distress in RC members? Discuss about the elements that catalyze the distress in RC structural elements.
5. What are all the agencies causing deterioration in RC structures? Explain the type of distress caused by the different agencies.
6. Explain clearly about the mechanism of freezing and thawing.
7. Explain various preventive measures in controlling distress RC structures.
8. Explain in detail plastic shrinkage and drying shrinkage.
9. Discuss briefly the strength and behaviour of concrete subjected to high temperatures.
10. Short notes on the following movements:
    (a) Sub–grade movement
    (b) Formwork movement
    (c) Foundation movement
11. Explain about the poor construction practices and errors in design of concrete structures.
12. Discuss briefly the effect of fire on the properties of concrete.
13. Write briefly about chemical attack on the concrete.
14. What is sulphate attack? Explain about the causes and effects of sulphate attack.
15. Explain about hydrolysis and leaching of concrete.
16. Explain the causes and effects of chloride attack and sulphate attack.
17. What is carbonation? Explain the various factors which affect the rate of carbonation.
18. Explain about the causes and effects of alkali reaction in concrete.
19. Explain about seawater attack on concrete.
20. Clearly explain about the chemical reactions involved in the corrosion reaction of steel of concrete.
21. Define corrosion. Tabulate the IS codal provisions for cover of various exposure conditions.
22. Explain clearly the various stages of corrosion.
23. Explain the factors affecting corrosion of steel in concrete.
24. What are the factors influencing corrosion of reinforcement? Explain the damages in RC structures due to corrosion of reinforcement.
25. Explain the basic principle of corrosion. Explain the various corrosion protection techniques.

PART II

DAMAGE ASSESSMENT

CHAPTER 4
Destructive Testing Systems

4.1. INTRODUCTION

Damage takes place due to natural or man-made causes. Even structures built in recent past have shown extensive damage and distress due to various causes, giving rise to problems both investigation and repair. Studies on durability of various construction materials and on methods of repairing and renovating the damaged structures are increasingly attracting the attention of engineers and researchers. Research efforts have resulted in the development of investigation procedures & instruments and repair materials & techniques. Concrete is the ace card for the construction industry. Civil and Structural engineers design concrete structures for an anticipated life of 100 years. But due to various factors durability of concrete became a matter of serious concern and today repairs to concrete structure are given importance and different methods are being developed for repairs.

Before going into details of this method it is necessary to discuss the probable causes of decay of concrete and its testing to evaluate the degree of deterioration before applying this method. Diagnosis of the damage is of greatest importance. Before remedies can be correctly prescribed the illness must be diagnosed and before an accurate diagnosis is possible the doctor has to have a thorough knowledge of the disease, its various symptoms and treatments. Similarly concrete repair expert needs similar knowledge in this field, so that the repairs he prescribes for the distressed/troubled structure are long lasting and successful. He must know the cause for the deterioration and distress of the existing structure rather than just the symptoms of each. A very methodical approach is required for this purpose. Data collection and analysis at each of the following stages will lead to methods and stages of repairs.

4.1.1 Purpose of assessment

During a disaster an engineer may be tasked with a variety of functions, some of which may not be within the purview of his/her expertise. One of these could be the assessment and monitoring of the structural stability of a reinforced concrete structure. It is the premise of this instruction that this can be performed in a cursory fashion by a non-structural engineer (in the absence of a competent structural engineer), within a relatively static environment. Therefore, the purpose of this instruction is to provide a fundamental understanding that will allow a nonstructural engineer to perform a cursory safety assessment/monitoring of a reinforced concrete structure.

Reinforced Concrete Structures:

Reinforced concrete is comprised of two basic materials, steel and concrete. The two materials work in a synergistic fashion when constructed properly to provide composite components which have very strong structural characteristics. Reinforced concrete is commonly used in structures designed for heavy use and long life, such as governmental and institutional buildings and public works structures. Concrete has a great capacity to support compressive loads. However, concrete has only a limited capacity to support tensile loads. Steel has a great capacity to carry both compressive and tensile loads. However, it is expensive to use as a single structural element and is prone to degradation (*e.g.*, rusting) in certain environments. Therefore, in order to construct structural elements which are long lasting, economical, and have both compressive and tensile capacity, steel is combined with concrete to harness the compressive strength of concrete and the tensile strength of steel. Beams are horizontal structural components that support floors, ceilings, roofs, or decks (i.e., bridge and parking decks). The loads carried by a beam are primarily perpendicular to the longitudinal axis of the beam. As load is applied to a beam it generally tends to cause bending of the beam, with the center of the unsupported section being forced away from the applied load.

This bending creates compressive forces (compression) in the upper depth of the beam and tensile forces (tension) in the lower depth of the beam. A typical reinforced concrete beam is designed to allow the compressive forces to be carried by the concrete material and the tensile forces to be carried by the steel. Columns are vertical structural components that support beams and other structural elements. The compressive loads carried by columns are primarily parallel to the vertical or longitudinal axis of the column. As a uniform concentric load is applied to a column, compressive loads occur within the column's cross-section, and are distributed between the concrete and longitudinal steel reinforcement. When an eccentric load is applied to a column bending will occur in the column, with outer sections of the column being in tension and inner sections being in compression. The tension and compression loads within the column are carried by vertical reinforcement and concrete. Columns are usually designed with a ring of structural steel around the outer perimeter to provide confinement of the concrete in case of column failure. Concrete structures are built using cast-in-place, precast, or a combination of either procedure. Structures built prior to the 1975 were built using non-ductile concrete methods and hence have more structural damage due to unforeseen events. Cast-in-place buildings are erected using temporary forms that concrete is placed into and then forms removed. Precast buildings elements are generally formed off-site and then shipped to be assembled on-site using some type of anchoring device. Cast-in-place structural failures will occur throughout the concrete member and/or its connections while failures in a precast building member will typically occur at the joints and/or connections.

4.1.2 Rapid structural safety assessment

The objective for the rapid structural safety assessment is to quickly inspect and evaluate the concrete structure and determine if the damaged structure is unsafe for personnel within the building and rescue personnel accessing the building. Two primary concerns need to be considered when performing this assessment of the structure that has sustained structural damage. This includes a quick evaluation of the building structural components (*e.g.*, beams, columns, decking, etc.) and of the building nonstructural components (*e.g.*, structural debris, partitions, ceilings, glass, pipe anchoring, electrical/mechanical equipment anchoring, etc.). If there are any visual signs of structural and/or nonstructural damage, then the specific building area needs to be isolated, secured, and marked as UNSAFE. The on-scene commander should be informed and the area remained in this UNSAFE condition, until a structural engineer proves otherwise. The rapid structural damage assessment would note the major failures within the structure including the major structural elements of beams, columns, and roof and floor decks. Typical failures would be found at the connections of the major structural elements, or at elements that no longer have adequate vertical support (*e.g.*, unsupported roof and floor decks that are now cantilever elements.) Indications would include cracking, spalling (i.e., loss of concrete from an exterior surface), and/or complete loss of all or part of a structural element. The on-scene commander should be notified immediately of the risk, and the area secured and marked UNSAFE.

The rapid nonstructural damage assessment would note the major failures within the building structure envelope including such items as structural debris, partitions, ceilings, glass, piping, and electrical/mechanical equipment. Concrete is a brittle material and, therefore, has a tendency to fragment into small, dense, hard pieces with rough edges. Many of these fragments may be precariously lying near or hanging from the exposed building edges and some fragments may be barely attached to exposed reinforcement steel. Settlement or shifting of the damaged structure may cause fragments to fall resulting in serious harm to personnel and/or additional damage to the remaining structure. The issue of potential harm to personnel from concrete fragments and other building materials (such as glass, and other items that may have been in and around the structure) is exacerbated by activities that may be occurring in and around the remaining structure by rescue personnel operating throughout the building, along with rescue and/or press rotary-wing aircraft operating around the building. These activities may tend to cause accidental (*e.g.*, physical interaction with something causing debris to fall) or inadvertent (*e.g.*, vibration or rotor wash from rotary-wing aircraft causing debris to fall) mishaps. Typical failures in other nonstructural elements would be found where they originally were attached and/or secured. Failures would include anchoring tensile or shear failure, nonstructural element damage, and potential of future damage due to gravity loads and/or inadequate support bracing. The on-site commander should be notified immediately of the risk, and the area secured and marked UNSAFE.

Depending on the structural and nonstructural elements within the building matrix and their relation to rescue/recovery activities, there may be varying importance on the ability of these elements to provide adequate support. For example, failure of a column at ground level may cause failure of the remaining structure above that columns, or failure of a beam above an area where rescue/recovery workers are operating may at least cause collapse of a floor (and its contents) into the rescue/recovery area. Therefore, the rapid structural safety assessment should focus on those remaining structural elements of greatest importance to the remaining structure and the safety of rescue/recovery workers. The ability to perform a rapid structural safety assessment is likely to require a good flashlight (or headlamp), because adequate lighting often is not available in or around the remaining structure. In addition, a notebook, pen or pencil, spray paint (for making annotations on or around the compromised structure) and a camera (if available) will be needed for the assessment. The locations of any structural and nonstructural elements should be documented via a written description and an associated sketch or photograph (digital) for use in more detailed assessment/analysis following the rapid structural safety assessment. Personnel performing the rapid structural safety assessment in and around the remaining structure should wear appropriate personal protective equipment (PPE). At a minimum PPE should include a hardhat, steel-toed boots (with steel shanks), gloves, and respiratory protection from dust. The specific site conditions may dictate further PPE. **The rapid structural assessment should be performed in the following order:**

(a) Review the entire outside of the structure.

(b) Enter the building only if necessary to determine extent of damage.

(c) Determine what degree of damage found in the structural and nonstructural elements.

(d) Secure all areas that need to be isolated and post UNSAFE signage.

4.1.3 Monitoring

A monitoring program can be established during the rapid assessment effort. Monitoring of potentially dangerous debris, unsecured equipment, structural cracks and failures until any mitigation/removal effort is complete may be simple periodic surveillance of the affected areas. The primary mode of failure for nonstructural elements within a compromised structure is due to the potential future movement of these elements. All unsecured elements need to be documented and recorded so that any future movement of these elements can be clearly established. Two primary modes of failure for structural elements within a compromised structure are shear and bucklings. Shear failure implies failure generally along a plane (*e.g.*, the failure of a deck slab that is no longer properly supported, could fail in a clean shear break). Buckling failure is a compressive failure that results in the element collapsing (*e.g.*, the lateral displacement of a column).

Failure events may occur suddenly (*e.g.*, the clean break of a cantilevered floor or roof section with no structural steel in the upper depth) or could be subtle as remaining structural elements adjust to the loading from the remaining structure. This adjustment may be radically different from the original design loads of the structural elements, as well as their potentially different structural composition. For example, they may have cracking and/or spalling affecting the carrying capacity of the element and/or the characteristics of the materials comprising the remaining structural elements may have changed. Therefore, monitoring is critical to expose subtle changes that may have the potential to result in structural failure. Monitoring individual structural elements for possible shear failure9 can be performed simply by marking a straight line perpendicularly across cracks of concern using a pencil or fine-tip permanent marker. Any slippage along the crack (possible shear failure) would then be indicated by a disjunction of the straight line. The length of cracks also should be monitored by drawing an arrow on the affected structural element with the tip of the arrowhead placed at the clearest discernible termination point of the crack. The date and time that the arrow was placed also should be clearly noted on the structural element such that its association with the arrow is unambiguous. This will document any extensions of cracks which may indicate possible weakening of the affected structural element with the potential end result being buckling failure. This data could then be utilized to make decisions on necessary shoring or bracing efforts by engineers with structural experience.

Monitoring of the overall structure can be accomplished by establishing a matrix of witness marks or survey marks on the exterior of the remaining structure. A line of markings can be made vertically through a line of columns and/or horizontally through a line of beams. This can be accomplished with the use of a transit, and bright-colored spray paint. It is desirable to have clear visibility of all markings from a common transit station that can be permanently established, without having to be periodically broken down during the monitoring effort. The intent of the placement of the witness marks is to allow monitoring of the various sections of the remaining structure relative to the others. This allows a relatively easy, fast, and safe means of determining if a given section is moving relative to the rest of the remaining structure and may illuminate significant and possibly imminent structural issues. The monitoring effort should be constant with data logging at regular time intervals. The interval will be dependent upon available resources, the stability of the remaining structure, and the dynamics of the environment and rescue/recovery efforts within and around the remaining structure.

4.2 INVESTIGATION OF DAMAGE

The investigation of reasons for damaged to structures is largely a matter of gathering information by observation, studying records and asking questions, supplemented if necessary by a certain amount of testing, and then interpreting the information thus obtained. A systematic investigation of concrete structures is highly important and essential in order to:

(a) Find out the cause of damage
(b) Assess the condition of the structure in its damaged state and
(c) Formulate recommendations for repair and restoration

The general procedure for an investigation can be briefly summarized as given below:

(a) Documentation of damage
(b) Visual observation
(c) Measurements on geometrical parameters such as deflections & deformations in vertically & horizontally
(d) Experiments for evaluating material properties & behaviour, these includes type of testing
(e) Interpretation and analysis of test results
(f) Analysis of the building its damaged state taking in to account the test results obtained during investigation
(g) Formulation of repair measures
(h) Post – repair evaluation

4.2.1 Observation

The texture of a concrete surface may suggest the possibility of chemical attack by a general softening, leaching of the matrix or in the case of sulphate attack, whitening of the concrete. Rust stains often indicate corrosion of reinforcement but they may be caused by contamination of the aggregates with iron pyrites. If cracked concrete is broken out, the state of the crack surface gives useful information. General flaking of an exposed concrete surface suggests frost damage. In fire damaged structures, the colour of the concrete gives an indication of the maximum temperature reached. When concrete is cracked the crack pattern can be informative. A mesh pattern suggests drying shrinkage, surface crazing, and frost attack or in rare cases, alkali–aggregate reaction. Exudations from cracks may be the result of water passing through the concrete and washing out calcium salts, or they may be gets from alkali–aggregate reaction. Relatively straight cracks usually indicate excessive but fairly uniform tensile strain, and they should be studied in relation to the probable stress pattern in the member concerned.

Cracks caused by unidirectional bending will be widest in the zone of maximum tensile stress and will taper along their length, while cracks caused by direct tension will be of roughly uniform width. Continuing movement at a crack often produces crumbling at the edges. Popouts in concrete are usually associated with particles of coarse aggregate just below the surface. Highly absorptive particles may expand if serve frost occurs while they are saturated. Occasionally, coarse aggregates are contaminated with particles of lime or clay that expand with weathering.

4.2.2 Questioning

Records of mix proportions, sources of materials, cube test results, weather conditions etc. may be available, particularly for recent built structures but, even with these, reliable information may be difficult to obtain. It is always useful to ask questions of as many as possible of the people who were concerned with the design or construction. Their recollections may not be completely accurate and their accounts of events may conflict but, by questioning, one can often find out what actually happened as distinct from what should have happened. This applies not only to work on site during construction but also to subsequent use or misuse of the structure.

4.2.3 Assessment procedure

The first key step of investigation is to make a diagnosis of the defect or the failure. It is basically an inductive procedure and it requires abundant caution. The figure 4.1 indicates the flow chart diagram of damage assessment procedure. **Some important steps are:**

(a) Visual inspection
(b) Study of available documentation
(c) Estimation of actual loads and environmental effects
(d) Diagnosis

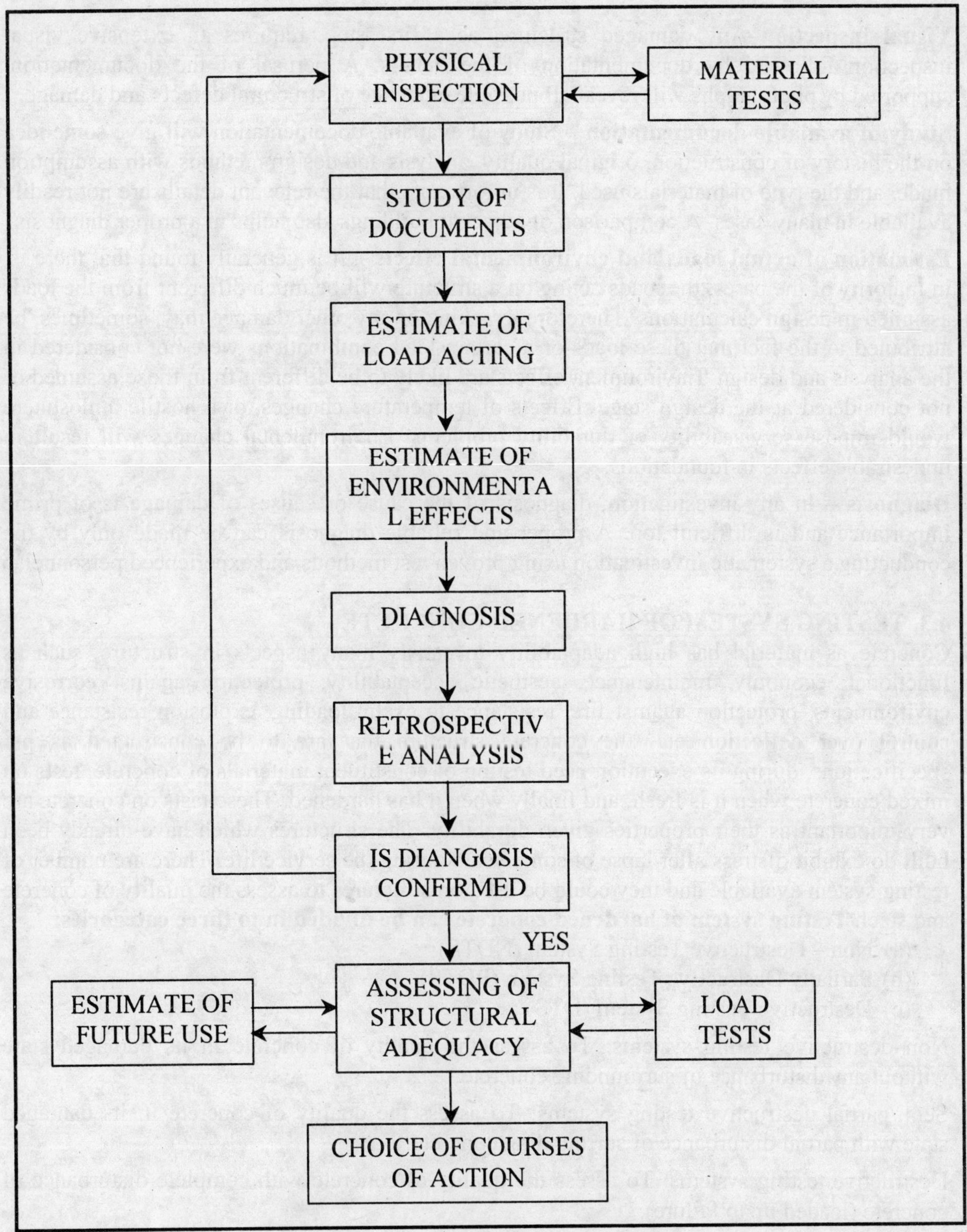

Figure 4.1. Assessment procedure for damage

Visual inspection–Any damaged structure, as a first step, requires an extensive visual inspection followed by documentation of the details. A perusal of the documentation, supported by photographs will reveal all possible evidence of structural defects and damage.

Study of available documentation – Study of available documentation will give some idea on the history of construction, original quality, analysis and design methods with assumption made, and the type of materials used. It is unfortunate that the relevant details are not readily available in many cases. A comparison of adjacent buildings also helps in a proper diagnosis.

Estimation of actual loads and environmental effects – It is generally found that there is, in majority of the cases, the loads acting on a structure will be much different from the loads assumed in design calculations. Therefore, cracking or any other damage may, sometimes, be attributed to the fact that these loads or a certain load combinations were not considered in the analysis and design. Environment effects are likely to be different from those assumed or not considered at the design stage. Effects of temperature changes, or a hostile atmosphere would impose serviceability or durability problems. Environmental changes will result in undesirable effects in foundations.

Diagnosis – In any investigation, diagnosis of the cause or causes of damage is of prime importance and is difficult too. A proper and reliable diagnosis can be made only by the conducting a systematic investigation using proven test methods and experienced personnel.

4.3. TESTING SYSTEM OF HARDENED CONCRETE

Concrete as material has high adaptability to satisfy many aspects in structure, such as functional, economy, maintenance, aesthetic acceptability, protection against corrosive environments, protection against fire, resistance to cyclic loading, explosion resistance and control over deflection etc. the concrete structure that are to be constructed as per specifications, during its execution need testing of constituent materials of concrete, tests on mixed concrete when it is fresh, and finally when it has hardened. These tests on concrete are very important as their properties given durability. The structures which have already been built do exhibit distress after lapse of some years during the service life. There are number of testing system available and they could be used on structures to assess the quality of concrete and steel. **Testing system of hardened concrete can be divided in to three categories:**

(a) Non – Destructive Testing System (NDTS)
(b) Partially Destructive Testing System (PDTS)
(c) Destructive Testing System (DTS)

Non–destructive testing systems: To assess the quality of concrete in its damaged state without any disturbance of surrounding concrete.

Semi/partial destructive testing systems: To assess the quality of concrete in its damaged state with partial disturbance of surrounding concrete.

Destructive testing systems: To assess the quality of concrete with complete disturbance of concrete (loaded up to failure).

The range of instruments is typically considered to be two parts. The first are non-destructive field tests of compressive strength. The second are tensile field tester systems to either determine the tensile strength of an overlay or bond material, or tensile strength of anchors embedded in the concrete. The first group is pure Non-Destructive Testing where the strength of the material is determined by correlation to another parameter more easily available and readily apparent. This is typically the hardness of the concrete or the resistance to penetration by either a pin or probe. The Windsor Probe, Windsor Pin and of Rebound Hammers all fall within this category. These are widely used standard tests and as such have seen use throughout the world.

The second set of instruments is our concrete tensile testers. These have been optimized to both test the strength of the anchors and repair overlay material. They can be used to test until failure or to simply verify that the material will not be affected by a specific amount of force. A number of considerations were taken into account when designing this line of products; include viscous damping of the resultant failure backlash, portability, and ruggedness.

4.4 EVALUATION OF CRACKS

The cracks in a structure are broadly classified in two categories: superficial cracks and structural cracks. The structural cracks may be active and dormant. A crack where a movement is observed to continue is termed active, whereas the crack where no movement occurs is termed dormant or static. **The following information may help in diagnosing the cracks:**

(a) Whether the crack is new or old,
(b) Type of crack, i.e. whether it is active or dormant,
(c) Whether it appears on the opposite face of the member also, pattern of the cracks,
(d) Soil condition, type of foundation used, sign of movement of ground, if any,
(e) Observations on the similar structures in the same locality,
(f) Study of specifications, method of construction used and the test results at the site, if any,
(g) Views of the designer, builder, occupants of the building, if any, and
(h) Weather during which the structure has been constructed.

From the above discussion it is evident that the cracking is a complex phenomenon.

As in the case of a medical practitioner prescribing medicine without thoroughly examining the patient, it is difficult for a repair engineer to advocate any repair technology without making a thorough investigation. Before proceeding with repair, the investigations should be made to determine the location and extent of cracking, the causes of damage, and the objectives of repair. Calculation can be made to determine stresses due to applied loads. For detailed information, the history of the structure, structural drawings and specifications, and construction and maintenance records should be reviewed. The objectives of repair include restoration and enhancement of durability, structural strength, functional requirements and aesthetics. **The evaluation of cracks is necessary for the following purposes:**

(a) To identify the cause of cracking.
(b) To assess the structure for its safety and serviceability.

(c) To establish the extent of the cracking.
(d) To establish the likely extent of further deterioration.
(e) To study the suitability of various remedial measures,
(f) To make a final assessment for serviceability after repairs.

Apart from visual inspection, tapping the surface and listening to the sound for hollow areas may be one of the simplest methods of identifying the weak spots. The suspected areas are then opened up by chipping the weak concrete for further assessment. The comparative strength of concrete in the structure may be assessed to a reasonable accuracy by non-destructive testing and by the tests on the cores extracted from the concrete. The commonly used non-destructive tests are the rebound-hammer test and ultrasonic-pulse-velocity test.

4.5 DESTRUCTIVE TESTING SYSTEMS

The most common destructive test is load test and is used to assess the strength of concrete structural elements. Load testing is sometimes used as an alternative method of assessing structural capacity. **Load tests are usually carried out for one of the following reasons:**

(a) There are still doubts about the satisfactory performance of the structure under load after a survey and local testing.
(b) It is difficult or impossible to determine adequate information about the structure and its materials.
(c) Verification of structural analysis in cases where the complexity of the structural form does not lend itself to rigorous analysis

Deficiencies in detail, material or construction are suspected and such deficiencies would mean that the normal procedures or assumptions on which structural analysis is based were not appropriate. After confirming that the reinforcement did not go in to plastic, it was decided to conduct full scale load tests on the beams, so as to determine the load carrying capacity of the cracked beams and compare it with original design strength before any remedial measures are suggested. Accordingly load tests were conducted on most distressed beams. **The following aspects were set for during the testing programme.**

(a) To conduct the load test on beams up to 1.25 times the designed live load.
(b) To monitor the deflections and recovery of the beams during incremental loading and unloading
(c) To compare the actual deflections with that of theoretical deflections

4.5.1 Assessment of existing concrete structures

It has been learnt that concrete structures require a closer inspection, not only immediately after construction but also periodically at a regular interval. The quality control measures during construction, generally, consist of workability tests on fresh concrete and cube compressive strength of concrete samples, after some specified days of curing. It is a well known fact that the results of the above tests do not reflect the true quality of the concrete, existing in the concrete structure because the quality of a concrete structure depends on many factors such as method of mixing, transporting, placing, compacting and curing of concrete. While concrete members with certain amount of imperfections can satisfy the requirements relating to strength and serviceability, such concrete may not satisfy durability requirements.

Assessment of quality of concrete is necessary to ensure that the quality of execution is satisfactory and to identify any deficiencies so that they can be rectified. This can be achieved only by conducting some in-situ tests on the structures besides visual inspection. The in-situ tests are non-destructive tests and partially destructive tests. These tests measure indirectly the strength of concrete except in the case of core test, where direct evidence on the condition of concrete and a measurement on compressive strength are possible. These testing methods are in use for evaluating existing concrete structures with regard to their strength apart from assessment and quality control of hardened concrete.

4.5.2 Direct load test

This is a method of assessing the strength of the in-situ concrete member. In most cases this test is performed for the proof of structure capacity, not for suspect or critical location. The principal aim in this testing is to demonstrate satisfactory performance under an over load, above the design working value. This is usually judged by measurement of deflections under this load which may be sustained for a specific period. The selection of specific members or portion of a structure to be tested will depend upon the general features of convenience as well as the relative importance of strength and expected load effect at various locations. Selection of member may often be assisted by non-destructive methods coupled with visual inspection to locate the weakest zones or elements.

There are certain inherent problems in this test. This method is very expensive and time consuming. In in-situ test whether the member under test is actually subjected to the assumed test load during fire is difficult to predict to predict due to load sharing effect. All the direct methods of assessing residual strength of concrete provide regarding average potential cube compressive strength and are not indicative of strength profile across the section. Direct load testing again may not be possible where the loading may damage the structure permanently or where the structural system in too stiff to show any deflection.

4.5.3 Load test on structural element

A live load of 1t/sqm was taken for imposed load calculation. Floor finishing load and partition wall load was also considered. Thus imposed load was taken as 1.25 times live load as per IS 456: 2000 Code recommendations. The load was imposed on the beam as udl spreads over an area of 55.5 sqm. Thus a total load of 84 t was arrived at and imposed in 12 stages of 7t each at an interval of 15 minutes. The load of 7t was imposed with the help of 200 sand bags weighing 35 kg each as first step and increased in steps up to 84t. Dial gauges and electrical strain gauges were fixed at different locations. Separate platform was provided to read and record the deflections from the dial gauges, strain from strain indicator during loading and unloading.

Deflections for the above mentioned loading steps were recorded. Total time for reaching maximum load was 26 hrs and the maximum load was kept for more than 43 hrs during which creep effect was monitored and the time taken for unloading was 32 hrs during which deflection recovery was monitored. The recorded deflections and theoretical deflections were compared.

CHAPTER 5

Non–Destructive Testing Systems

5.1. INTRODUCTION

Non-destructive evaluation is widely employed for inspecting the condition of structures. Non destructive techniques, which are less time consuming and relatively inexpensive, can be used for the following purposes:

(a) Test on actual structures.
(b) Test at several locations.
(c) Test at various stages.
(d) Assess the quality control of actual structures.
(e) Assess the uniformity of the concrete.
(f) Assess the materials used and workmanship with specification.
(g) Assess the poor construction practices.
(h) Assessment of the extent of cracks, voids, honeycombs.
(i) Confirmation of suspected distress due to poor design
(j) Assessment of partial durability
(k) Integrity testing of piles
(l) Monitoring of progressive changes in structure

The results of non-destructive tests are most useful when supplemented by a limited number of destructive test procedures. There are more testing techniques with different principles and applications are available to evaluate the properties of concrete. The concrete material is so complicated that the efficiency and quality cannot be established just by one single test. Most of the tests, which are used, for estimating the parameters of concrete provide an excellent means of establishing and evaluating the uniformity of concrete.

5.2 NON–DESTRUCTIVE TESTING METHODS

Potential and limitations of various non-destructive techniques cited below and are briefly discussed in the following paragraphs to apprise users of their relevance in field application.

(a) Surface Hardness Method
(b) Ultrasonic Pulse Velocity Method
(c) Resonant Frequency Method
(d) Dynamic or vibration method
(e) Pulse Attenuation Method
(f) Pulse Echo Method
(g) Radioactive Method
(h) Nuclear Methods
(i) Magnetic Methods

(j) Electro magnetic methods
(k) Electrical Methods
(l) Acoustic Emission Technique
(m) Radar Technique
(n) Radiography Methods

5.2.1 Surface hardness test

The surface hardness method consists of impacting the concrete surface in a standard manner. Activating a mass by a given energy and measuring the indentation or rebound achieve this. The most commonly and widely used instrument is a **"Rebound Hammer".** There are several types of hammers having varying impact energy from 0.07 kg m to 3 kg m. the high impact energy is used for mass concrete, road pavements and airport runways. The low impact energy hammers (0.07 to 0.09 kg m) are used for small and low strength materials. A typical rebound hammer is shown in figure 5.1.

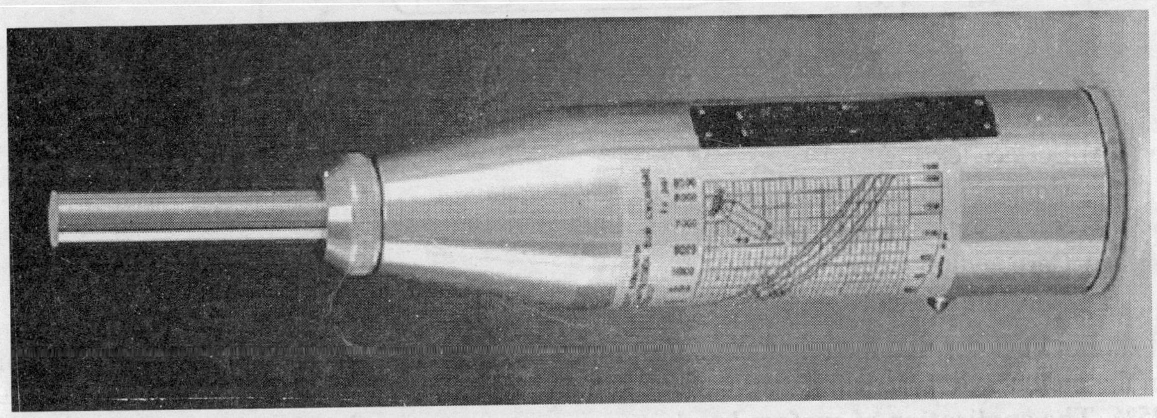

Figure 5.1. A typical rebound hammer

Test procedure:

The test procedure consists of applying the hammer on the concrete surface and observing the rebound reading indicated by a rider over a scale. Before applying the hammer, the surface of the concrete is cleaned and smoothened. A minimum of 10 readings is compared and each reading should not differ by more than 7 units. The average of remaining readings is determined for evaluating the strength. If more than two reading differ from the average by 7 units, than the entire set of readings are taken afresh. The testing procedure in hammer is as shown in figure 5.2.

The procedure for determining the rebound values has been specified in ASTM C 805-85, BIS-13311 Part 2 and also in the latest ASTM specification. Estimation of concrete compressive strength from rebound number is determined from standard calibration curve based on the laboratory results. The calibration curve should be established for each type of concrete. A typical calibration curve is shown in figure 5.3. **The main factors that affect the readings are:**

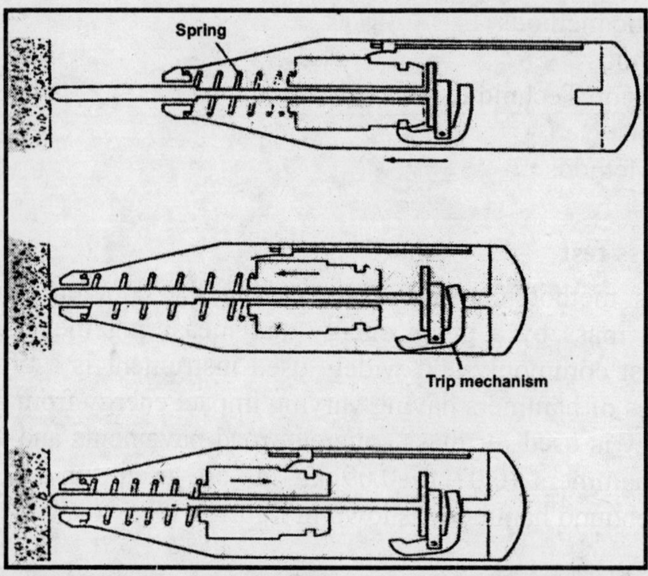

Figure 5.2. Testing procedure of rebound hammer

(a) Size and age of concrete
(b) Surface texture
(c) Concrete mix characteristics
(d) Stress state & temperature
(e) Carbonation level in concrete &
(f) Moisture content

It should be noted that the rebound values reflect the concrete quality up to a depth of 50 mm in the member. In practical situation, the strength prediction can be made to an accuracy of ± 25%. **The application of surface hardness measurements can be used for:**

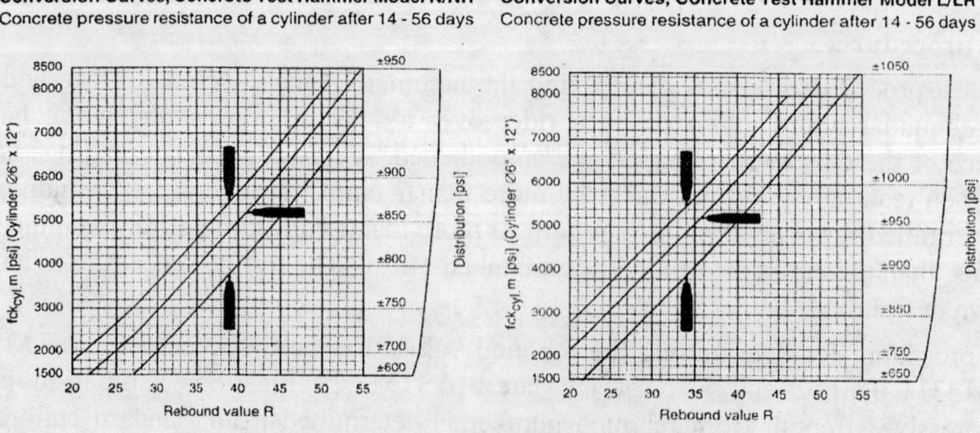

Figure 5.3. Typical calibration curves

(a) Checking the uniformity of concrete
(b) Comparing a given concrete with a specified requirement

(c) Approximate estimation of strength by using laboratory calibrated graphs
(d) Abrasion resistance classification

The following table 5.1 shows the quality of concrete cover from rebound number.

Table 5.1. Quality of cover concrete from rebound number

Average rebound number	*Quality of concrete*
Greater than 40	very good hard layer
30 to 40	good layer
20 to 30	fair
less than 20	poor concrete
0	delaminated

5.2.2 Ultrasonic pulse velocity (UPV) test

The object of this method is basically to measure the velocity of the pulses of longitudinal vibrations passing through concrete. The first report of the measurement of the velocity appeared in USA in the mid of 1940s. It was found that the velocity depended upon the elastic property and geometry of the material. With the development of reliable measurement techniques, the ultrasonic methods have become widely accepted around the world and commercially produced lightweight equipments. If properly used by an experienced operator, a considerable amount of information about the interior of the concrete member can be obtained. Recommendations for the use of this method are given in BS-4408 Part 5, ASTM C 597–71, and BIS-13311 Part 1.

Theory of pulse propagation through concrete:

Three types of waves are generated by an impulse applied to a solid mass. They are longitudinal (compression), shear (transverse) and surface (Reyleigh) waves. Longitudinal waves are fastest and provide more useful information than others. For an infinite, homogenous isotropic elastic medium, the compression wave velocity is given by

$$V = \sqrt{k\, E_d/\rho},$$

Where,

V = Compression wave velocity km/sec,
$k = (1-\gamma)/[(1+\gamma)(1-2\gamma)]$,
E_d = dynamic modulus of elasticity,
ρ = mass density, and
γ = dynamic Poisson's ratio.

Test procedure:

The velocity of an ultrasonic pulse is influenced by properties of concrete, which determine its elastic stiffness and mechanical strength. Thus the variations in the pulse velocity values reflect a corresponding variation in the state of concrete under test. In order to evaluate the strength of concrete, calibration charts should be established based on laboratory tests. The calibration chart is influenced by a number of factors such as type of cement, cement content, admixtures, type & size of aggregate, curing conditions and age of concrete. Hence, adequate care and caution should be exercised while translating the pulse velocity values in terms of strength. The test consists of transmitting longitudinal vibrations produced by an electro – acoustical transducer from one side of the concrete, receiving the signal from the other side, and measuring the transit time (T) of the pulse using electronic time circuits. The path length (L) of the pulse is measured and velocity (V) is calculated as V = L/T.

The measuring equipment consists of an electrical pulse generator, a pair of transducers, an amplifier, and an electronic timing device for measuring the time interval elapsing between the onset of a pulse generated at the transmitting transducer and the onset of its arrival at the receiving transducer. Transducers of 50 to 60 kHz are found to be useful for most of the applications in concrete testing. A typical measuring instrument is as shown in figure 5.4.

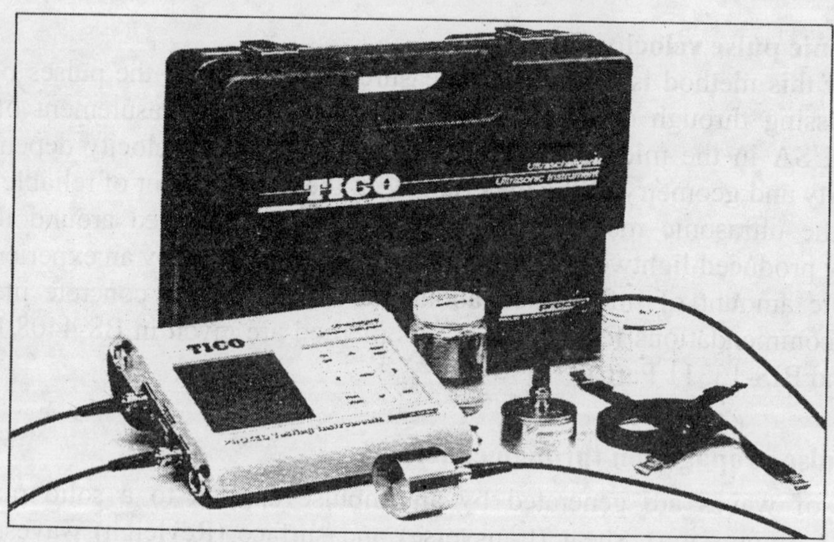

Figure 5.4. UPV measuring instrument

The pulse velocity measurements may be used to establish:

(a) The homogeneity of the concrete
(b) The presence of cracks, voids and other imperfections
(c) Changes in the structure of the concrete which occurs with time
(d) The quality of the concrete in relation to standard requirements
(e) The quality of one element of concrete in relation to another
(f) The values of elastic modulii of concrete

There are three possible ways of measuring pulse velocity are shown in figure 5.5:

(a) Direct transmission
(b) Indirect or Surface transmission
(c) Semi–direct transmission

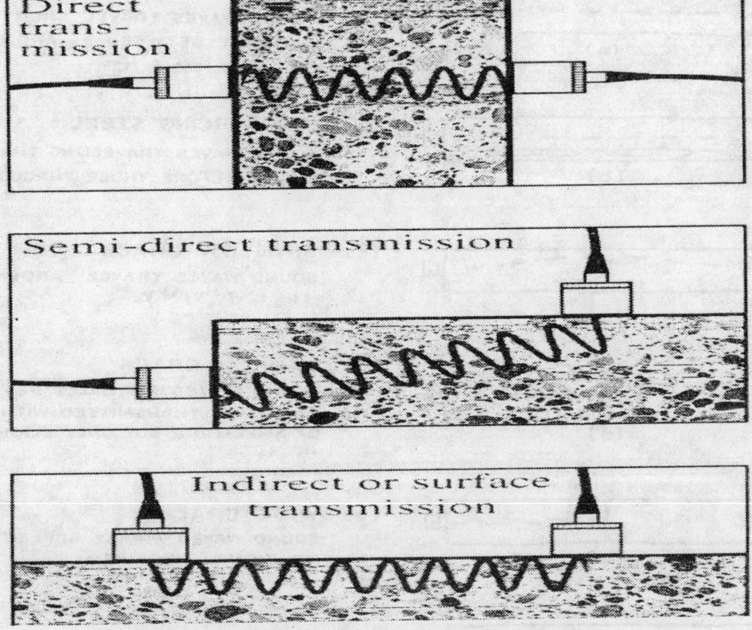

Figure 5.5. Methods of transmission system

The direct transmission method is generally preferred since the maximum energy of the pulse is being directed at the receiving transducer and this gives maximum sensitivity. The indirect transmission arrangement is the least sensitive and for a given path length, produces a signal which has only about 2% or 3% of that produced by direct transmission. This arrangement is used when only one face of the concrete is accessible, when the depth of the surface crack is to be determined or when the quality of the surface concrete relative to the overall quality is of interest. The semi direct transmission arrangement has a sensitivity intermediate between those of the other two arrangements. In this method, there is uncertainly regarding path length. It is generally found to be sufficiently accurate, if the length is measured from centre of transducer faces.

Some of the factors that influence pulse velocity measurements are:

(a) Surface condition
(b) Moisture content
(c) Temperature of concrete
(d) Path length
(e) Shape & size of specimen
(f) Reinforcement

The following figure 5.6 shows the behaviour of ultrasonic pulses in concrete medium under different conditions.

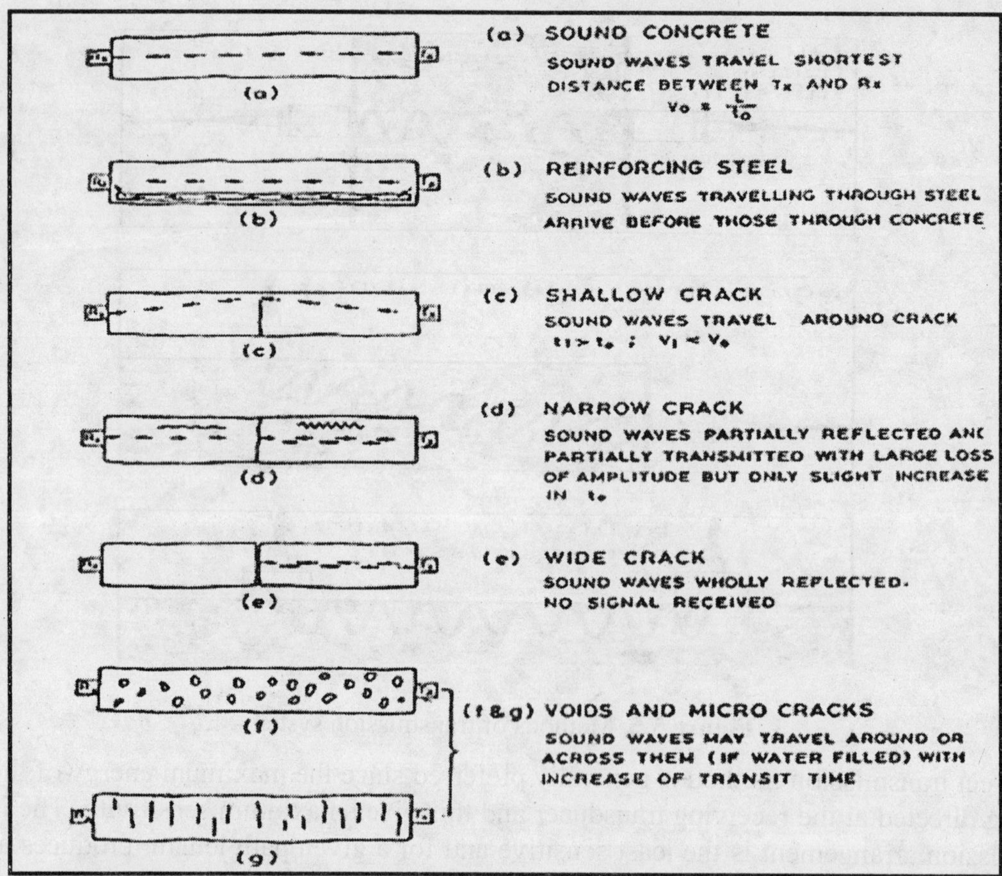

Figure 5.6. Behaviour of ultrasonic pulses in different concrete medium

The following table 5.2 gives the criteria are adopted for certification qualitatively of concrete under UPV results.

Table 5.2. Quality of concrete from UPV

UPV value km/sec (V)	*Concrete quality*
V greater than 4	Very good
V between 3.5 & 4	Good
V between 3 & 3.5	Poor
V between 2.5 & 3	Very poor
V between 2 & 2.5	Very poor & low integrity
V less than 2	Large voids suspected

Advanced ultrasonic testing equipments:

This category comprises the range of instruments that use sound or stress waves in order to determine the properties of concrete and other materials non-destructively. The first and most widely used System is **V-Meter**, which utilizes the ultrasonic pulse velocity method for evaluating construction materials in the field. Transducers are available for a variety of frequencies from 24 KHz to 500 KHz. This unit has also been modified to suit the special needs of ceramics users and can be found as the ultra pulse.

The **Portable Impact-Echo System (PIES)** is an advanced instrument for non-destructive detection of flaws and defects in a variety of civil infrastructures ranging from bridges, parking structures and buildings to dams, piles, tunnels, tanks and marine structures. The **E-Meter System** represents the state of the art in bench top laboratory resonant frequency analysis of materials. This unit comes standard with a test bench designed to handle a variety of modes of vibration, including longitudinal, torsional and flexural. This line of products represents the most effective and efficient way to evaluate concrete and other materials in the field and utilizes the latest in technology in order to guarantee accurate results.

5.2.3 Resonant frequency method

This method is based up on the determination of the fundamental resonant frequency of vibration of a specimen. The equipment used for this is usually known as **SONOMETER.** Resonant frequency methods are mostly used in the laboratory. The size of specimens in these tests is usually limited to 150 mm × 300 mm cylinder or 75 mm × 75 mm × 300 mm prisms. **The resonant frequency test is used for the following purposes:**

(a) For studying the deterioration effects of concrete subjected for repeated cycles of freezing and thawing.
(b) To study the effects due to acidic and alkali reactions
(c) For determining the damage due to fire
(d) To calculate the dynamic young's modulus of elasticity of concrete.

5.2.4 Dynamic or vibration method

These methods are important non-destructive methods used for testing concrete strength and other properties. The fundamental principle of this method is in the propagation of sound velocity through a solid material. A mathematical relationship could be established between the velocity of sound through the specimen and its resonant frequency. From this relationship, the modulus of elasticity of the material is determined. For deriving this relationship the solid mediums are considered to be homogenous, isotopic and perfectly elastic. However these relations are also applied to heterogeneous materials like concrete. The velocity of sound V in a solid material is a function of the square root of the ratio of its modulus of elasticity E and its density ρ. Thus

$$V = f\,[gE/\rho]^{1/2}$$

Where g is the acceleration due to gravity

5.2.5 Pulse attenuation method

This is a wave propagation method in which electromagnetic waves, typically in the frequency ranged 500 MHz to 1000 MHz are allowed to propagate through solids and the methodology is found to be useful for subsurface investigation in civil engineering structures and in particular concrete structures. The basic principle is that the attenuation properties of the electromagnetic waves are influenced by the electrical properties of the solid material tested. The dominant properties are the electrical permittivity, which determines the signal velocity, and the electrical conductivity, which determines the signal attenuation. Reflections and refractions of the radar wave will occur at inter faces between different materials and the signal returning to the surface antenna can be interpreted to provide an evaluation of the properties and geometry of subsurface features. Limited use of radar has been seen in the investigation of tunnel linings in India. **It is increasingly recognized that there are many structural applications for the radar system as listed below:**

- Determination of major construction features
- Assess element thickness
- Locate reinforcing bars
- Locate moisture
- Locate voids, honey combing/cracking
- Locate chlorides
- Size of voids
- Estimate chloride concentration
- Locate rebar corrosion

5.2.6 Pulse echo method

Considerable developments have taken place in marking commercially available instrument on pulse echo technique and this has become popular through extensive research and

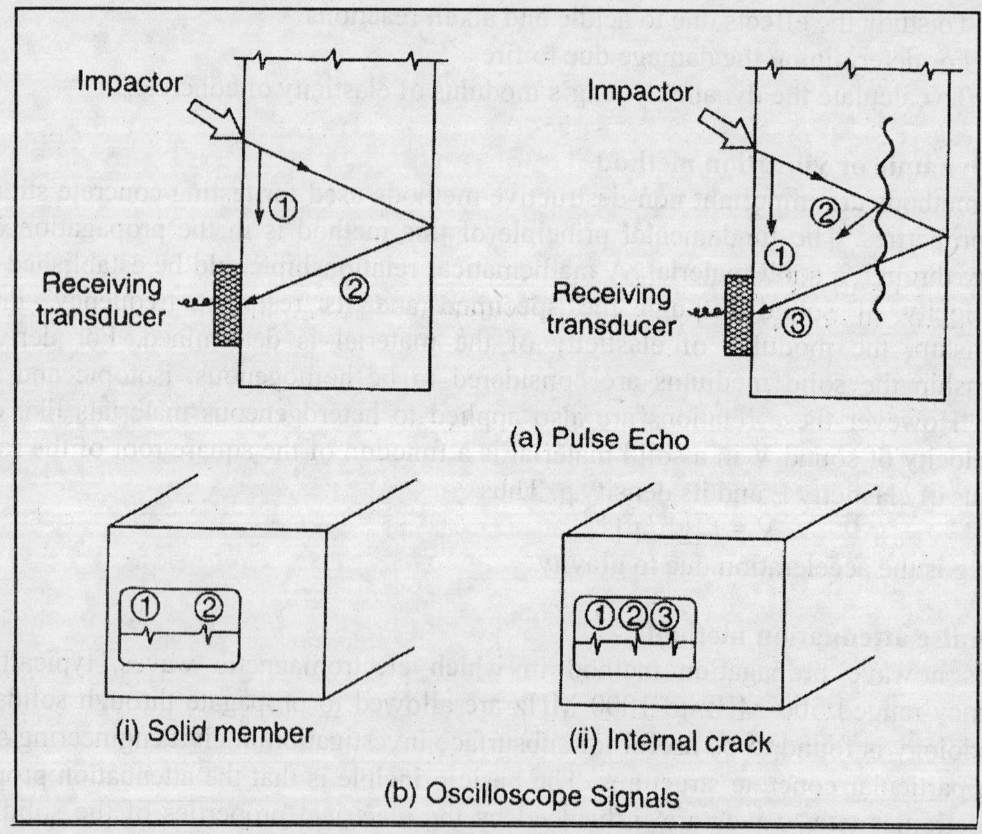

Figure 5.7. Pulse echo method for crack location

practical applications. Originally, pulse echo technique was developed for pile integrity testing and now it is popular for concrete structures and structural elements. An instrument system commercially known as **DOCTER** marked in Denmark is available with a field computer, related software and transducers. The pulse echo and oscilloscope signal techniques are shown in figure 5.7.

The system found to be useful for:

(a) Testing for thickness & flaws in concrete
(b) Testing of wave speed on the surface
(c) Testing for depth of surface opening cracks

5.2.7 Radioactive method

The use of X – rays and gamma rays as non – destructive method for testing properties of concrete is relatively new. X – rays and gamma rays both components of the high-energy region on the electromagnetic spectrum penetrate concrete but undergo attenuation in the process. The degree of attenuation depends on the kind of matter traversed, its thickness, and the wavelength of the radiation. The intensity of the incident gamma rays and the emerging gamma rays after passing through the specimens are measured. These two values are made use of for calculating the density of structural concrete members. Gamma – rays transmission method has been used to measure the thickness of concrete slabs of known density. Gamma radiation source of known intensity is made to pass and penetrate through the concrete. The intensity at the other face is measured. From this thickness of the concrete is calculated.

5.2.8 Nuclear methods

Use of nuclear methods for non – destructive measurement of some properties of concrete is of recent origin. Two principle techniques have been reported, namely **Neutron Scattering Method**, for determining the **moisture content** of concrete and **Neutron Activation Analysis** for the determination of **cement content**. These methods are not suitable for finding out the strength of concrete.

5.2.9 Magnetic methods

Battery operated magnetic devices that can measure the depth of reinforcement cover in concrete and detect the position of reinforcement bars is now available. The apparatus is known as **Cover meter/Pachometer.** This can be used for measuring the cover given in the lightly reinforced sections. Pachometer or covermeter is an electromagnetic device used for determining the location of reinforcing steel embedded in concrete, and to determine the amount of cover over steel. The instrument consists of a search head connected by a cable to a metering unit, which may have a digital or analogue read-out. The purpose of locating the reinforcing steel, prestressing strand, cladding ties etc, in concrete, is to avoid them during core sampling, direct pulse velocity surveys etc, and to make connections for electrochemical measurements. The knowledge of cover depth is essential if it is desired to obtain samples of the concrete at the level of the rebar for chloride ion analysis, pH measurements and the presence of carbonation.

It is also useful in determining the potential for corrosion and subsequent concrete deterioration since it has been well established that structures in corrosive environments with inadequate cover get subjected to early deterioration. By this test, bar diameter, cover to reinforcement, spacing of the reinforcement, number of bars and any discontinuity in the bars can be detected. This test is performed using covermeter, which is based on electro – magnetic theory.

Some constituents of concrete may affect covermeter readings and so it is always advisable to carry out a calibration check for the actual situation. This can be done by taking readings on bars at a range of different depths and either drilling down or breaking out to determine the actual cover. A calibration curve can then be constructed. When measuring the cover on a member with reinforcement in two directions at right angles to each other such a slab, it is probably best to quickly determine the locations of the bars by moving the head across the surface and marking the location of the minimum reading. Some covermeters can also be used for assessing the diameter of the reinforcing bars. These instruments have different scales for different scale diameters. The bar diameter is assessed by locating a bar using any arbitrary scale and then recording the indicated reading using each bar diameter scale.

5.2.10 Electro magnetic method

This category of products comprises the range of instruments used to locate and determine size and depth of steel reinforcing bar in concrete. The first type most widely used utilizes a low frequency electromagnetic field to locate ferrous objects within a structure. The second and latest technology utilizes Ground Penetrating radar to locate steel and other objects

Figure 5.8. HR Rebar Locator **Figure 5.9.** Datascan **Figure 5.10.** Datascan MK II

within a structure. The **HR Rebar Locator** is shown in figure 5.8 and the Datascan both utilize a low frequency electromagnetic field to locate ferrous objects in a structure. By closely monitoring changes in the electromagnetic field the proximity of steel reinforcing bar can be determined. The HR is most economical model. It uses a simple analog meter to determine the strength of the electromagnetic field and correlate this to either the size of the rebar or the distance to the rebar. The **Datascan** is shown in figure 5.9 units utilize a more sensitive digital system to monitor the strength of the electromagnetic field. This allows the user more sensitivity in locating steel at greater depth and more closely spaced. The

Datascan MK II is shown in figure 5.10 utilizes the state of the art in ground penetrating radar technology. By sending out a high frequency pulse and observing the reflected electromagnetic waves features inside a concrete structure can be determined. The unit implements a number of different ways to visualize the data received, including a three dimensional image of the objects within the structure. This unit represents the current cutting edge technology in object location. The line of products represents the most effective and efficient way to evaluate steel reinforcing bar in the field and utilizes the latest in technology in order to guarantee accurate results.

Rebar locator:

A Rugged Field Instrument for finding the location, depth and size of reinforcement bars in place. **Features:**

(a) Single unit construction; no physical strain for the operator and no probe cord to snag or break

(b) Post-mounted sensor with meter mounted separately for convenient viewing by operator in stand-up position.

(c) Pinpoint accuracy for precise rebar location from direct reading, stable analog meter.

(d) Detection up to 10 inches (250mm) of cover with easy to read calibrated meter.

(e) Light operating weight- less than 3.5 pounds (1.6 kg)

(f) Up to 8 hours continuous use between battery charges; improved electronic circuitry.

Technical Specifications:

Detections and Orientations of Rebars - The exact position and orientation of rebars can be measured quickly and accurately. Rebar-free areas can be identified for coring, grinding, resurfacing or insertion of new machinery mountings and under inspection is as shown in figure 5. 11. HR Rebar Locator can be used to inspect new structures for compliance with specifications as well as old structures under modification.

Measurement of concrete cover - The amount of concrete cover over reinforcement bars is an important parameter on bridge decks, highways, columns and slabs. Concrete cover can de determined with an accuracy of +/- 1/8 inch. Concrete thickness can be measured in tanks, pipes and other structures where the inner lining is steel and within 10 inches of the surface. Reinforcement bar size can also be estimated.

Location of Ferrous Metals - Locates any ferrous base material such as pipe, flues,

Figure 5.11. Rebar locator under process

wire and sheets embedded in concrete, masonry or wood. Identification is possible to a depth of 10 inches.

Location of Prestressed Cables - Locates the position of cables and lost tendon splices in pre or post tension concrete products.

Special Mounting: The HR Rebar Locator can be mounted on an adjustable telescopic tube with handle. The sensor is mounted at the end of the tube and the meter placed within convenient operator viewing range. This enables areas to be scanned with the operator in a standing position.

Rebar Datascan:

Rebar Datascan is a sophisticated reinforcement bar locator with an easy to read Liquid Crystal Display. Features:

(a) Reinforcement bar cover can be detected to a maximum depth of 12 inches (300 mm) by reading directly from the high legibility liquid crystal display.
(b) Analog meter on probe facilitates rapid and easy location of rebars.
(c) Fast scanning of large areas is enhanced by an audible output via headphones.
(d) Built in memory stores over 1000 cover readings for late downloading to a personal computer.
(e) Calibration for rebars and high tensile steel tendons is available.
(f) The probe meter can be preset to react to any rebar within the value, ignoring those outside the range.

5.2.11 Electrical methods

Recently some electrical methods have been employed for determining the moisture content of hardened concrete, tracing of moisture permeation through concrete and determining the thickness of concrete pavements. The accurate determination of the moisture content of hardened concrete is required in connection with creep, shrinkage and thermal conductivity studies. The fact that a dielectric property of hardened concrete change with changes in moisture content is made use of in this method. Electrical resistivity methods have been used to find out the thickness of concrete pavements. The method is based on the principle that the material offers resistance to the passage of an electric current. A concrete pavement has a resistivity characteristic that is different from that of the underlying subgrade layers. A change in the slope of the resistivity versus depth curve is used to estimate thickness of concrete pavements.

5.2.12 Acoustic emission technique

The acoustic emission technique has been used to study the initiation and growth of cracks in concrete (locating delaminated concrete). Sounding of concrete with a hammer provides a low–cost, accurate method for identifying delaminated areas. When striking areas of delaminated concrete, the sound changes from “ping” to a hollow sounding “puck”. Boundaries of delaminations can easily be determined by sounding areas surrounding the first puck until pings are heard. Hammer–sounding of large areas generally proves to be extremely time consuming. More productive sounding methods are available when working with horizontal surfaces. Chain dragging accomplishes the same result as hammer – sounding. As the chain is dragged across a concrete surface, a distinctly different should is heard when it crosses a delaminated area.

Automated electronic system is also available, eliminating the need for an operator to listen for differences in acoustic emissions. However, these methods have proved to give only a general picture of the areas of delamination. Consequently, they should be used only for general assessment, not for the detailed layout required for reconstruction.

In fared **Thermography** is a useful method of detecting delaminations in bridge decks. This method is also used for other concrete components exposed to direct sunlight. The method works on the principle that, as concrete heats and cools, there is substantial thermal gradient with in the concrete because concrete is a poor conductor of heat. Delamination and other discontinuities interrupt the heat transfer through the concrete. These defects cause a higher surface temperature than surrounding concrete during periods of heating, and a lower surface temperature than the surrounding concrete during periods of cooling. The equipment can record and identify areas of delamination and indicate the depth of delaminations below the surface.

5.2.13 Radar technique

Ground–probing or subsurface interface profiling radar can be employed in a number of different ways in the investigation of concrete structures. The technique has been used for measuring the thickness of members. For determining the spacing and cover to reinforcement, and for detecting the position and extent of voids. The technique is similar to the methods used in seismology except that radio frequency pulses are used rather than sound. The equipment generates electromagnetic pulses, which are transmitted to the member under investigation by an antenna close to its surface. The pulses travel through the member but are partially reflected at any interfaces where there are distinct changes in dielectric characteristics. Reflected pulses are received by a second antenna and processed by the equipment.

The surface of the member is marked out with transit lines with regular distance markers. The antenna is moved along the transit lines and a signal is inserted in to the system each time a distance marked is passed. The radar signal is attenuated, or diminished in strength, as it passes through the member under investigation. The amount of attenuation is dependent on the conductivity of the material and the frequency of the signal used. Attenuation is increased in conductive materials and therefore it may be difficult to penetrate saturated or salt contaminated concrete. The velocity of propagation of the radar pulse is dependent on the dielectric constant of the concrete. This can vary quite widely depending on aggregate type, moisture content and other factors. There is a need, therefore, for depths determined by radar to be checked and calibrated by direct physical measurements at a few locations.

5.2.14 Radiography method

Radiography employing X–rays or Gamma rays is some times used to investigate defects such as honey combing or poor compaction inside concrete members. The technique may also be used to determine the location of the reinforcement in congested sections or other sections, where the use of cover meter is not possible or is inappropriate. British Standards 1881: Part 205 gives recommendations for the use of technique. Gamma rays sources are usually employed for concrete sections up to 500mm thick. High energy X–rays are appropriate for greater thickness. The sources consisting of cobalt 60 or iridium 192 for Gamma rays or a linear accelerator for X–rays is placed on one side of the member under investigation. A film, sensitive to X-rays, is placed in light–tight cassette in contact with the other side of the member. Intensifying screens of lead foil mounted on card may be placed on either side of the films. The foil emits electrons when it is subjected to irradiation and this causes an intensification of the image, thus reducing the explosive time. The British Standards suggest thickness of both front and back screens for X-ray and Gamma rays sources. Sources to film distances are also given for various source diameters to minimize lack of sharpness due to the configuration.

5.3 RECENT DEVELOPMENTS ON NDT INSTRUMENTS

For rehabilitation and repair of concrete structures various test techniques are used for evaluating the input parameters required for the remedial measures for undertaking repairs. One of the major problems associated with concrete structures is the evaluation of the strength and serviceability of a structure during and after construction. As of now the acceptance of concrete quality is essential based on the compressive strength of specimens (cylinders or cubes) obtained from standard tests. Situations, however, arise to test the concrete in structure. Such situations most frequently arise as a result of

(a) Doubt about the validity of standard test results
(b) Failure of these test results to meet the specifications
(c) Damage and deterioration of concrete due to various causes such as corrosion, fire, and improper construction, and
(d) Expansion of an existing structure

Further, it is often realised that the standard test results are not easily traced after a period of completion of a structure. Under these circumstances, it is necessary to test the concrete in a structure and evaluate is condition for taking further action. Assessment of quality of in–situ concrete during and after construction and periodical monitoring of structures call for several field testing techniques. Several testing methods have been developed the recent past; a few like rebound hammer and core test, have been in use for a number of years and others like ultrasonic pulse velocity test and penetration and internal fracture test are more recent developments. These tests have been developed with a primary objective of quickly evaluating the condition of in–situ concrete in structural members. While some of them are non–destructive, others are partially destructive. Each method has its own merits and demerits. **Various test instruments deployed all over the world as follows:**

5.3.1 Schmidt concrete test hammer

Non–destructive quality testing of concrete and other building materials in completed structures and prefabricated structural sections.

5.3.2 Digi–Schmidt tester

Testing the strength of concrete, assessment of uniformity of concrete and structural components and detecting damaged zones.

5.3.3 Tico – Ultrasonic concrete tester

Portable instrument for accessing the quality of concrete either in–situ or precast, provides information on wide variety of parameters such as uniformity of concrete, cavities, cracks, honey combing, defect due to fire and frost, modulus of elasticity and concrete strength.

5.3.4 Profometer 4 rebar locater

Digital metal detector for reinforcement bar spacing, concrete cover depth measurement and reinforcement bar size determination. High accuracy depth measurement with spot probe up to 130 mm and depth probe up to 300 mm. Measure bar diameter to an accuracy of ± 1 mm. Path measuring device with probe makes rebar visible through integrated **cyber scan software.**

5.3.5 Canin corrosion analyzing instrument

Non-destructive detections of corrosion in the reinforcement bars of concrete building elements. Large area measurements can be made using road and wheel electrodes and printouts can be taken directly to scale with out **PC via RS232.**

5.3.6 Resistivity meter

Measurement of the electrical resistance of reinforced concrete components for assessing the risk of corrosion.

5.3.7 Dyna pull off tester

Measures the surface strength of concrete. It also measures the adhesive strength of all kinds of applied coatings (mortars, plasters and plastic coatings) flexible and themo plastic coatings, paint finishes and coatings are metals and concrete etc.,

5.3.8 Dyna extraction tester

It measures pull out force/displacement on anchored bolts and plugs.

5.3.9 Torrent permeability tester

Fast reliable and non destructive measurements of the permeability of concrete structures

5.3.10 New concrete workability meter

Rapid, accurate measurement of the workability (water cement ratio and slump) and temperature of fresh concrete for improved on–site quality control.

5.3.11 Multi–cell surveyor

Monitoring steel corrosion in concrete using eight probes to record eight individual readings simultaneously.

5.3.12 Half – cell surveyor

A single probe instruments that measures and records electrical activity in reinforced concrete.

5.3.13 Resistivity logger

A simple push and press instrument that measures resistivity of concrete, masonry and other materials and also stores the gathered data.

5.3.14 Logging cover meter

Effectively locates steel reinforcement bar and measures the depth of concrete cover over them, it also stores the data it has gathered.

5.3.15 Bremor test

For finding out cement content of fresh mortar.

5.3.16 New rebar checker:

Precession instrument that makes the detection of concrete cover over steel reinforcement bars rapid, simple and extremely accurate

5.3.17 Portprobe

Most accurate rugged and portable instrument for density and moisture measurement of soil, pavement, concrete and other construction materials by nuclear method.

5.3.18 AC – 2R Asphalt content gauge

Rapid verification of asphalt content of paving mixtures sampled at asphalt plant, lab or paving site prior to placement and compaction.

5.3.19 Foundation pile diagnostic system

For integrity testing. Pile driving monitoring, dynamic load testing, statnamic load testing, structural vibration monitoring, hammer monitoring, sway monitoring, inclination monitoring, sheet pile monitoring and pile driving prediction.

5.3.20 Doctor – impact echo system

For flaw detection or measurement of thickness of construction material such as concrete.

5.3.21 Electrical time domain reflectometry (ETDR)

ETDR stress/strain sensing technique has been successfully used in geotechnical applications to detect rock deformation and longwall movement. Since durable sensor media that can survive harsh construction environment and has long service life can be used, the ETDR sensing method appears to be a practical technique for health monitoring applications of civil concrete structures. In this paper, the applicability of using an embedded coaxial ETDR cable to sense structural deformation and detect load-induced crack damages of a concrete structure is investigated. Bending test was performed on small-scale concrete specimens with embedded coaxial ETDR sensors. The test results show that the deformation patterns of the concrete specimens are clearly represented and the location and magnitude of load-induced crack damages that pass the sensing line are evidently shown in the ETDR signal response of the sensor. The results of the current study demonstrate the great potentials of the ETDR technique for the health monitoring application of large civil concrete structures.

CHAPTER 6

Semi-Destructive Testing Systems

6.1. INTRODUCTION

Partially destructive test are used for assessing the in–situ concrete strength. These tests cause localized damage, which do not cause any loss in the performance of the structure. All tests in this group are surface zone test which require assess to only one exposed concrete face. Compared to core test the accuracy may not be as good but estimation of strength is immediately available and the testing is less disruptive and damaging. Some of the testing methods are available to assess the quality of concrete–damage of concrete under partial disturbance of the surrounding concrete. **The common methods in construction industry are given below:**

(a) Penetration Techniques
(b) Pull-out and Pull-off Tests
(c) Core sampling and testing
(d) Break off test
(e) Permeability Test
(f) Half–cell potential survey
(g) Resistivity survey
(h) Carbonation and pH value test
(i) Chloride content test
(j) Abrasion resistance test.

In addition, some of the tests are used to find the quantity and quality of concrete ingredients and are used to help in the damage assessment of existing concrete. **These types of tests are given below under chemical testing of concrete:**

(a) Determination of cement content
(b) Determination of water content
(c) Determination of water cement ratio
(d) Determination of free chloride content
(e) Determination of pH value of concrete

In addition, some of the special tests are available to assess the quality of concrete and rebar in corrosion-damaged structures and fire damaged structures. These are also discussed in this chapter.

6.1.1 Penetration Techniques

The measurement of hardness by probing techniques was first reported during 1954, two techniques were used. In one case, a hammer known as **Simbi** was used to perforate concrete and depth of borehole was correlated to compressive strength of concrete cubes. In the other technique, the probing of concrete was achieved by blasting with **Spit Pins** and the depth of penetration of the pins was correlated with compressive strength of concrete. The accuracy of the test was found to be ± 25%. Simbi and Spit pins were more affected by the arrangement of coarse aggregate, than the test using rebound hammers.

During 1964 and 1966, a technique known as the **Windsor probes** was advanced for testing concrete in the laboratory and in situ. The Windsor probe is a hardness tester of the surface of the concrete. It is an equipment consisting of a powder activated gun, hardened alloy probes, loaded cartridges, and depth gauge for measuring penetration of probes. The probe is driven in to the concrete by firing of a precision powder charge cartridges. The exposed length is measured by calibrated depth gauge and this is correlated to the strength of concrete cylinder.

The **penetration resistance method** of assessing strength uses a hardened steel probe, which is fired into a concrete surface using a special gun and a standardized explosive charge. The surface is cleared of loose material and a steel plate is positioned adjacent to the probe. A steel cap is screwed onto the probe and the length of probe protruding from the surface is measured using a spring-loaded depth gauge shown in figure 6.1 and the depth of penetration can be determined. It is clear that the forces acting on the probe and the processes occurring within the concrete are extremely complex. As the tip of the probe enters the concrete, it causes local crushing of the surface and creates a shock wave, which results in local spalling. As penetration continues, the forward motion of the probe is resisted by fraction along its sides, concrete is crushed at the tip and the concrete ahead of the probe is compressed in a zone of indeterminate shape and size. The initial kinetic energy of the probe is dissipated by

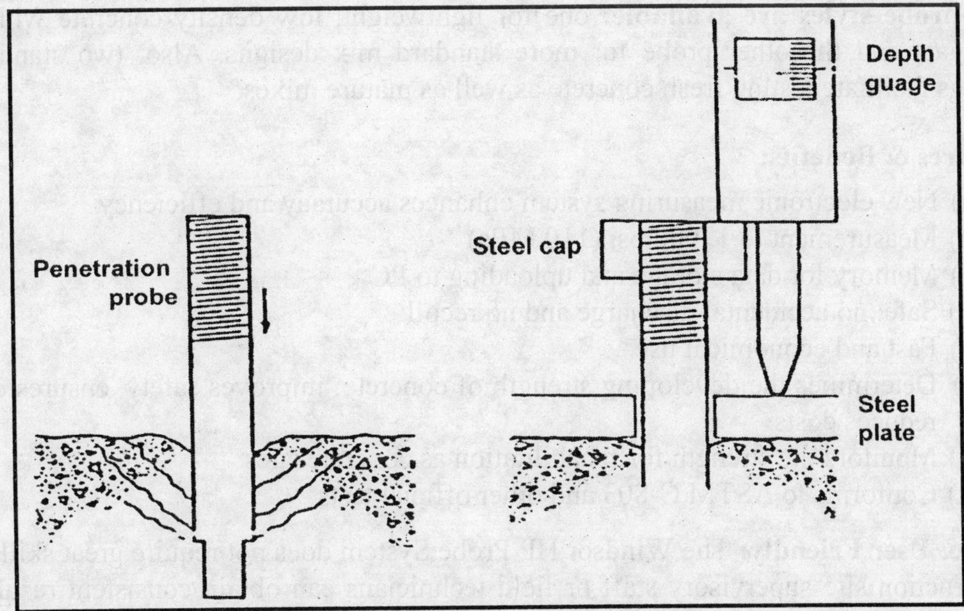

Figure 6.1. Penetration resistance test

all of these processes and it is asserted that the depth of penetration can be calibrated against compressive strength. Cartridges are available at a choice of two power levels according to the strength of the concrete to be tested.

The use of the penetration resistance method is described in ASTM C 803. The standard notes that penetration resistance testing can be used to obtain an indication of the development of strength in concrete, to investigate areas of deterioration or poor quality, and to assess the uniformity of concrete within a structure, but that the method is not intended to take the place of strength determination by conventional methods. The standard also notes that a relationship may be established between penetration resistance and strength for a particular concrete mix and particular test equipment and used to assess strength in structures. However, the relationship may be affected by curing conditions and the type and size of aggregate. The estimated strength values are unlikely to have an accuracy of less than ± 20 per cent at confidence levels of 95 per cent.

Windsor HP Probe system: For In-Place Testing of Normal and High-Performance Concrete. Windsor HP Probe Measures the compressive strength of concrete accurately and effectively, on site in the field.

The Windsor HP Probe system rapidly and accurately determines the concrete compressive strength of a structure by driving a probe into the concrete with a known amount of force. Improved and enhanced over thirty years, this modern system is capable of measuring concrete with a maximum compressive strength of 110Mpa. It has been followed for use in the construction environment, yet refined to provide the user with a simpler system to operate. An electronic measuring unit has been added to help ensure proper test results, which can be recorded for later review or uploading to a personal computer.

Two probe styles are available: one for lightweight, low density concrete with air filled aggregate and the other probe for more standard mix designs. Also, two standard power settings facilitate testing fresh concrete as well as mature mixes.

Features & Benefits:

(a) New electronic measuring system enhances accuracy and efficiency
(b) Measurement to 17000 psi (110 MPa)
(c) Memory for data storage and uploading to PC
(d) Safe: no accidental discharge and no recoil
(e) Fast and economical use
(f) Determines the developing strength of concrete; improves safety, ensures quality and reduces costs
(g) Monitors the strength for rehabilitation as concrete ages
(h) Conforms to ASTM C-803 and other official tests

Safe & User Friendly: The Windsor HP Probe System does not require great skill to use and construction site supervisory staff or field technicians can obtain consistent results. In fact, among its users are contractors, engineers, architects, testing laboratories, ready mix concrete producers, owners, managers and government authorities. This system has widespread use in testing concrete in-situ, on conventionally placed, sprayed or precast concrete; on horizontal or vertical slabs; on floors or overhead; on fresh or mature concrete. The system is safe to use. The driver cannot be discharged unless it is fully depressed with some force against the actuating template, which rests against the surface being examined.

Probes: There is two-power settings available, low and standard power. The low power is used where concrete strength is less than 19.4 MPa. The newly designed silver probes can be used for high performance concrete with strength up to 110 MPa.

Procedure: A systematic test procedure is shown in figure 6.2.

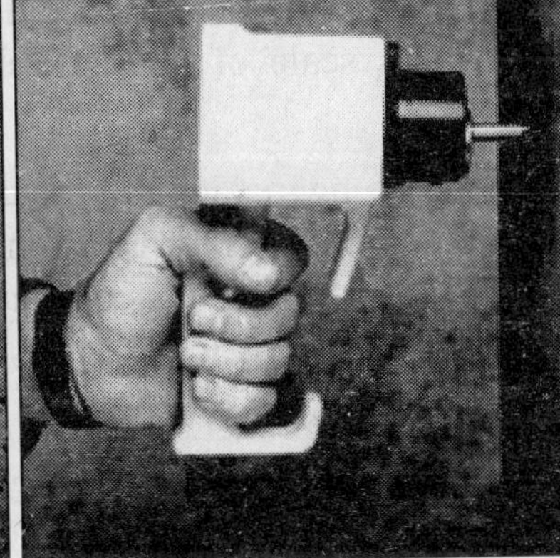

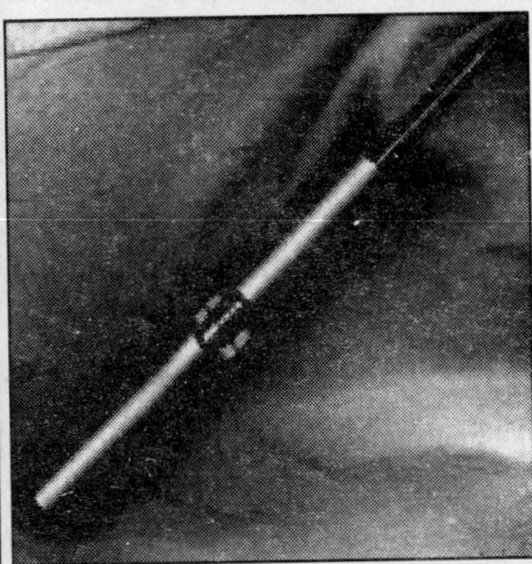

Figure 6.2. Wnidsor HP probe test procedure

A. Actuating

Load the driver with a power load and probe suited for the type of concrete being examined. Place the driver firmly on the actuating template and fire. The locating template is then used to locate the probes at the corners of a fixed triangle. Normally, three measurements are required for consistent and statistically reliable results.

B. Measuring

The electronic measuring device is menu driven and programmed for selection according to the following parameters.

- Aggregate hardness
- Light weight, normal, or HP concrete
- American or Metric units

The three individual tests are automatically averaged and displayed on the LCD in accordance with ASTM procedure. This data together with time and date of the test are stored in the memory for subsequent uploading to a PC. An extractor is supplied to facilitate probe removal after the test. The Standard Windsor Probe System includes a depth gauge and "Strength Chart" for determining the concrete strength. This is an economical alternative for many concrete test environments.

Technical: The Windsor HP Probe System is designed to evaluate the compressive strength of concrete in place. It is non-destructive and can be used with equal effectiveness on fresh and mature concrete. Equally accurate results are obtained on horizontal or vertical surfaces provided that the probe is perpendicular or at right angles to the test surface.

A hardened steel alloy probe is propelled at high speed by an exactly measured explosive charge into the concrete and its penetration measured. Each power load is guaranteed to have an energy level to give an exit muzzle velocity tolerance within ±3%. The compressive strength of the concrete is directly related to the resistance to penetration of the crushed aggregate and cement matrix: the distance required to absorb the specific amount of kinetic energy of the probe determines this. The compressive strength of the concrete is empirically related to the penetration that varies with the hardness of the aggregate. This relationship is recognized by determining the Moh's scale of hardness of the aggregate and applying a correction factor to the penetration.

The combined contributions of both the aggregate and the cement paste to concrete strength are examined by the test. The accuracy of the inferred strengths has been examined in many independent tests and trials. The Windsor HP results correlate well with the concrete strength determinations obtained by conventional means.

For most accurate test results ASTM recommends that a correlation be developed for the particular mix design being tested. Exact duplication of cylinder test results should not be expected. The probes measure the strength of the actual concrete in a structure rather than that of a sample compacted and cured under strict and somewhat artificial conditions, which do not necessarily represent those of the structure itself.

6.1.2 Pullout and pull off test

Several tests are available which give estimates of compressive strength by measuring the force required to pull out embedded anchors or to pull off discs glued to the surface. Many of the pull out tests use anchors, which are placed in to the concrete at the time of construction. Pull out test–in this test, either an insert is cast in the concrete or fixed in to a hole, which is drilled in to the concrete. Force required to pull out the insert is measured which is correlated with the compressive strength. Although the results relates to the surface zone only, the advantage is that a more direct measure of strength and a greater depth, compared to the surface, hardness test, and is available. Pull off test–it is based on the measurement of the in–situ tensile strength of concrete by applying a direct tensile force. The method is specially useful in measuring the bond between the overlays. In this test, a metallic disk is glued either to the concrete surface to the surface of partial core. The force required to pull of the disk, causing tensile failure of concrete, is measured and correlated to the strength of concrete. The test requires the care in preparing surface and cause difficulty with damp surfaces.

A pullout test measures the force required to pullout from the concrete a specially shaped rod whose enlarged end has been cast into that concrete. The stronger the concrete, the more is the force required to pullout. The ideal way to use pullout test in the field would be to incorporate assemblies in the structure. These standard specimens could than be pulled out at any point of time. The force required denotes the strength of concrete. Another way to use pullout test in the field would be to cast one or two large blocks of concrete incorporating pullout assemblies. Pullout test could than be performed to assess the strength of concrete.

Internal fracture test:
It is the pull out test, which has been most widely used in the field. It was first developed for testing high alumina cement concrete components but further work has been carried out to produce correlation curves for Portland cement concretes. To carry out the test 6mm hole is drilled to a depth of 30–35 mm and the hole is blown out. Holes for individual tests should be at least 100 mm apart, should not be close to an edge and should be drilled away from reinforcement. It is usual to carry out six tests on a member. The average of the six maximum readings is used to read off the mean compressive strength from a standard calibration curve. It is suggested that the method allows concrete strength to be estimated to with in ±30 % at the 95% confidence level. The test causes internal fracture of the concrete and their should be little damage to the surface. After the test the bolt can be sawn of f and the surface made good.

The capo test:
The capo test is a modification of the **Lok test** which was developed in Denmark. The lok test uses cast- in inserts and confirms to ASTM C 900. The capo test names based on the initial letters of the phrase 'cut and pull out'. To carry out the test, on 18 mm hole is drilled to a depth of 45 mm. The hole is reamed out to form a 25 mm diameter groove at a depth of 25 mm. A ring insert is placed in the hole and expanded in the groove. The insert is connected to the hand–operated hydraulic jack and load–indicating gauge. Connecting to the test equipment is made by a threaded coupling rod 7.2 mm in diameter. The hydraulic jack produces a tensile force in the rod and bears on a reaction ring of internal diameter 55 mm on the concrete surface. The failure surface generated by the test is in the form of the frustum of core from the outer edge of the expanding insert to the inner edge of the reacting edge. The test can be stopped and the load released as soon as the peak is reached. This results in only a fine crack on the surface. Alternatively, the test can be continued until a plug of concrete is removed so that the disc can be recovered and reused.

Limpet test:
The limpet test is a pull off test. An aluminum disc, 50mm in diameter is glued to the concrete surface at the test position using an epoxy resin or other suitable adhesive. The disc is connected to a mechanical loading device by a threaded rod. Load is applied by turning a handle on the side of the machine. Applied load measured by integral strain gauge equipment with digital read–out. The applied force pulls the disc away from the surface with an attached thin layer of concrete. The concrete fails principally in direct tension. It is normal to carry out three tests on a member and calculate a mean pull–off force. The compressive strength can be assessed using suitable calibration curves. Since the layer of concrete detached with the disc is usually only a few mm thick, the results can be strongly influenced by surface effects such as laitance, poor curing or carbonation. This difficulty can be over come by core drilling

using a 50 mm diameter barrel to a depth of 20 mm or greater, to penetrate concrete below the surface layer. The central section is not broken off and the aluminum disc is attached to it before pulling in the normal way. This technique is also useful for testing the bond of repairs or overlays. In this case, the core is drilled through the repair or overlay in to the parent concrete.

6.1.3 Core sampling and testing

The core tests are performed under the following situations:

(a) When the standard 28 days cube strength test gives lower strength than acceptable and the primary aim of the core test is to ascertain whether the structural element is of adequate strength.
(b) When it is required to estimate the load carrying capacity of the structure for its safety under change of loading contemplated for the structure.
(c) To detect the segregation, honey combing and to check the bond at construction joint or to verify the thickness of the pavement.

Core test serve as a good support test for any field investigation. Special electric or gasoline driven core drilling machines have been developed to cut cores from the structural members. Cores of 25 mm to 300 mm diameter can be cut including reinforcements. The cores can be tested for compressive strength, chemical analysis, petrographic examination and evaluating physical parameters. The strength testing of cores provide almost a direct evidence on the quality of concrete as it exits on the structure. The final results are influenced by the factors such as: diameter of the core, slenderness ratio, core location, presence of reinforcement and curing condition.

The general procedure includes:

Core location and size - Cores for test should be normally taken at points where minimum strength and maximum stress coincide, for example, from the top surface near mid span for simple beams & slabs or any face near the top of lifts–for columns or walls. In general, the accuracy decreases as the ratio of size of aggregate to core diameter increases, and 150 mm is regarded as the preferred size for aggregates of up to 40 mm, 100 mm cores must not be used if the maximum aggregate size exceeds 25 mm, and this should preferably be less than 20 mm for 75 mm diameter core.

Testing - Each core must be trimmed and capped before visual examination, assessment of voidage, and density determinations.

Trimming–Trimming with a masonry or diamond saw, should give a core of a suitable length with parallel ends, which are normal to the axis of the core. If possible, reinforcement and unrepresentative concrete should be removed.

Capping–Cores should be capped with high alumina cement mortar or sulphur–sand mixture to provide parallel end surfaces normal to the axis of the core. Capping should be kept as thin as possible.

The honey combing, cracks, aggregate distribution, drilling damages and other defects are easily seen on a dry surface. However, for the assessment of type of aggregate, size and then characteristics a wet surface is preferable. Voids between 0.5 to 3 mm are classified as small, 3 to 6 mm sized voids as medium as and greater than 6mm large voids. When the voids are interconnected, they are called honey combing. The precise locations of the reinforcement present in the core must also be recorded.

Density determination:

(a) Measure volume of trimmed core (Vu) by water displacement

(b) Establish density of capping material. (Dc)

(c) Before compression testing, weigh soaked/surface dry capped core in air and water to determine gross weight W_1, and gross volume V_1

(d) If reinforcement is present this should be removed from the concrete after compression testing and the weight Ws and Vs determined.

(e) Calculate saturated density of concrete in the uncapped core from

D2 = W1–Dc (V1–Vu)–Ws/Vu - Vs

(f) If no steel is present, Ws and Vs are both zero.

(g) Estimated excess voidage = [(Dp–Da)/(Dp–500)] * 100

When Dp = the potential density based on available values for 28 days old cubes of the same mix.

Da = the actual density from core samples.

Factors influencing the core compressive strength:

(a) Moisture and voids
(b) Length/diameter ratio of core
(c) Diameter of core
(d) Position of cutout concrete on structure
(e) Direction of drilling
(f) Effect of age
(g) Concrete characteristics
(h) Method of capping
(i) Reinforcement

Compression testing:

Compression testing will be carried out at a rate of $15N/mm^2/min$ in a suitable testing machine and the mode of failure noted. If there is cracking of the caps, or separation of cape and core, the result should be considered as being of doubtful accuracy. Ideally, cracking should be similar all round the circumference of the core, but a diagonal shear crack is considered satisfactory except in short cores or where reinforcement or honeycombing is present.

Estimation of an equivalent cube strength corresponding to a particular core result must initially account for two main factors. There are:

(a) The effect of the length/diameter ratio (correction factor)
(b) Conversion to an equivalent cube strength (conversion factor based on general relationship; cylinder strength = 0.8 × cube strength

Horizontal drilled cores;
Estimated cube strength $= 2.5\ f_{\lambda}/(1.5 + 1/\lambda)$
Vertical drilled cores;
Estimated cube strength $= 2.3\ f_{\lambda}/(1.5 + 1/\lambda)$
Where f_{λ} is the measured strength of a core with $l/d = \lambda$

Due to the presence of reinforcement, the measured strength of concrete is found reduced up to 10%. Thus as for as possible reinforcement from the core must be avoided. In case the reinforcement is present in the core, the measured core strength should be corrected as follows:

Corrected strength = Measured strength X [1 + 1.5(¢r h/¢c l)]

Where

¢r = diameter of reinforcement bar
¢c = diameter of the core
h = distance of bar axis from the nearest end of core
l = uncapped length core

In case core has multiple bars, then corrected strength

= Measured strength X [1 + 1.5(ε¢r h/¢c l)]

IS 516–1959 describes the method of preparation of core after drilling and the procedure of test. According to IS 456-1978 the concrete in the member represented by the core test shall be considered acceptable if the average equivalent cube strength of the core is equal to at least 85% of the cube strength of the grade of concrete specified for the corresponding age, and no individual core has the strength less than 75%.

Advantages of core testing:

(a) The structural integrity of the concrete across the full cross section may be affected.
(b) The ratio of length/diameter other than standard ratio 2 will have its effect on strength.
(c) Capping of both ends of the core is also liable to affect the strength of concrete.
(d) Existence of reinforcement will also have its effects on the strength of concrete.

Petrographic examination:

Petrographic is a section of **geology** dealing with description and classification of rocks, but petrographic techniques can also be used to inspect samples of concrete in the laboratory. A standard practice is described in ASTM C 856 and a description of microscopic methods is given in Concrete Society Technical Report No 32. **Areas in which microscopic methods have been found to be useful include:**

(a) Identification of mix ingredients including coarse and fine aggregate type, cements type, pulverized fuel ash and ground granulated blast-furnace slag.

(b) Assessment of mix proportions and air void content
(c) Investigation of various durability problems including carbonation, alkali-aggregate reaction, fire damage and chemical attack

Petrographic techniques can be used in a quantitative way to assess the original mix proportions. Two methods using either point counting or linear traversing of polished sections are commonly employed. The basic principles of both methods are described in ASTM C 457 in relation to the determination of air void content and bubble size and spacing. In each of the methods the polished specimen is placed on a hand or motor–driven stage which can be moved in two directions at right angles. Using either of these methods, the volumetric percentages of each of the components can be determined. With the measurement of some additional parameters the cement content, water/cement ratio and other details of the original mix can be assessed.

6.1.4 Break off test

A test, which measures the force required to break of a core, has been developed in Norway. The test was originally developed to monitor the early strength of concrete, and utilized a plastic cylindrical insert at the time of casting to from the cores. How ever, it can also be used on drilled cores. A core of 55 mm diameter is drilled to d depth of 70 mm. the annular hole on the surface is enlarged to form a circular group 10 mm wide and 10 mm deep to accept the loading device, as shown in figure 6.3. The load is applied transversely at a depth of 5 mm from the surface. The flexural, tensile stress developed at the base of the core causes it to rupture and the core breaks off.

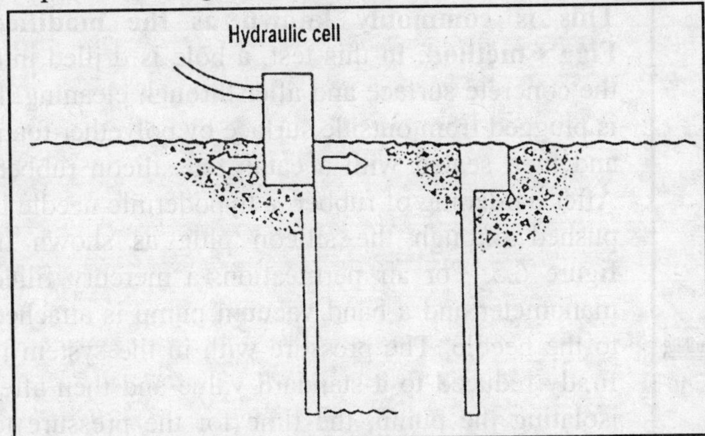

Figure 6.3. Breakoff test

6.1.5 Permeability test

Several types of apparatus have been developed for site use in measuring properties related to permeability. When the test is undertaken on a vertical surface, a means of keeping the cap firmly in contact with the surface has to be provided. Water is introduced into the cap to give a pressure head of 200 mm using a filter funnel. A second port in the cap leads to a horizontal capillary tube. The rate at which water is absorbed into the concrete surface is determined by closing the connecting to the reservoir and measuring the movement of water surface in the capillary tube during a fixed time period. On site the test is usually carried out

10 minutes after filling the cap with water. A diagram of the apparatus is given in Figure 6.4.

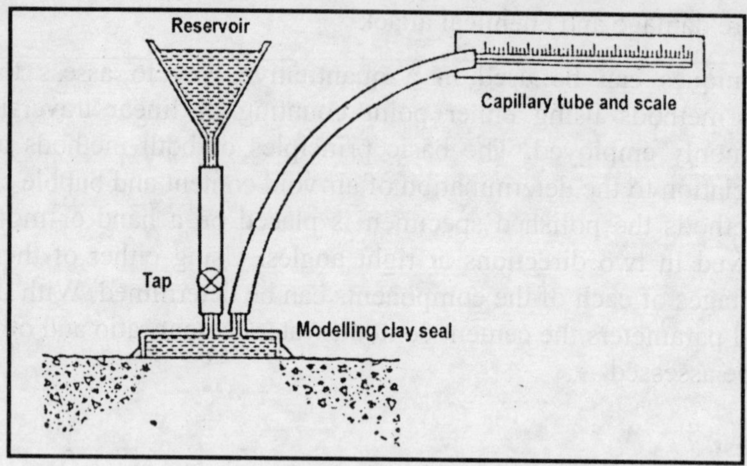

Figure 6.4. Initial surface absorption test

The second test was developed at the building research establishment, U.K in the early 1970 to enable measurements of either air permeation index or water absorption of concrete to be made. **This is commonly known as the modified Figg's method.** In this test, a hole is drilled into the concrete surface and after through cleaning. It is plugged from outside surface by polyether foam and then sealed with a catalyzed silicon rubber. After hardening of rubber, a hypodermic needle is pushed through the silicon plug as shown in figure 6.5. For air permeation, a mercury filled manometer and a hand vacuum pump is attached to the needle. The pressure with in the system is firstly reduced to a standard value and then after isolating the pump, the time for the pressure to raise a standard value is recorded. This time is taken as a measure of the **air permeation index** of concrete. For water absorption, a water head of 100 mm is used. Water is forced in to the assembly using a syringe and one minute after contact between water and concrete, the syringe is shut off. The time for the meniscus in a capillary tube to travel 150 mm is recorded and is taken as a measure of the **water absorption index** of concrete.

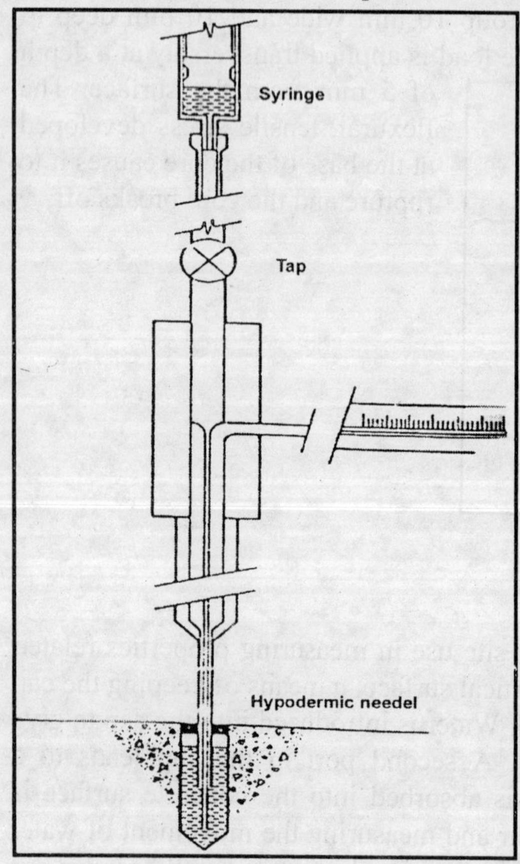

Figure 6.5. Figg's apparatus

A third portable instrument called the Clam uses a circular steel ring bonded to the concrete surface. The main body of the instrument is constant pressure is applied to the surface shown in figure 6.6. The pressure is maintained by means of a piston controlled by a micrometer screw gauge. Readings on the gauge are noted at the beginning and end of a test, which allows the volume of water, which has flowed into the concrete to be calculated.

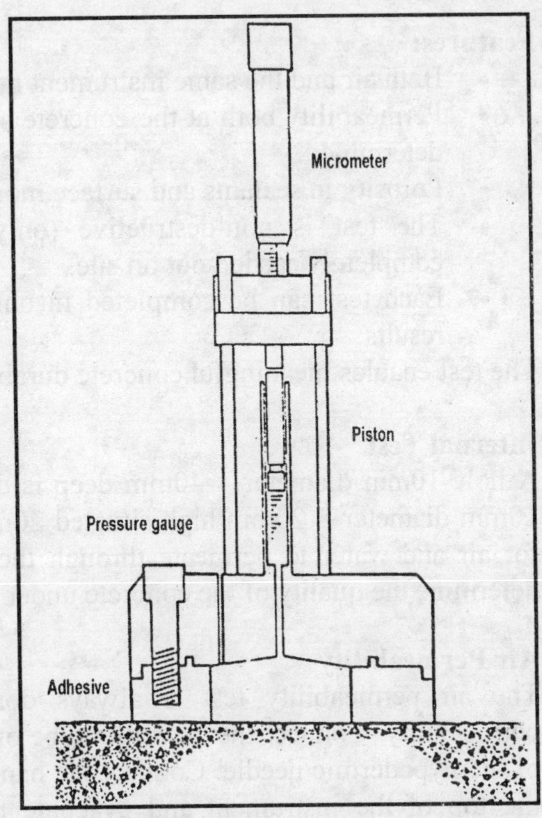

Figure 6.6. Calm apparatus

Advanced permeability tester:
Poroscope-Plus Concrete Air/Water Permeability Tester shown in figure 6.7. The field test for air and water permeability of concrete using the FIGG technique.

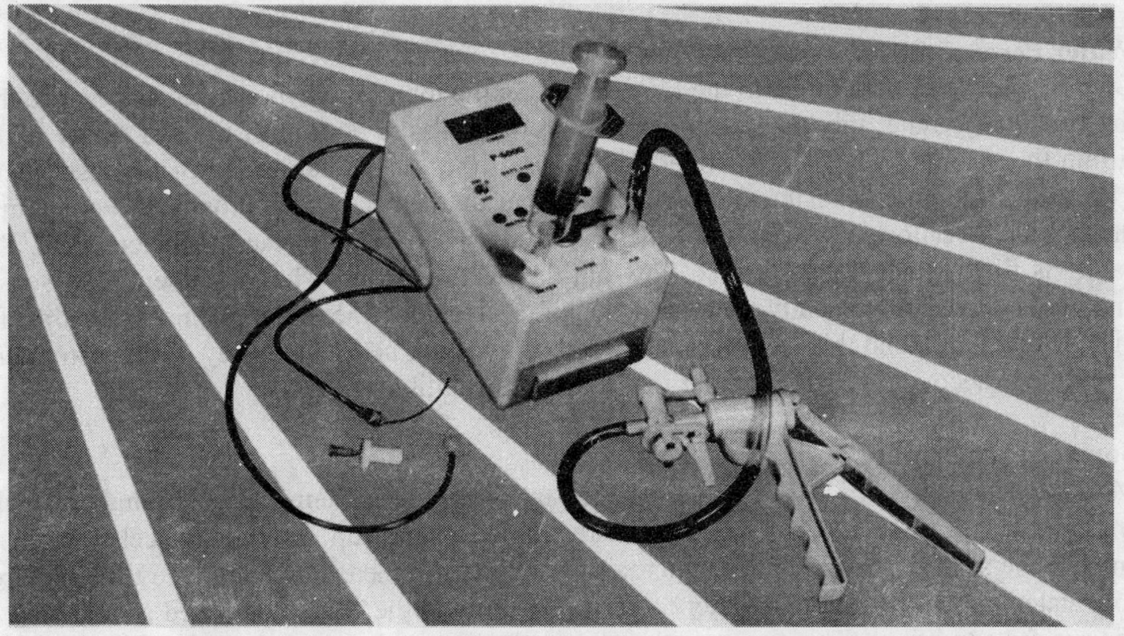

Figure 6.7. Poroscope-Plus Concrete Air/Water Permeability Tester

Features:

- Both air and the same instrument measures water permeability of concrete.
- Permeability both at the concrete surface as well as within the concrete mass can be determined.
- Porosity in sealants and surface mortars can be checked.
- The test is non-destructive (only a small plugged hole required) and can be completely carried out on site.
- Each test can be completed in only a few minutes and gives reliable reproducible results.

The test enables meaningful concrete durability prediction to be made.

Internal Test

A hole 10mm diameter × 40mm deep is drilled and plugged leaving a cylindrical test void 10mm diameter × 20mm high situated 20mm below the concrete surface. The time required for air and water to permeate through the test material to the void is used as an index to determine the quality of the concrete under test.

Air Permeability

The air permeability test is always done first since Moisture has a large effect on permeability. Connect the air outlet tube on the instrument to the Luer Connector on the top of the hypodermic needle. Connect the hand operated vacuum pump to the air connection on the top of the instrument and evacuate to greater than 55KPa. The instrument timer in seconds for the vacuum to fall from -55kPa to -50kPa. This time is the Figg number and is a measure of the air permeability of the concrete.

Water Permeability:

Connect the water outlet tube to the Luer socket on the top of the hypodermic and ensure that the fine plastic inner tube is of sufficient lengh to reach the bottom of the test cavitv. After filling the syringe with distilled water connect it to the water inlet on top of the instrument. The water is then forced into the cavity and the air displaced out through the outer tube through the overflow tube which is 4 inches(100 mm) above the surface of the concrete. The cavity is filled when water starts to flow out the overflow tube. The instrument flow sensor and timer then automatically measured the time taken for the water meniscus to travel a distance of 50mm and this time is seconds is displayed on the LCD display of the instrument. The time in seconds is FIGG number for water permeability.

Surface Permeability Test:

Measurements are carried out at the surface by clamping a Stained steel chamber on the smooth surface of the concrete. An exactly dimensioned cup grinding wheel is used to smooth the sealing surface of the concrete if necessary. A measurement of the time required for related amounts of air and water to permeate through the concrete is used as an index of the surface conditions. This time can then be used to determine the condition of any concrete sealant or surface mortar.

Surface Technical Specification:

A stainless steel surface chamber with the same surface area and exactly twice the volume of the hole used in the internal test is now used as the void for this test. The method of sealing the surface chamber to the concrete eliminates the possibility of variation in the test due to sealants seeping into the chamber, or voids along the sealing surface. The surface chamber is sealed to the concrete by grinding a smooth donut in the surface with the cup wheel provided. This cup wheel is sized to exactly match the dimensions of the surface chamber. A pair of o-rings mounted concentrically in the surface testers 'flange is then used to seal the chamber to surface The two o-rings eliminate the possibility of a surface void in the material being tested defeating the test. After clamping the surface chamber to the surface a strong seal is now provided with no variation in volume.

This surface chamber is now used as the void for testing porosity of the surface. Rather than the walls of the hole being the tested surface the surface that the chamber is sealed against is now the surface tested. This provided a check for water and air penetration through concrete sealants, surface mortars and any other methods used to seal construction material surfaces.

The surface chamber has been designed to easily accommodate attachment to the instrumentation. By first performing the air test as outlined in the internal test and then the water test the instrumentation will provide the time required for the chamber to lose 5kPa vacuum or once filled with water, 0.01ml water. With the surface tester attached to the instrument both a Figg number and direct indices for air and water surface permeability can be established.

Technical Information:

The ingress of air and moisture into the concrete can cause corrosion of the steel reinforcement and lead to a deterioration in concrete strength. Therefore, measure of the ease of movement of the liquids and gases through the surface layer of the concrete is a better method of assessing the soundness and expected life of concrete than strength alone. Permeability is recognized as being the most important parameter in assessing concrete durability.

The air permeability test involves measuring the time then for air to flow into a known volume of a sealed, evacuated chamber in the concrete reducing the vacuum from – 55KPa to – 5-KPa. This time is a measure of the air permeability of the concrete.

The water permeability test utilizes the same sealed chamber in the concrete, which is completely filled with water, and the total time in seconds for a volume of 0.01ml of water to escape is taken as a measure of the water permeability of the concrete.

The moisture content of the concrete has a major effect on permeability. For example, fully saturated concrete is almost impermeable to air and results in extremely long times in the water permeability test. For effective testing the concrete should be dry and near surface moisture content measured.

Permeability test results have shown that there is a good corrolation with both water/cement ratio and compressive strength of the concrete.

6.1.6 Half–cell potential survey

Corrosion being an electrochemical phenomenon, the electrode potential of steel rebars with reference to a standard electrode under goes changes depending on corrosion activity. A systematic survey on well-defined grid points gives useful information on the presence of probability of corrosion activity. The grid points used for other measurements, namely, rebound hammer and UPV can be used for making the data more meaningful. **The common standard electrodes used are:**

(a) Copper–Copper Sulphate Electrode
(b) Silver–Silver Chloride Electrode
(c) Standard Calomel Electrode

To perform the half–cell potential measurements, electrical connections have to be made with reinforcement and with concrete surrounding it. Electrical connection with the reinforcement is made simply by making a hole in the concrete to expose part of the reinforcement and connecting wire to it. Electrical connection with concrete is made by electrolyte in the $Cu/CuSO_4$ reference electrode which wets both the concrete surface and the conductor to which another wire can be attached. Thus for taking half–cell potential readings, the half–cell *i.e.* $Cu/CuSO_4$ reference electrode is contacted to the surface of the concrete via a sponge soaked in contact fluid and the wires attached to the reinforcement and reference electrode are connected to a high impedance millivoltmeter which gives half–cell potential readings. The measurement consists of giving an electrical connection to the rebar and observing the potential difference between the rebar and a reference electrode in contact with concrete surface. A simple field arrangement is shown in figure 6.8.

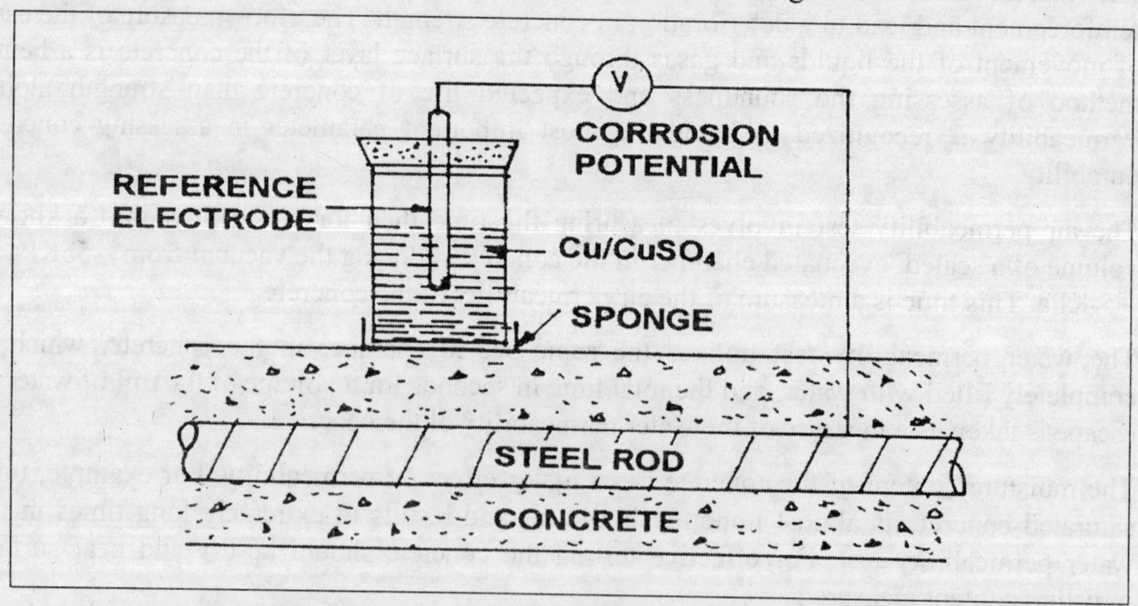

Figure 6.8. Typical setup for potential measurement in rebars

Generally, **the potential value become more and more negative as the corrosion becomes more and more active.** How ever, less negative potential values may also indicate the presence of corrosion activity, if the pH values of concrete are less. The following Table 6.1 shows the general guidelines for identifying the probability of corrosion based on half–cell potential values covered by ASTM C 876-1991.

Figure 6.1. Corrosion risk based on potential readings

Potential readings	*Corrosion*
More negative than –350 mv	> 90%
– 200 mv to –350 mv	50%
More positive than –200 mv	< 10%

However, these interpretation criteria should not be normally utilized under the following conditions:

(a) To evaluate reinforced steel in concrete that has been carbonated/and have highly variable moisture or oxygen content at the level of steel; and

(b) To evaluate galvanized reinforcements which have a range of active/passive potentials completely different of that to bare steel

Advanced testing systems:

The **Ohmcor** and **Cormap** systems represent more economical methods of evaluating steel reinforcement corrosion. The Cormap Systems use half-cell potential mapping in order to identify areas of probably corrosion. The Ohmcorr System is typically used in conjunction with the Cormap system in order to verify the presence of steel reinforcement corrosion. This older technology is still widely used in the field.

A. CorMap

A simple economical method for identifying areas of probable rebars corrosion.

(a) Easy to use

(b) Detachable electrode extension pieces facilities measurements in hard to reach locations.

(c) High impedance digital meter is designed for tough field conditions

(d) Economical

Technical specifications

Corrosion, which is an electrochemical process, occurs in concrete when oxygen and moisture are present. The actual corrosion is an exchange of energy within different sections of the uncoated reinforcing steel. The relative energy levels can be determined in relation to a reference electrode with a stable electrochemical potential. By connecting a high impedance voltmeter between the reinforcing steel and a reference electrode placed on the concrete surface, a measurement can be made for the half-cell potential at the location of the reference cell. This then is a measurement of the vicinity of the reference cell. The reference cell is copper in copper sulphate solution. By taking half-cell potential measurements a fixed distance apart a grid of half-cell potentials can be quickly made and thus areas delineated with a high probability of corrosion of the reinforcing steel. To analyze the results, the measurements made with CorMap can be plotted on a grid and lines of possible corrosion activity. For example the following guide is listed in ATM C876 using a Copper/Copper sulphate half-cell:

(a) For readings of – 350 mV and greater there is a 95% chance of active steel corrosion.
(b) For readings – 200 to – 350 mV there is a 50% chance of active steel corrosion.
(c) For reading less than – 200 mV there is only 5% chance of active steel corrosion.

The Method is particularly useful for:

- Bridge Decks
- Parking Garages
- Concrete Piers & Docks
- Substructure
- Tunnel Lining
- Foundations

B. Cor-Map II

Advanced System for corrosion potential data acquisition and analysis, allowing the user to quickly identify areas of probable corrosion in the field is shown in figure 6.9.

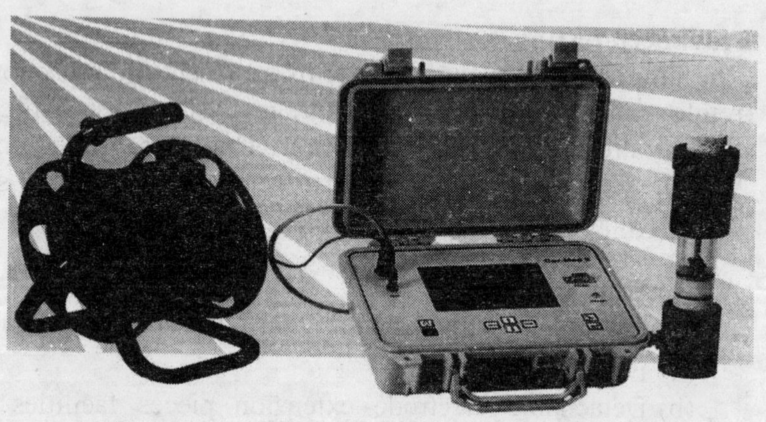

Figure 6.9. Corrosion analyzer equipment – Cor map II

Features and benefits

Ruggedized Electronics allows rapid analysis of data in the field or office. Conforms to ASTM C-876 Standard Method of half–Cell Potential of un-coated reinforcing steel in concrete. Electrode is designed for use on horizontal, vertical and inverted positions. Temperature and humidity sensors facilities inclusion of environmental condition in data analysis.

For steel reinforcing bars in concrete, corrosion is an exchange of ions from the steel to the concrete. This chemical exchange of ions produces rust (FeO_2). It also produces areas of concrete where there is a larger concentration of negative ions due to the corrosion process of the steel reinforcing bar than areas where there is no corrosion. This larger concentration of ions creates a small electric voltage potential. By measuring and mapping the voltage potential found in the concrete we are able to determine rapidly the presence of corroded steel reinforcement without costly and time consuming demolition of concrete.

This is done by recording the voltage between the rebar and half-cell, which is mapped across the surface of the concrete. Areas of rust with high corrosion will exhibited significantly lower voltage than areas without corrosion, thus areas of corroding steel reinforcing bar in concrete can be rapidly found. There is no need to know the exact position of the steel reinforcing bar or the amount of cover, the presence of the steel is that is required. However, the voltmeter has to be connected to an exposed piece of the rebar network, and because the concrete is being tested, any material on the surface should be removed.

Half-cell reference Electrode:

The Cor-Map II system comes complete with rugged half- cells designed for the tough construction environments. Porous ceramic tips are used in order to provide long life and eliminate problems from clogs in Cu/CuSo4 half–cell. The specially shaped tip has also been designed to allow the half-cell to take readings in the vertical, horizontal or inverted position. They also have a semi transparent full view window that allows liquid level observation without removal of sealed ends and still protects the half-cell from damage by sunlight. The Ag/AgCL electrode uses an easily replaceable cell. This also allows reading to be taken in the vertical, horizontal or inverted position. It also makes maintenance a simple matter.

Instrumentation Unit:

They fully integrated data acquisition and analysis unit has been designed for the rapid analysis of data in the field or office. A large amounts of data are normally generated, interpretation of this information can be very difficult. Employing the simple to use menu driven Cor-Map main unit, data can not only be collected quickly and easily, but it can also be analyzed directly in the field on the graphic display. The unit produces a symbolic map of the structure, where symbols represent various half–cell potential voltage levels previously acquired. This symbolic map can then be interpreted like a contour map where areas of high potential represent areas most likely to be corroding. Not only can this information be acquired and analyzed but the unit will also read the general environmental parameters of temperature and relative humidity. All this data can also be stored and uploaded to a P.C this allows the user to include the data in subsequent reports and spreadsheets for further analysis.

Specifications:

Instrument Weight:	6 lbs. (2.75 Kg)
Ship Weight:	15 lbs (6.8 Kg)
Instrument Dimension:	4.5” × 8.5” × 10.5” (115mm × 225mm × 267 mm)
Battery:	12 Volt 4-10 Hours Continues Operation
Display:	320 × 240 Pixels backlit for daylight use

Storage: over 5000 readings

Operating Temperature: 0-50 C
Temperature Reading Range: – 273 - +130 C
Temperature Accuracy: +/-0.5%
Humidity Reading range: 0 -100%
Humidity Accuracy: +/-5%

6.1.7 Resistivity mapping

The corrosion of a specific length of reinforcement is dependent on the algebraic summation of the electrical currents originating from the corroding sites on the steel and flowing through the moist surrounding concrete to non–corroding sites. Hence the **electrical resistance of concrete** plays an important role in determining the magnitude of corrosion at any specific location. This parameter is expressed in terms of **resistivity in ohm centimeter or kilo ohm centimeter** by four-probe resistivity meter shown in figure 6.10.

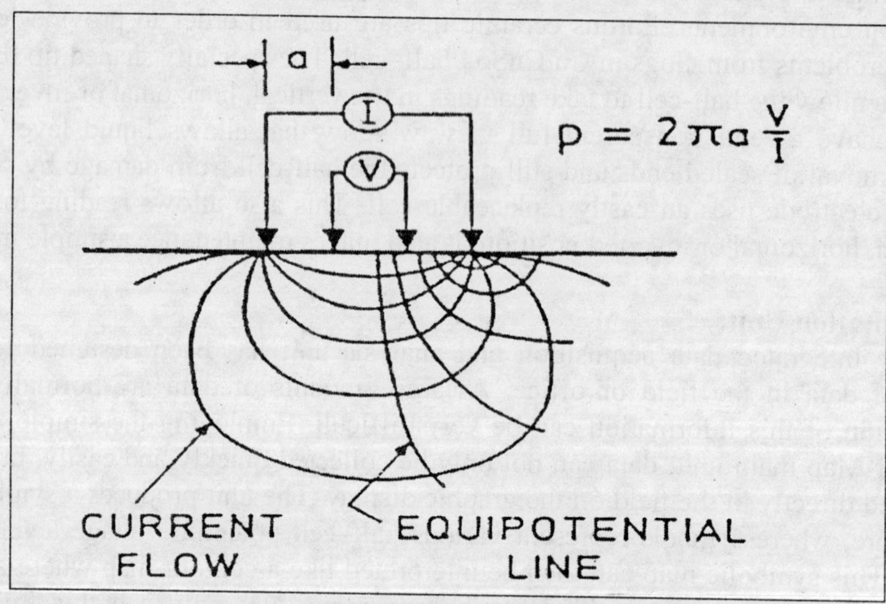

Figure 6.10. Resistivity testing for concrete

The method essentially consists of using a 4 probe technique and the probes are inserted into the concrete to a depth of 6mm with a spacing of 50mm between the probes in which a known current is applied between two outer probe and the voltage drop between the inner two elements is read off allowing for a direct evaluation of resistance R. Using a mathematical conversion factor, **resistivity is calculated as $\rho = 2 \pi R a$.** The factors, which govern the resistivity values, are:

(a) Constituents of concrete
(b) Chemical contents of concrete
(c) Type of pore structure of concrete

The below Table 6.2 indicates the general guidelines of resistivity values based on which areas having probable corrosion risk can be identified in concrete structures.

Table 6.2. Corrosion risk based on resistivity values

Resistivity ohm cm	*Corrosion probability*
Greater than 20,000	Negligible
10,000 to 20,000	Low
5,000 to 10,000	High
less than 5,000	Very high

Advanced resistivity meter:
RM-8000 OHMCORR - Resistivity Meter to Assess Corrosion Currents in Concrete as shown in figure 6.11.

Features

(a) Assesses damaging corrosion currents in concrete.
(b) Economic and easy to use.
(c) Direct digital readout of resistivity. Measuring from two small holes avoids the problems and errors of surface measurements.

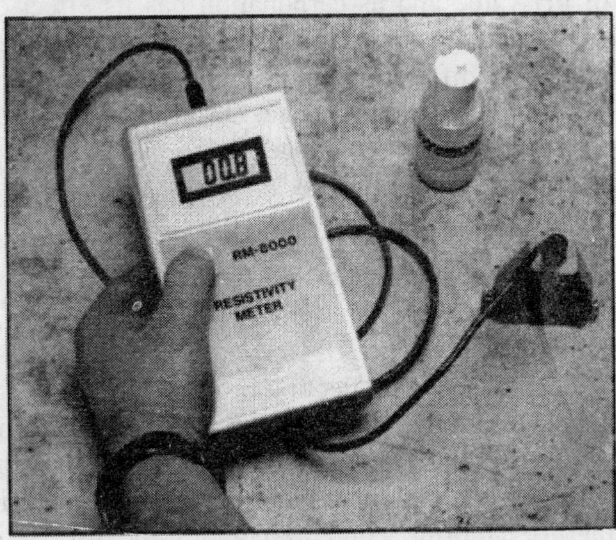

Figure 6.11. Ohmcorr resistivity meter

Used in conjunction with Cor-Map system to produce resistively plots. The OHMCORR, when used in conjunction with The CORMAP system provides an economic and sound means of diagnoses of corrosion in reinforced concrete

Technical Information
The electrical conductivity of concrete is an electrolytic process that takes place by the movements of ions in the cement matrix. This ionic movement will take place when

contaminants such as chloride ions or carbon dioxide are introduced into the cement mortar matrix. A highly permeable concrete will have a high conductivity and low electrical resistance. Because resistively is proportional to current flow, the measurement of the electric resistance of concrete surface should be avoided. The resistively meter, OHMCORR, has two probes spaced 5cm (1.97 inches) apart which Are placed in two holes drilled to a depth of 8mm (3/8 inch) and filled with conductive gel. The concrete resistively is displayed on an LCD. When the control switch is activated. The following table 6.3 correlates arrange of Values vs. the possible rate of corrosion of the Reinforcement bars.

Table 6.3. Risk of corrosion due to resistivity level

Resistivity Level (K Ohms cm)	*Possible Corrosion Rate of Reinforcement Rebars*
< 5	Very High
5 to 10	High
10 to 20	Moderate to Low
> 20	Insignificant

6.1.8 Carbonation test and pH value

Carbonation test can be carried out using **simple phenolphthalein test**. Phenolphthalein spray is an indicator of pH value and as carbonation has its effect on pH of concrete, the change in colour indicates the extent of carbonation. The test is conducted by drilling a hole on the concrete surface to different depths up to cover concrete, spraying with phenolphthalein, and observing the colour change. Uncarbonated surface exhibits pink colour while carbonated concrete exhibits no change in colour. The depth of carbonation is estimated based on the change in colour profile.

The pH value can be determined by collecting broken samples or from core samples by analyzing in the laboratory from water extracts and pH meter.

6.1.9 Chloride content test

Chloride content can be determined by collecting broken samples or from core samples. The test consists of powdering the sample, obtaining the water extracts, and conducting standard titration experiment for determining the water-soluble chloride content, which is expressed by weight of concrete or by weight of cement if the mix ratio is known. One recent development for testing for chloride content includes the use of chloride ion sensitive electrodes. This is commercially known as **"Rapid Chloride Test Kit"** and the test consists of obtaining powdered sample by drilling, collecting the sample from different depths (every 5 mm), mixing the sample (of about 1.5 gm weight) with a special chloride extraction liquid, and measuring the electrical potential of the liquid by chloride–ion sensitive electrode, with the help of a calibration graph relating electrical potential and chloride content, the chloride content of the sample can be directly determined.

The following Table 6.4 gives qualitative guidelines for identification of corrosion prone locations based on pH values and chloride content.

Table 6.4. Corrosion prone based on pH chloride content

Test results	*Interpretations*
1. High pH value greater than 11.5 & very low chloride content	No corrosion
2. High pH value & high chloride content (0.4 to 0.6% by wt of cement)	Corrosion prone
3. High pH value & high chloride content greater than 0.6% by wt of cement	Corrosion prone

Advanced equipments:

The **Poroscope** and **Chloride Field Systems** allow the user to rapidly evaluate mechanisms that cause corrosion. The Poroscope quickly and simply evaluates concrete permeability. The Chloride Field System will rapidly evaluate field chloride levels without expensive and time-consuming laboratory tests.

Features

(a) Fast – Results within minutes at the site.
(b) Economical – Low coat per sample compared to laboratory testing.
(c) Accurate – Results are comparable to laboratory testing.
(d) Covers wide range from 0.002% to 2% chloride by weight.
(e) Automatic compensation for changes in ambient temperature.
(f) Digital display for direct reading of lbs/cu. yd. and percentage of chloride by weight.
(g) Correlates to ASTM C114 and meter conform to AASHTO-T260.

Technical Specifications

The determination of the chloride ion concentration in concrete is essential in assessing the need for maintenance on, for example, bridge decks and parking structures. The test can also be used to ensure that materials used in new construction are free from potentially harmful chloride ion levels. With this method, the concentration of acid soluble chlorides is measured. In most cases, this is equivalent to total chloride concentration. A sample of powder is obtained by drilling and careful quartering. Then an accurately weighed 3gr sample is dissolved in 20ml of extraction liquid which consists of a precise, measured concentration of acid. For sampling wet concrete a 3gr sample of mortar (*i.e.* without coarse aggregate) is used. The chloride ion reacts with the acid of the extraction liquid in an electrochemical reaction. An electrode, with integral temperature sensor, is insured into the liquid and the electrochemical reaction measured. A uniquely designed instrument converts the voltage generated by the chloride concentration. The instrument automatically applies the temperature correction and correlates to ASTM C114. Once the sample is obtained, Test results can be determined and read in less that five minutes. To avoid contamination, the electrode should be thoroughly washed with deionizer Water after each test. Replacement packs, containing twelve bottles of extraction liquid each for one time use, are available. Five calibration liquids each with known concentrations are supplied with each pack. These liquids are used to establish the Calibration curve, and establish the Calibration curve, and to check that the system is functioning correctly. Calibration is not required for each use and the instrument will indicate when calibration is required. The calibration Liquids are colored to avoid confusion between then and with the extraction liquid. All equipment necessary to complete the chloride test is supplies in standard size carrying case.

6.1.10 Abrasion resistance test

ASTM Standard C779-89a prescribes three test procedures for abrasion resistance of horizontal concrete surfaces. In procedure A, the revolving disk test, the abrasive action is provided by the flat faces of three (60.3mm diameter) cold-rolled steel revolving disks, each attached to motor-driven vertical shafts which also revolve about a vertical axis. The abrasive grit used in this procedure is silicon carbide. The comparison of measurements of average depth of wears of representative surfaces (specimen size 305x305x95mm) at 30 and 60 min, exposure to abrasion will indicate the relative abrasion resistance of these surfaces. The other two procedures are Dressing Wheels Abrasion Test and Ball Bearing Abrasion Test.

6.1.11 Alkali silica reaction test

Alkali silica reaction damage is assessed by **ASR Detect** equipment. Simply apply each of the two regents to the broken surface of a concrete core drilled in a suspect structure and rinse off the excess. On ASR contaminated concrete, the resultant stain reveals the presence of ASR. The stain's distribution shows the extent of ASR in the concrete, and their proximity to different components of the aggregate gives clues to the source of trouble. The two gels

Figure 6.12. Simple coloured dye field test to detect alkali silica reaction

that are identified-one staining yellow, the other pink–indicate the stage of ASR's progression. Yellow signals that degradation has begun. Pink warns that degradation is advancing. Typically, ASR occurs in cracks and these cracks often cut through the aggregate and usually do not follow the aggregate-paste boundaries. ASR tends to fill air voids. The ASR detect is shown in figure 6.12.

Features:

(a) Test can be carried out completely on site.
(b) Minimal operator training and no special equipment required.
(c) Utilizes only two environmentally safe days.
(d) Identifies ASR in concrete and differentiates ASR from other causes of degradation.
(e) Results obtained in less than five minutes are easy to interpret. Economic, fast and easy to use.

Advantages:

In contrast to the two established methods of ASR detection–petrography analysis and uranyl acetate analysis, ASR detect has numerous benefits. Because the reagent stains are clearly visible even before the treated sample dries, a complete diagnosis is possible in less than 5 minutes. ASR detect systems are relatively inexpensive. Petrographic analysis requires shipment to a laboratory, adding time and raising the costs to hundreds of dollars per sample, also the uranyl acetate reagent is almost prohibitively expensive. ASR detect is simple enough to use in the field. Its reagent stains are visible to the naked eye and are distinctive enough to be recognized and interpreted by anyone with minimal training. Petrographic analysis requires specially trained technicians working in a well equipped laboratory. Therefore, the use of ASR detect can considerably reduce the cost of diagnosis by reducing drastically the need for petrographic analysis.

The ASR Detect reagents present minimal danger to either human health or the environment. Uranyl acetate is radioactive and contains a heavy metal and therefore, has the potential to cause health and disposal problems. ASR detect provides information not only about the presence but also the severity of ASR. Its affordability allows an engineer to analyze enough samples to obtain an accurate diagnostic picture of an entire structure. The high cost of a petrographic analysis permits only a limited number of samples to be examined. Concrete tested with pink gel, yellow gel and pink yellow gel are shown in figure 6.13 and showing beginning and advanced stages of ASR degradation.

Applications:

ASR detect is both a practical and a scientific tool. Its principal application is analyzing existing concrete structures. By identifying ASR deterioration in its earliest stage, ASR detect facilitates the problem being identified when remediation techniques can be applied; for example, treating the concrete with a lithium bearing solution to inhibit further deterioration is advanced, ASR detect provides a clear picture of the extent and depth of the damage. As a scientific tool, ASR detect can be applied to improving the understanding of where, how and why ASR occurs. That understanding is basic to developing ASR preventatives that allow high-alkali cements or poor-quality aggregates to be used in concrete mixes without risking the development of ASR.

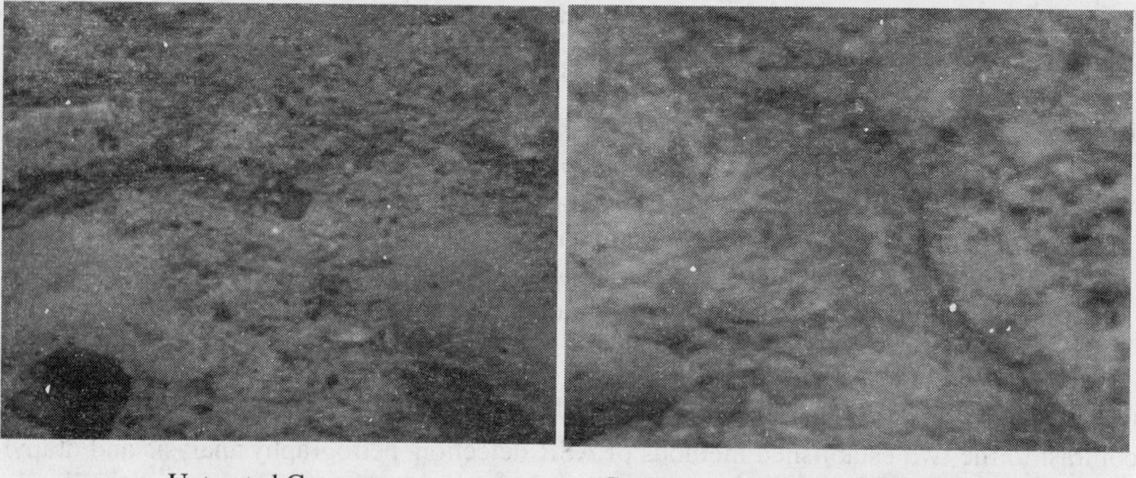

Untreated Concrete

Concrete tested with pink gel only showing advanced ASR degradation

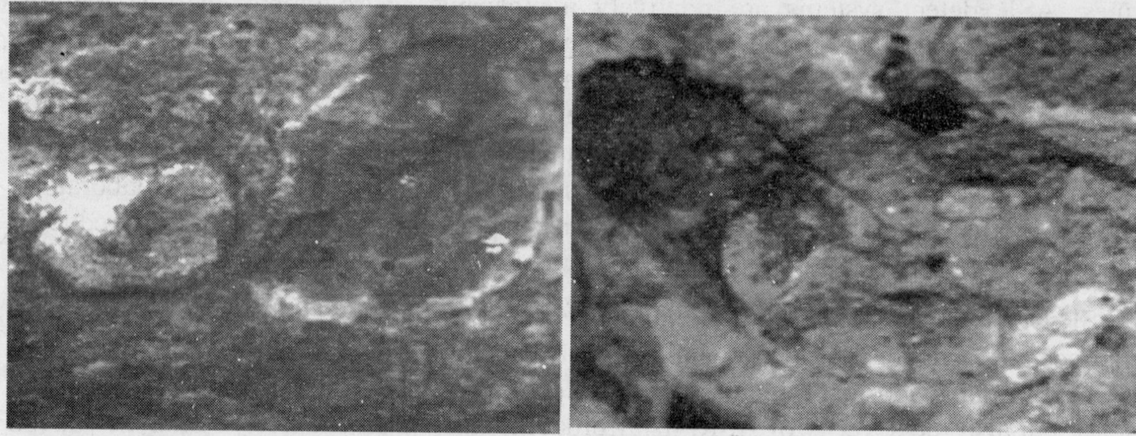

Concrete tested with yellow gel only showing beginning stages of ASR degradation

Concrete tested with pink and yellow gels showing both beginning and advanced stages of ASR

Figure 6.13. Beginning and advanced stages of ASR degradation

Technical specification:

One of the primary causes of premature concrete deterioration is alkali-silica reaction (ASR). ASR causes concrete to deteriorate when sodium and/ or potassium from the cement attacks silica rich components in the aggregate, producing gels that expand and eventually crack the structure. Most gels contain cations (positively charged atoms or,molecules) that readily exchange with other cations in solution. ASR detect 's two reagents react with cations found in the two gels associated with ASR. The first reagent exchanges sodium with the potassium found in some ASR gels and then reacts to a form a bright yellow precipitate. The second reagent reacts with calcium-rich ASR gel to form a bright pink satin. In concrete containing

ASR, the result is brightly colored surface showing the presence of the targeted gels; concrete with no ASR is unaffected.

6.2 CHEMICAL TESTING OF CONCRETE

Chemical testing of hardened concrete is mainly carried out to identify the causes of deterioration, and in the context of rebar corrosion, mainly consist of the determination of original water and cement contents, chloride content, carbonation depth and pH of concrete. Chemical testing of in–situ concrete can be carried out in the laboratory through proper sampling of concrete from the structure under consideration.

Determination of cement content:

The cement content of the powdered sample of concrete can be determined following the standard procedure recommended in ASTM: C 1084–87. In this procedure percentage silica in the concrete or percentage calcium oxide in the concrete is determined by soluble silica sub–procedure or calcium oxide sub–procedure. The cement percentage by weight is expressed as:

$$C = Cs \text{ or } Ce$$

Where

Cs = (% silica in the concrete/% silica in the cement) × 100 (assume 21%, if not known) and

Ce = (% calcium oxide in the concrete/% calcium oxide in the cement) × 100 (assume 63.5 % if not known)

Knowing the cement % (C) by weight, the cement content of concrete can be obtained as:

Cement content = C × D/100

Where D is density of concrete

The cement of concrete has important bearing on its corrosion resistance properties and it affects the rebar corrosion because high alkalinity of the concrete, which resist the depassivation of rebar, is mainly due to the cement. It has been demonstrated by fried land that concrete with cement content less than 235 kg/m^3 can have a highly enhanced rate of rusting even when the w/c ratio is proper, and increasing the cement content to 300kg/m^3 improves the product nature of concrete; however beyond this cement content the improvement in productive nature of concrete is not proportional. Various codes of practices have recogonised the above fact and IS: 456–2000 restricts the minimum cement content for reinforced concrete as 250–360 kg/m^3 for mild, moderate and severe exposure respectively.

Determination of original water content and w/c ratio:

Determination of original water content and w/c ratio of the hardened concrete is covered in BS 1881-Part-124; 1988. The original water content is the amount of water present in the concrete mix at the time of setting. It is the sum of the capillary porosity of the concrete (Q) originally filled with water at the time of setting; and the combined water of hydration present in the prepared concrete sample. The capillary porosity of concrete can be determined using unfractured core sample. The sample is oven dried at 105'c for at least 16 hours cooled to room temperature in the desiccators, weighed, and immersed in a liquid of known density in vacuum desiccators. The pressure is reduced, causing the air from the capillaries to be evolved. The vacuum is then released and the sample kept immersed in a liquid further 5 min. the weight of liquid observed can then be calculated, and the capillary porosity can be determined as: Q = Wt of liquid observed liquid density (1.33) × Wt of dry sample × 100.

If aggregate control samples are not available as in case of diagnosis of old RC structures, it may be assumed that the combined water of hydration of concrete is typically 0.23 × C, where C is the percentage cement content. Further, if the aggregate absorption value is not known, then only the original total w/c, which includes the water absorbed by the aggregates at the time of setting unlike the free w/c, which excludes the water absorbed by the aggregates at the time of setting, can be quoted and is given:

The original total w/c = (Q/C) + 0.23

The w/c of concrete plays different roles in the process of rebar corrosion in different exposure of concrete. When the RC structures are immersed in some aggressive solutions, it is the permeability of concrete, which is a function of a water cement ratio, affects the corrosion of rebar.

Determination of free chloride content:

The researchers have furnished recently a rapid, accurate method for determining the water–soluble chloride (free Cl–ions) in hardened concrete samples obtained from the site. This rapid method involves the extraction of water soluble free chloride ions, present in concrete samples, by boiling. After boiling the powdered sample for ten minutes in water, un dissolved mater is filtered out, and the filtrate acidified to a pH value of 3 before titration against standard mercury nitrate solution using diphenylcarbasone as an indicator. This titration gives an accurate estimate of the free chloride, percent by weight of concrete. IS 456-1978 limits the total amount of chloride in the concrete at the time of placing to 0.15% by mass of cement.

6.3 DIAGNOSTIC METHODS OF CORROSION DAMAGE

Diagnosis of the distressed structures is the first step in the process of their repairs and rehabilitation to identify the real causes of distress. Its importance lies in the fact that rehabilitation carried out without comprehensive diagnosis, will not be able to cure the real causes of distress; and thus rendering the repaired structure susceptible to damage due to same causes again. Hence, if a reinforced concrete structure shows signs of cracking, spalling or rust staining, or any other sign that might indicate corrosion distress, the first task should be to find out how serious it is, and what has caused it, and how it can be set right.

The corrosion of reinforcing steel in R.C. structures is one of the most commonly occurring causes of distress in many structures even in places situated far away from marine environment or sea shore. It is a well-known fact that the corrosion of reinforcing bar in R.C. structures occurs mainly due to two reasons:

(a) Reduction of alkalinity of concrete at rebar level due to carbonation and/or penetration of other acidic gases such as sulphur-dioxide
(b) Presences of aggressive anions such as free chloride ions which can lead to the depassivation of rebar.

In an urban environment such as one encountered in industrial zones of Delhi, the reduction of alkalinity of concrete may be either due to carbonation owing to atmospheric carbon dioxide or due to penetration of polluting industrial acidic gases. The penetration of external chloride into the concrete is more prevalent near the sea-shore; however at places situated far away from marine environment, presence of free chloride ion in concrete is likely to be through addition of $CaCl_2$ during the time of construction as an accelerating admixture or through impurities in aggregates and /or mixing water.

A large number of cases corrosion distress have been reported in recent years. Some examples are: corrosion distress of RCC bridge across little Rann of Kutch, corrosion distress of prestressed concrete Thana Creek bridge, and corrosion distress in buildings located in various parts of Northern India, most of which were constructed in the sixties and seventies, whereas a few cases refer to relatively recent construction. The cause of corrosion in R.C. structures under marine or chloride bearing industrial environment has been well understood which is due to presence of large amount of chloride in concrete made available through penetration from outside, but the cases of corrosion distress in the R.C. structures away from marine or chloride bearing industrial environment, such as those encountered in Northern India, draw attention towards the need of pinpointing the real cause of rebar corrosion which may possibly be more than one.

6.3.1 Investigation strategies

A visual survey of the affected structures provides valuable information to ensure that, whether the corrosion of rebar is really a cause of distress or there is some other cause of distress. This survey consist of a careful investigation of the structures for any sign of distress, such as, cracking, cracking and spalling, cracking and rust staining, and simply rust staining. If visual inspection of structure suggests that the cause of distress is the corrosion of rebar only, the next step is to make a careful examination of the structure and carryout detailed test which will positively identify the cause and extends of distress, and allow some predictions to be made about the future life of the structure. The prediction of the future life of the structure needs measurement of the rebar corrosion rate and assessment of the loss of sections from the reinforcement or prestressing steel due to rusting. Measurement of the corrosion rate is useful for life prediction in terms of the residual time during which the cover concrete would crack or spall due to rusting; where as the assessment of the loss of section from the rebar or prestressing steel is useful to predict the life in terms of the residual time during which the strength of structure is impaired.

After the initial sign of corrosion has been confirmed from visual observation, a thorough preliminary investigation may be carried out to locate the rebars and to estimate the cover depth. This may be followed by half–cell potential mapping and concrete resistivity mapping to isolate the corroding and non–corroding regions. The causes of rusting can then be ascertained by chemical testing of concrete both in–situ and in the laboratory using cored samples obtained from corroding regions. The rate and extent of damage may be estimated through electrochemical testing techniques and through chemical cleaning of randomly collected samples of rusted rebars.

6.3.2 Detailed test and inspection techniques

A considerable range of test and inspection techniques is available for use in–situ during inspection, or on samples removed from the structures for laboratory investigations. **Test and inspection techniques can be subdivided as follows:**

(a) Those dealing with structural integrity and location of reinforcement, example, Pachometer or Covermeter service;

(b) Those dealing with prestressing steel or reinforcing steel serviceability and condition, example, electrochemical and non-electrochemical measurements;

(c) Those dealing with concrete quality and composition namely chemical testing of samples obtained by crushing of cores collected from site, for determination of chloride content. pH, cement content and original w/c ratio etc, additionally the core samples obtained from the site can be tested for estimation of potential cube compressive strength and absorption test etc; in order to ascertain the durability quality of concrete.

6.3.3 Determination of structural integrity and location of reinforcement

Visual surveys

The first visible of reinforcing steel distress in a conventional RC structure is frequently the appearance of delamination of the outer concrete layer and rust–staining on the concrete surface. Obviously, this only occurs where the surface can be visually examined. A later visual stage occurs where cracks which may be have developed at the subsurface, reach the surface and spalling of the concrete ultimately occurs. Hairline cracks in concrete running parallel to the rebar may be due to shrinkage or they may be the first sign of corrosion distress. That cracks developed, whether due to rusting or due to shrinkage, can be differentiated on the fact that the delamination with rust staining on concrete surface appears due to rusting, rather than shrinkage, but this evidence is not always apparent. There are many tools which may be used to detect delamination or surface fracture planes parallel to the concrete surface. These devices range from simple chain drags or light-weight striking hammers, involved in surrounding technique, to more sophisticated devices such as Delamtect. Almost any surrounding device can be used to locate hollow areas are delamination caused by corrosion of rebar. Ultrasonic reflective method to determine the existence and extent of sub-surface delamination and the depth of visible surface cracks can be alternatively used. The suitability of various delamination detection techniques/devices in different situations is presented in Table 6.5.

Table 6.5. Suitability of delamination detection techniques/devices

S.No	*Delamination detection technique/devices*	*Suitability*
1	Simple chain drag	For locating delaminated areas during repair operations
2	Automated Delamtect	For surveying large numbers of bridge decks or other horizontal surfaces such as parking garage floors if record of the area of delamination is required.
3	Surrounding technique using Light-weight striking hammer	For upper surfaces of bridge decks and also on vertical surfaces
4	Ultrasonic reflective technique using Soniscope	May be modified so as to be of use even on submersed surfaces

Visual survey should be carried out experienced personnel capable of making careful observations and record findings in a systematic manner, paying attention both to areas exhibiting distress and areas which are visually sound for comparison. The visual survey is normally used as the basis of confirmation of the reinforcement corrosion distress and for any on site testing, non-destructive testing and sampling.

B. Pachometer or Covermeter surveys:

Pachometer or Covermeter is used to locate reinforcing steel embedded in concrete, and to determine the amount of cover over steel. The purpose of locating the reinforcing steel, prestressing strand, cladding ties etc., in concrete, is to avoid them during core sampling, direct pulse velocity surveys etc., and to make connections for electrochemical measurements. The knowledge of cover depth is essential if it is desired to obtained samples of the concrete at the level of the rebar for chloride ion analysis, pH measurement and the presence of carbonation. It is also useful in determining the potential for corrosion and subsequent concrete deterioration since it has been well established that structures in corrosive environments with inadequate cover get subjected to early deterioration.

6.3.4 Determination of steel serviceability and condition

Measurement of reinforcement corrosion in concrete structures needs a method which can determine simply, accurately and non–destructivity not only whether corrosion of reinforcement is taking place bit also its intensity and the rate of damage. The most widely used in–situ methods are half–cell potential mapping, and the resistivity testing of concrete. Of the other methods, the linear polarization or polarization resistance method is considered to be the most suitable for determining the rate of corrosion. Furthermore, the gravimetric test and visual observation can also be performed on rebar samples, obtained from site, to assess the percentage weight loss of rebar and its geometrical features after rusting.

A. Half–cell potential mapping

The half–cell potential is also called as open circuit potential or rust potential or corrosion potential. The test procedure is discussed in the previous headings. Half–cell potential data are normally presented as isopotential contour plots known as half–cell mapping which can be used to delineate anodic and cathodic areas of reinforcement. It is important to note that the potential mapping can provide information on the extent of corrosion and its probability but rates of corrosion cannot be inferred from absolute values of potential. In general, where corrosion is slow, potential difference rarely exceed 100 mV, where as in reinforced concrete undergoing significant corrosion potential differences over 200 mV are common. This method is widely used when assessing maintenance and repair requirements, with the advantages of being non destructive, inexpensive, simplest to perform, and able to survey the whole structure quickly. It is particularly valuable in isolating the regions in which corrosion may cause future difficulties, from those in which it has already occurred but with no visible evidence at the surface of concrete. This helps in restricting the corroded regions of structure for detailed investigation. However, this method is difficult to perform when contaminants are present on or in concrete.

B. Resistivity testing of concrete

The procedure of that test is explained in the previous headings. Like half–cell potential mapping, resistivity measurements, particularly using surface contact probes, can be a rapid scan technique that provides useful information of rebars corrosion. However, resistivity testing is normally used only in conjunction with detailed half–cell potential mapping. For in–situ usage, variations in moisture conditions are likely to have the major influence upon results. Furthermore, it should be prone in mind that resistivity measurements generally relate to the concrete cover to the steel and not to the resistance of the concrete immediately next to the steel where corrosion is occurring. It is generally accepted however that for practical purposes the depth of zone of concrete affecting the measurement will be equal to the electrode spacing and a value of 50mm is commonly adopted. The method is therefore only indicative and other factors, such as oxygen availability at the cathode or anode-cathode ratios, may also influence the corrosion rate of the reinforcement.

C. Corrosion rate measurement

In reinforced concrete structures, determination of actual rate at which the reinforcement is corroding assumes larger importance. One method is known as Linear Polarisation Resistance (LPR) method for the on–site study of corrosion rates of steel in concrete. The fundamental principle of linear polarization is based on the experimentally observed assumption that for a simple model corroding system, the polarization curve for a few millivolts around the corrosion potential obeys a quasi-linear relationship. The slope of this curve is the so-called polarization resistance (R_p):

$$R_P = \left(\frac{\Delta E}{\Delta I}\right)_{\Delta E-0}$$

From this slope, the corrosion rate can be determined

$$I_{corr} = \frac{B}{R_P}$$

Where *B* is a constant which is a function of the Tafel slopes and βa, βc are determined from the formula given below:

$$B = (\beta a \cdot \beta c) / 2.3\,(\beta a + \beta c)$$

The value of B usually lies between 13 and 52mv depending on the passive and active corroding system. For on-site measurements, the testing system consists of a potentiostat, counter electrode, reference electrode, and the reinforcement as working electrode. This system is schematically illustrated in figure 6.14. A typical plot of linear polarization curve is shown in figure 6.15. It is necessary that, for measurements in concrete, the potentiostat should have electronic ohmic compensation or otherwise, the value is to be obtained by calculation or separate experiments. The application of commercial instrument known as GECOR-6, for measuring the corrosion current in reinforcement in an actual structure. This works on the principle of LPR technique.

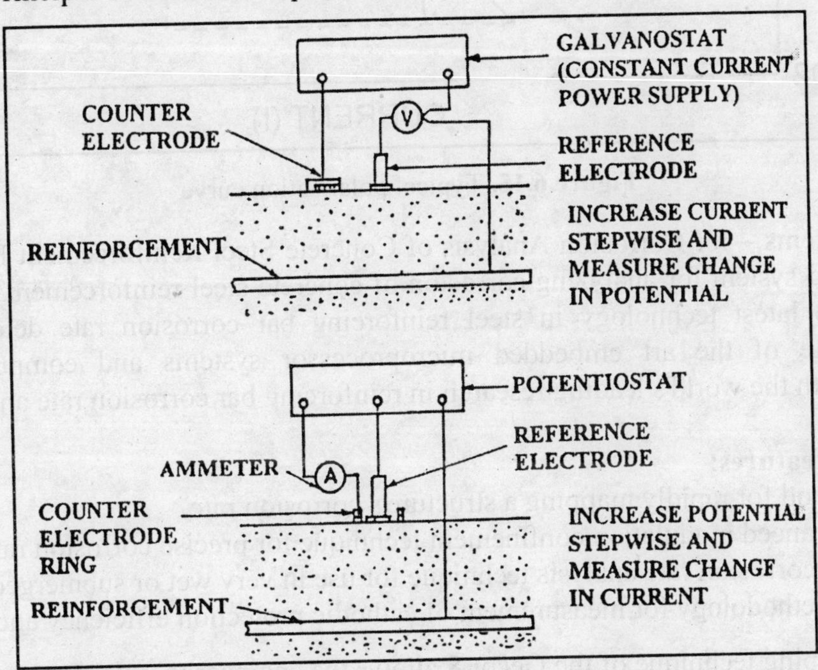

Figure 6.14. Typical setup of linear polarisation resistance method

This category comprises the range of instruments that evaluate parameters related to the corrosion of the concrete or the steel reinforcing bar within the concrete. The leading instruments in this field are Gecor Systems. These represent the latest research and technology in determining the corrosion rate of steel reinforcing bar in concrete. The Gecor 8 is most sophisticated system using a closed loop modular confinement system in order to confine the linear polarization test from outside parameters. The Gecor 6 is a more economical method of determining corrosion rate measurement. These will allow the end user to accurately locate areas of high corrosion with out expensive demolition work and to estimate the service life of the structure.

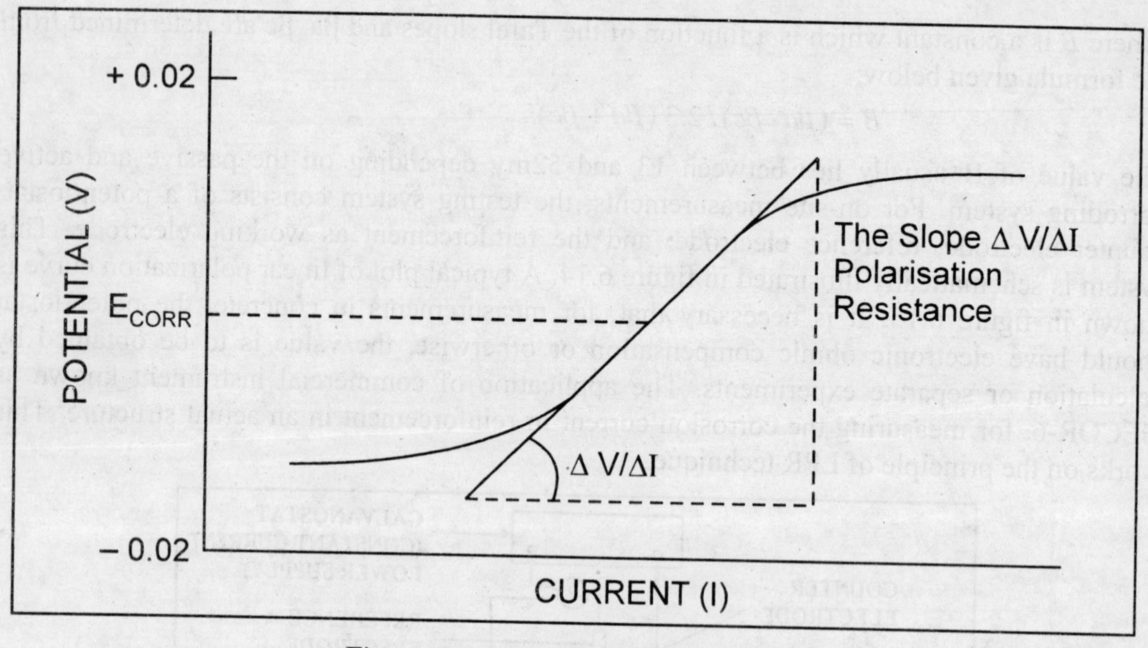

Figure 6.15. Typical polarisation curve

GECOR Systems – for Corrosion Analysis of Concrete Steel Reinforcement Bars - World's most advanced system for analyzing corrosion of concrete steel reinforcement bars. Gecor 8 represents the latest technology in steel reinforcing bar corrosion rate determination. It combines state of the art embedded microprocessor systems and computerized flash technology with the world's leading research in reinforcing bar corrosion rate analysis.

The Gecor 8 features:

(a) A method for rapidly mapping a structure's corrosion rate.
(b) An advanced modulation confinement technique for precise corrosion rate analysis.
(c) A new corrosion rate analysis technique for use in very wet or submerged structures.
(d) New methodology for measurement of cathodic protection efficiency and more.

The rapid mapping technique of the Gecor 8 allows the engineer to quickly classify areas of a structure while the unit's built-in programming can graphically analyze this data and produce contour maps of suspected areas. And, our advanced modulation confinement technique accurately measures the true polarization resistance of the steel reinforcing bar. Combined with the latest advances in interface systems and database storage technology, the Gecor 8 is an easy to use, reliable, automatic and intelligent corrosion rate analysis system. A system that can save an engineer time, money and effort in evaluating structures for subsequent rehabilitation.

D. Gravimetric test and determination of geometrical features of rebar sample

Gravimetric test and measurement of the reduced diameter of rusted rebar can be carried out by cleaning the rebar samples using Clark's solution as per ASTM G1-81. Composition of the Clark's solution is as follows; HCl of sp.gr.1.19 one litre, Antimony trioxide 20gms and stannous chloride 50gms. After proper cleaning the difference in weight and reduced diameter of bar can be measured. The difference in weight can be used to determine the average corrosion rate of rebar. No differential or instantaneous corrosion rates can be measured, but only a mean during the period of test. When pitting is present, it must be accompanied by visual observation. The measurement of the reduced diameter of the rebar can help in determining the residual strength of a structure.

6.3.5 Determination of concrete quality and composition

A. Testing of drilled cores:

The test procedure of core drilling and sampling is already explained in the previous headings. The only way to get detailed information about the in–situ concrete is by obtaining and testing drilled cores from the distressed structures. However, obtaining the cores is relatively expensive and time consuming. The cores drilled from concrete structures can be tested in compression in addition to chemical testing to measure the compressive strength of concrete; they can be tested in splitting to measure the tensile strength of concrete; they can be weighed to measure its density; sliced to measure its permeability; and examined petrographically. After these physical tests on cores, they can be crushed to analyse the concrete chemically to determined cement content and type, chloride content, water cement ratio, pH, aggregate type and grading. And if a core is cut through the reinforcement, its condition can be examined too by electrochemical and non–electrochemical methods. However, normally coring through reinforcement should be avoided because it can seriously weaken the structure besides rendering it unsuitable for compression testing. Corrected absorption value of 3% for 75 mm dia core sample indicates good concrete, where as the same between 3–5 % indicates moderate concrete and absorption higher than 5 % indicates a poor concrete.

B. Chemical testing of concrete

These tests include:

- Determination of cement content
- Determination of original water content and w/c ratio
- Determination of free chloride content
- In–situ determination of depth of carbonation
- Determination of pH value of concrete

The above tests procedures are already explained in the previous headings.

6.4 SYSTEMATIC ASSESSMENT OF FIRE AFFECTED STRUCTURES

The assessment of damage caused by fire in a building is of almost importance as it is a key to any subsequent repair work and its economic and practical viability. Damage caused by fire can be of three categories named damage beyond repair, which is more difficult and requires close examination of each structural member. The strategy generally adopted in this situation is to determine the extent of damage and residual material strength of the structural member either directly or through indirect methods. Direct estimation of the material strength of the member is usually by means of non–destructive involve estimation of peak temperature and the peak temperature profile attained in the material during the time of fire; from search known temperature history of the material, a reasonable estimate of their residual strength can then be made from available charts etc. Yet another way of estimating the temperature history of material is through theoretical calculation. By estimating the probable value of the air temperature attained during fire in the room, from heat transfer calculation and using charts, one can estimate the temperature profile in structural sections. The steps in the procedure generally adopted, estimation of residual strength and degree of damage, and lastly stabilize check of the building.

6.4.1 Preliminary investigation

A prompt and careful preliminary investigation of fire damaged structure is necessary to discover any urgent requirements for temporary propping before cleaning up and safe detail investigation. The evidence of the preliminary survey should also enable the engineer to identify the structural element to be examined first. The investigation helps in selecting investigational techniques and test methods which would be best suited to the damaged structure in question on the basis of preliminary investigation, a damage classification of structure can be done as given in table 6.6.

Table 6.6. Damage classification of fire damaged concrete

S.No	*Damage level*	*Investigation requirement*
1	Repairable damage	Requires close examination of affected structural member by means of the best suitable investigational technique and test
2	No damage	Decision regarding this part of structural member can be made during preliminary investigation just by visual observation
3	Damage beyond repair	Decision regarding this part of structural member can be made just by observation such as deflection etc of the member concerned

6.4.2 Detailed investigation

Detailed inspection and investigation may be taken up once the structure has been propped up and found reasonably safe. This process involves a range of technique starting from simple thorough visual inspection to highly sophisticated methods. Each individual method has its own advantages and disadvantages and a number of methods are used together in conjunction to arrive at worth–while conclusion. The detail investigation strategy may be summed up as shown in figure 6.16.

A. Detail visual inspection

This includes visual survey of colour, texture and damages of surface concrete and collection of evidential debris, resulting in identification of zones based on extent of damage,

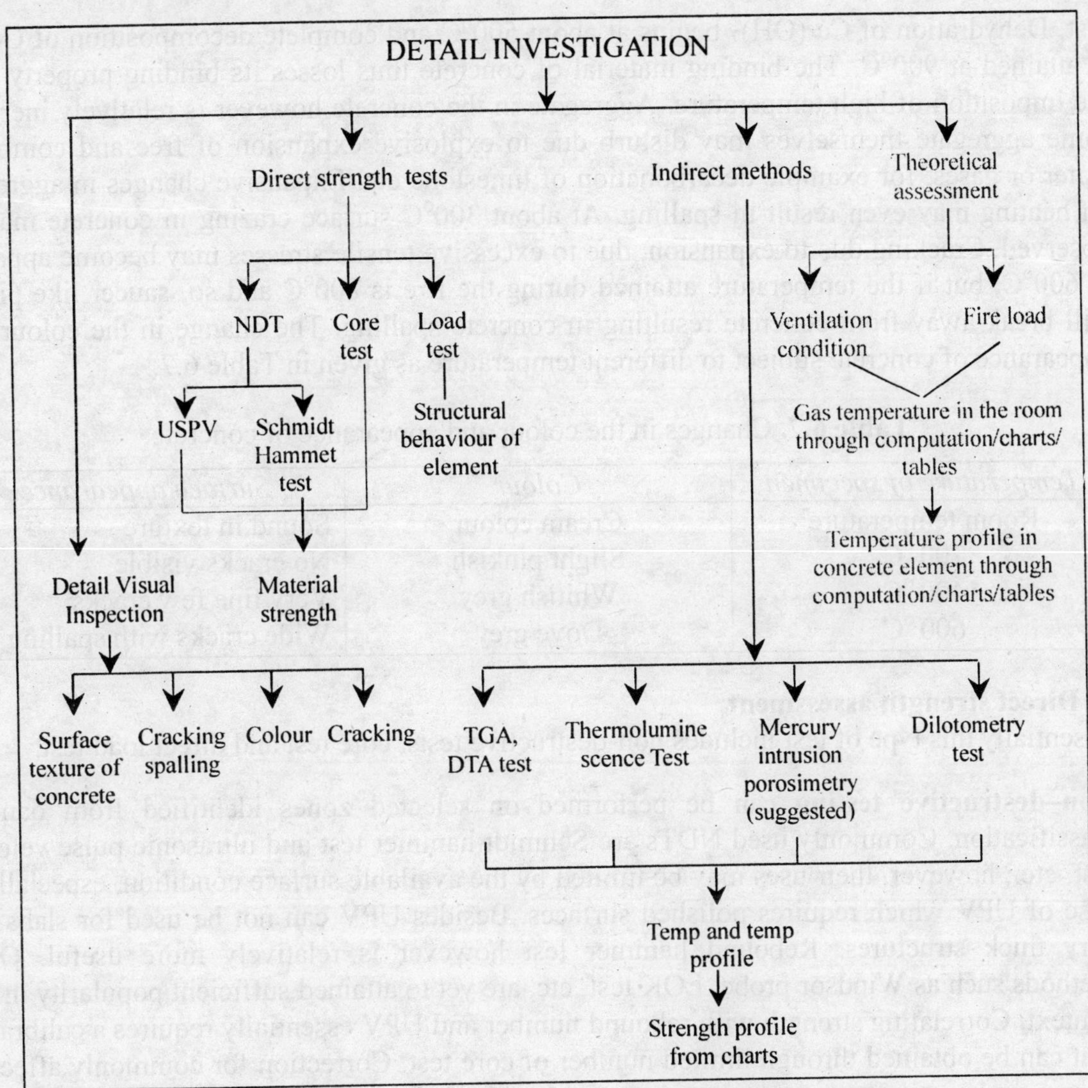

Figure 6.16. Flowchart diagram of detail investigation

approximate assessment of gas temperature in the room and temperature attained by the surface concrete. Concrete is known to undergo a set of irreversible chemical and physical changes when subjected to high temperature regime. Concrete made from non-igneous aggregates usually undergoes a permanent colour change through oxidation on heating. It has been stated that below 300°C of temperature colour of concrete is normal, progressively between 425°C-600°C it may be pink hue depends to red. Between 600°C–1000°C colour changes back through grey buff. However, this colour change depends upon type of aggregate for example, calcareous and igneous crushed aggregate are less susceptible to this effect.

When concrete is subjected to high temperature, the evaporable water present in concrete is lost first, followed by the commencement of dehydration and breakdown of cement gel about 180°C. At about 300°C inter layer water at C-S-H and some chemically combined water is lost. Dehydration of Ca $(OH)_2$ begins at about 500°C and complete decomposition of C-S-H is attained at 900°C. The binding material of concrete thus losses its binding property with the imposition of high temperature. Aggregate in the concrete however is relatively inert but some aggregate themselves may disturb due to explosive expansion of free and combined water or gases, for example decarbonation of limestone etc. Expansive changes in aggregate on heating may even result in spalling. At about 300°C surface crazing in concrete may be observed. Cracking due to expansion, due to excessive tensile stresses may become apparent at 600°C, but if the temperature attained during the fire is 800°C and so, saucer like pieces will break away from concrete resulting in concrete spalling. The change in the colour and appearance of concrete subject to different temperature as given in Table 6.7.

Table 6.7. Changes in the colour and appearance of concrete

Temperature of specimen	*Colour*	*Surface appearance*
Room temperature	Cream colour	Sound in texture
200 °C	Slight pinkish	No cracks visible
400 °C	Whitish grey	Very fine few cracks
600 °C	Dove grey	Wide cracks with spalling

B. Direct strength assessment

Essentially this type of test includes non-destructive tests, core test and direct load test.

Non–destructive testing can be performed on selected zones identified from damage classification. Commonly used NDTs are Schmidt hammer test and ultrasonic pulse velocity test, etc.; however, their uses may be limited by the available surface condition, especially in case of UPV which requires polished surfaces. Besides UPV can not be used for slabs and very thick structures. Rebound hammer test however is relatively more useful. Other methods such as Windsor probe, LOK test, etc, are yet to attained sufficient popularity in this context. Correlating strength with rebound number and UPV essentially requires a calibration that can be obtained through limited number of core test. Correction for commonly affecting factors such as moisture content, carbonation, etc. must be duly applied.

Core test is one positive way to confirm suspicious about the quality of cast in–situ concrete is to take core from it. The relevant properties are determined by the standard test procedures, where detailed results of compressive strength, modulus of elasticity, hardness etc. are required. Cores which intersect reinforcement steel provide an opportunity of examining the steel and its condition of bond with cement paste. In order to see the compari-son between the undamaged and damaged concrete, cores are taken from undamaged concrete also. Core test can also be helpful in obtaining the calibration for other NDTs. This method provides an in depth investigation of the damage occurred to concrete during fire. However, taking out more and more cores will make the already weaken structure more weak.

Direct load test method is very expensive and time consuming. In in-situ test whether the member under test is actually subjected to the assumed test load during fire is difficult to predict due to load sharing effect. All the direct methods of assessing residual strength of concrete provide information regarding average potential cube compressive strength and are not indicative of strength profile across the section. Direct load testing again may be not be possible where the loading may damage the structure permanently or where the structural system in to stiff to show any deflection.

C. Indirect test

For indirect assessment of residual strength, its relationship with the temperature attained must be known well ahead of the test. Thermogravimetric test (TG), Dilatometric test, Differential thermal analysis (DTA), Thermoluminescence test (TL), etc. belong to this category.

Thermogravimetric test (TG):

This test together with dilatometric test was used to determine the temperature attained during fire. As the concrete undergoes irreversible chemical changes during fire there is weight loss due to the process of dehydration of C–S-H and dehydroxylation of calcium hydroxide and is indicated by thermogravimetry test. This process can be considered as a process of chemical stabilization. Thus the concrete which has not been subjected to fire at all is likely to show weight loss at a slow rate initially below 100°C may become rapid beyond 300°C upto a temperature of 800°C when the dehydration process is complete. A concrete which on the other hand has been subjected to a temperature of say 500°C if tested with in a reasonable period from the date of fire is unlikely to show any weight loss till 500°C beyond which weight loss patterned will be identical as that of unaffected concrete. The unaffected concrete can be treated as a reference specimen with which thermogravimetry curves of fire affected concrete may be compare to ascertain the temperature. The quantity of test samples required is of order of 500mg of ground material, therefore if concrete sample is collected from different depths, approximate temperature profile can be established through this test.

Dilatometric test:

The principle underlying this test is similar to that mentioned above for TG test. During the process of dehydration irreversible shrinkage of concrete takes place and which may be detected through °C test. In this test again unaffected concrete is used as reference where the shrinkage takes place rapidly beyond a temperature of about 150°C after initial expansion, till about 850°C and line expansion vs temperature represents dilatometric curves. The fire damaged concrete on the other hands demonstrates thermal expansion monotonically till the temperature to which it has been subjected during fire, beyond which the shrinkage follow reference curve for unaffected concrete. Thus, by comparison of the dilatometric curve of fir affected concrete with that of un affected concrete, one can established the probable temperature to which concrete was subjected to. The sample of material originally used was 7.62cm long × 1.3cm × 1.3 cm square representing the test specimen. The rate of temperature rise at which the test was performed as 5°C/min. Details of this test together with TG test and their use for assessing the temperature of fire damaged concrete. It must be recognized that this method provide only indicative value of temperature and error can be anything of the order of ±15 to ±25%.

Differential thermal analysis (DTA):

This technique is based up on measurement of temperature change of small sample of powdered concrete accompany the irreversible physiochemical transformations at a temperature when subjected to a heating region in a furnace. Usually the rate of change is compared with an inert sample whose rate of temperature rises in controlled to be as nearly uniform as possible. The sign of the measured temperature change increase or decrease in relation to the reference determines whether the change is exothermic or endothermic and indicated by peaks in the measurement. The exothermic/endothermic peak temperatures are characteristics of particular physiochemical changes such as loss of water of crystallization, dehydro oxylation and decarbonation etc. The DTA can be performed for an unaffected sample and expected peaks can be identified. The absence of peak at a desired temperatures for the concrete subjected to fire excludes the possibility of the expected physiochemical changes there by giving an indication that this temperature was already attained by the sample during the fire. DTA curve therefore can throw light on the temperate history of the sample and an approximate assessment of the temperature profile of the concrete section can be made when drilled sample from different depths are analysed. Sample weight may be of the order of 16–20 mg depending up on the instrument, and rate of temperature increases in the furnace may be of the order of 20°C/min. However literature on application of this method to post fire assessment of temperature of concrete is not many.

Thermoluminescence test (TL):

The test is useful in finding out the temperature history of concrete exposed to a temperature of 300°C-500°C. Like the other indirect tests this test also requires very small amount of sample and is sensitive to the thermal exposure experienced by concrete rather than just maximum temperature attained. TL is the emission of visible light which occurs when certain minerals are heated. The intensity of emission of light vs temperature curve for a particular material depends up on its thermal and radiation history. Essentially TL is a result of presence of deflects in crystal in materials like quartz, feldspars, etc. When energy is supplied, the vibration of lattice causes escape of electron or ion trapped as defect in the crystal which become free to defuse and may be recombined with another ion resulting in radioactive light emission called thermoluminescence. If the sample has already been subjected to heating up to temperature under consideration then this emission will be absent at that particular temperature during the test. This forms the basis of use of the test. Temperature Vs TL curve of the fire affected samples may be compared with that of unaffected samples, and from the comparison, the duration of exposure to the given temperature and the temperature its self for the sample can be estimated. Using cored sample from the cross–section of the structural element, the profile of temperature thus can be obtained. The equipment required for this method is some what complex and drilled sample used may be washed in concentrated acid solution to remove mineral which can give spurious results. Usually it is TL of quartz up to 500°C. However this test is sensitive to thermal exposure rather than maximum temperature and is sensitive to exposure of concrete to ultraviolet radiation.

OBJECTIVE TYPE QUESTIONS

1. The problems rises due to damage,
   (a) In investigation and repair
   (b) In repair and rehabilitation
   (c) In damage assessment
   (d) All the above

2. A systematic investigation consists of,
   (a) Find out the cause of damage
   (b) Assess the condition of structure
   (c) Formulate repair and restoration
   (d) All the above

3. The first key of investigation is
   (a) Visual inspection
   (b) Diagnosis
   (c) Estimation of loads
   (d) Evaluation of strength

4. The testing system of hardened concrete
   (a) Non destructive testing systems
   (b) Partial destructive testing systems
   (c) Destructive testing systems
   (d) All the above

5. One of the non destructive testing method is
   (a) Permeability Test
   (b) Windsor Probe
   (c) Resonant Frequency Method
   (d) All the above

6. One of the partial destructive testing method is
   (a) Pulse echo method
   (b) Internal Fracture Test
   (c) Radioactive method
   (d) All the above

7. Name of the destructive testing method is
   (a) Load tests
   (b) Strength tests
   (c) Stability tests
   (d) All the above

8. Non destructive testing methods conducted
   (a) Without any disturbance of surrounding concrete

(b) Partial disturbance of surrounding concrete
(c) Complete disturbance of concrete
(d) None of the above

9. Semi destructive testing methods conducted
(a) Without any disturbance of surrounding concrete
(b) Partial disturbance of surrounding concrete
(c) Complete disturbance of concrete
(d) None of the above

10. Destructive testing methods conducted
(a) Without any disturbance of surrounding concrete
(b) Partial disturbance of surrounding concrete
(c) Complete disturbance of concrete
(d) None of the above

11. The principle of rebound hammer test is
(a) Impacting the concrete surface
(b) Impacting the body of concrete
(c) Impacting the edge of the concrete
(d) None of the above

12. The impact energy in rebound hammers are
(a) 0.7 kg m to 3.0 kg m
(b) 0.07 kg m to 3.0 kg m
(c) kg m to 3.0 kg m
(d) None of the above

13. The high impact energy hammer is used in
(a) Mass concrete
(b) Road pavements
(c) Air port runways
(d) All the above

14. The low impact energy hammer is used in
(a) Ordinary concrete
(b) Light weight concrete
(c) Normal concrete
(d) All the above

15. The code recommended the test procedure of rebound hammer is
(a) BIS–13311 Part 2
(b) BIS–13311 Part 1
(c) BIS–13311 Part 3
(d) BS–13311 Part 2

16. To estimate the concrete compressive strength from rebound numbers from
   (a) Workability curves
   (b) Calibration curves
   (c) Grading curves
   (d) None of the above

17. The factors affect the readings of rebound numbers
   (a) Age of concrete
   (b) Surface texture
   (c) Concrete mix characteristics
   (d) All the above

18. The rebound values reflect the concrete quality up to a depth of
   (a) 70 mm (b) 20 mm
   (c) 50mm (d) 40 mm

19. The strength prediction can be made to an accuracy of
   (a) ± 20% (b) ± 25%
   (c) ± 15% (d) ± 10%

20. The application of rebound hammer test is
   (a) Checking the uniformity of concrete
   (b) Estimation of strength
   (c) Abrasion resistance classification
   (d) All the above

21. The fundamental principle of resonant frequency method is
   (a) Frequency due to vibration
   (b) Vibration due to frequency
   (c) Frequency due to reverberation
   (d) All the above

22. Name of the instrument used under frequency test is
   (a) Multimeter
   (b) Sonometer
   (c) Covermeter
   (d) All the above

23. Identify the correct statement of the following:
   (a) Concrete core test gives better idea of tensile strength of concrete
   (b) The size of aggregate used in the flexure test specimen has no influence in its strength
   (c) The strength results obtained by one point loading and two point loading method are same
   (d) The tensile strength of concrete is more sensitive to improper curing than its compressive strength

24. The core testing of concrete should be performed.
    (a) When the primary aim of core testing is to ascertain the adequate strength of the structural member
    (b) When it is required to detect the segregation and bleeding of the concrete
    (c) When it is required to estimate the load carrying capacity of the structural member
    (d) All the above.

25. Identify the incorrect statement of the following :
    (a) The core should be taken at the nearest non critical section
    (b) The core should be taken at locations where minimum strength and maximum stress and walls
    (c) The core should be taken occupying the top 25% portion in case of columns and walls.
    (d) The reinforcement of concrete should be avoided from the core

26. Identify the correct statement of the followings:
    (a) As per code the minimum number of cores should be 3
    (b) For core drilling rotary drill having cutting cool with diamond bits and water supply arrangements is sufficient
    (c) To minimize uncertainty in correlating the core strength to cylinder or cube strength the ratio of height/diameter should be kept 2.0
    (d) All the above.

27. Identify the incorrect statements of the following:
    (a) A saturated core has been found to have 10 to 15% less compressive strength than a dry core.
    (b) It has been observed that as the l/d ratio of the core increases, the measured strength also increases
    (c) The core drilled vertically has been found to have more strength than that taken horizontally
    (d) The in situ concrete gains little strength after 28 days

28. The disadvantages of core testing are
    (a) The structural integrity of concrete across the full cross section may be affected
    (b) The length/diameter ratio other than 2.0 will have effect on the test results
    (c) Capping of the specimen on both sides may affect the results
    (d) All the above

29. The results of rebound method of concrete testing are affected
    (a) Age of concrete
    (b) Compaction of concrete
    (c) Smoothness of surface
    (d) All the above

30. The rebound test can be used
    (a) To check the uniformity of concrete quality
    (b) To estimate the approximate strength of concrete
    (c) To compare a given concrete with a specified quality concrete
    (d) All the above

31. The resonant frequency test can be used
    (a) To study the deterioration effect of concrete subjected to repeated freezing and thawing
    (b) To study the damage due to fire
    (c) To study the effect of alkali and acidic reaction
    (d) All the above

32. The transducers may be arranged on concrete surface
    (a) On opposite faces
    (b) On adjacent faces
    (c) On the same face
    (d) All the above

33. Identify the correct statement of the followings
    (a) The value of pulse velocity is affected by the contact surface under test
    (b) The variation of ambient temperature affects the pulse velocity of concrete
    (c) With the increase in moisture the pulse velocity increases
    (d) All the above

34. The pulse velocity of concrete is used
    (a) To establish the uniformity of concrete
    (b) To estimate the strength of concrete
    (c) To estimate the fire hazards
    (d) All the above

35. The specimen size used for resonance frequency test is
    (a) 150 mm × 300 mm cylinder
    (b) 75 mm × 300 mm cylinder
    (c) 150 mm × 150 mm cube
    (d) All the above

36. The object of UPV test is
    (a) To measure the pulses of longitudinal vibrations
    (b) To measure the velocity of longitudinal vibrations
    (c) To measure the velocity of horizontal vibrations
    (d) All the above

37. Name of the wave generated in UPV instrument is
    (a) Compression waves (b) Tension waves
    (c) Torsion waves (d) All the above

38. To evaluate the strength of concrete by UPV test
    (a) Using pulses chart
    (b) Using calibration chart
    (c) Using velocity chart
    (d) All the above

39. The determination by ultrasonic pulses are
    (a) Elastic stiffness and mechanical strength
    (b) Plastic stiffness and mechanical strength
    (c) Elastic stiffness and metallurgical strength
    (d) None of the above

40. The pulse velocity measurements may be used to establish
    (a) The cement content of the concrete
    (b) The water content of the concrete
    (c) The homogeneity of the concrete
    (d) None of the above

41. Transducers of --------------- are found to be useful for most of the applications in concrete testing.
    (a) 15 to 60 kHz
    (b) 150 to 160 kHz
    (c) 50 to 60 kHz
    (d) 5 to 10 kHz

42. The principle of pulse attenuation method is based on
    (a) The electrical properties of the solid material tested
    (b) The magnetic properties of the solid material tested
    (c) The homogeneity properties of the solid material tested
    (d) None of the above

43. The dominant properties used in pulse attenuation method are
    (a) The electrical resistivity and conductivity
    (b) The electrical conductivity and dielectrics
    (c) The electrical permittivity and conductivity
    (d) None of the above

44. Pulse echo technique was developed for
    (a) Pile integrity testing
    (b) Tunnel testing
    (c) Carbonation testing
    (d) None of the above

45. Name of the instrument used in pulse echo method is
    (a) Sonometer
    (b) Multimeter

(c) Doctor
(d) None of the above

46. The type of rays used in radio active methods
(a) X–rays & Beta rays
(b) X–rays & Gamma rays
(c) X–rays & Delta rays
(d) None of the above

47. The use of radio active method is to find
(a) Density of concrete members
(b) Strength of concrete members
(c) Cracks of concrete members
(d) All the above

48. The principle techniques have been reported in nuclear methods
(a) Nuclear Scattering Method
(b) Electron Activation Analysis
(c) Neutron Scattering & Neutron Activation
(d) None of the above

49. The purpose of nuclear method for determining the
(a) Moisture content of the concrete
(b) Cement content of the concrete
(c) Surface humidity of the concrete
(d) All the above

50. Name the instrument used by magnetic system
(a) Pachometer
(b) Tachometer
(c) Sonometer
(d) Magneto meter

51. The determination by electrical methods
(a) The moisture content
(b) Tracing of moisture permeation
(c) The thickness of concrete
(d) All the above

52. The property involving in the electrical methods
(a) Dielectric properties of fresh concrete
(b) Dielectric properties of hardened concrete
(c) Trielectric properties of hardened concrete
(d) All the above

53. Electrical resistivity methods have been used to find out
(a) The thickness of concrete pavements

(b) The moisture content
(c) Tracing of moisture permeation
(d) All the above

54. The finding of acoustic emission technique is
(a) Tracing moisture of concrete
(b) Locating delaminated concrete
(c) Finding strength of concrete
(d) None of the above

55. The factor influence the final results of cores
(a) Slenderness ratio
(b) Presence of reinforcement
(c) Curing condition
(d) All the above

56. The properties find out from core test
(a) Density
(b) Water absorption
(c) Elastic strength
(d) All the above

57. Cores for test should be normally taken at points where
(a) Minimum strength and maximum stress coincide
(b) Maximum strength and minimum stress coincide
(c) Minimum strength and minimum stress coincide
(d) All the above

58. One of the procedures in core testing is
(a) Palling & scaling
(b) Trimming & capping
(c) Screening & topping
(d) None of the above

59. The properties assessed in visual examination
(a) Honeycombing
(b) Cracks
(c) Drilling damage
(d) All the above

60. From the core test, small voids are
(a) 0.5 mm–3 mm
(b) 1.5 mm–3 mm
(c) 2.0 mm–6 mm
(d) None of the above

61. From the core test, medium voids are
(a) 0.5 mm (b) 3 mm
(c) 3.6 mm (d) 6 mm

62. From the core test, larger voids are
(a) More than 3 mm
(b) More than 5 mm
(c) More than 6 mm
(d) None of the above

63. The rate of loading for core compressive test
(a) $15 N/mm^2/min$
(b) $12 N/mm^2/min$
(c) $16 N/mm^2/min$
(d) $20 N/mm^2/min$

64. The reasons for higher permeability in concrete
(a) Long term shrinkage
(b) Transition zone
(c) Volume change
(d) All the above

65. Name the apparatus used for permeability test
(a) Lechatlier apparatus
(b) Figg's apparatus
(c) Slump cone apparatus
(d) None of the above

66. The in situ testing for concrete corrosion
(a) Chloride content test
(b) Half cell potential survey
(c) Resistivity test
(d) All the above

67. Carbonation test can be carried out using
(a) Phenolphthalein test
(b) Ammonia test
(c) Phenol test
(d) None of the above

68. The commercial name for chloride test equipment is
(a) Rapid Corrosion test kit
(b) Rapid Chloride test kit
(c) Rapid Carbonation test kit
(d) None of the above

69. The standard electrodes used in half–cell potential survey
   (a) Copper–Copper Sulphate Electrode
   (b) Silver–Silver Chloride Electrode
   (c) Standard Calomel Electrode
   (d) All the above

70. The potential value become more & more negative as the corrosion becomes
   (a) More & more inactive
   (b) Less & more active
   (c) More & more active
   (d) All the above

71. Which property plays an important role in the resistivity of concrete?
   (a) Electrical resistance
   (b) Mechanical resistance
   (c) Electrical inductance
   (d) None of the above

72. The technique used in concrete resistance measurement
   (a) 2 probe technique
   (b) 3 probe technique
   (c) 4 probe technique
   (d) All the above

73. The factors that govern the resistivity value
   (a) Constituents of concrete
   (b) Chemical contents of concrete
   (c) Pore structure of concrete
   (d) All the above

ANSWERS

1. d	2. d	3. a	4. d	5. c	6. b	7. a	8. a	9. b	10. c
11. a	12. b	13. d	14. d	15. a	16. b	17. d	18. c	19. b	20. d
21. a	22. b	23. d	24. d	25. c	26. d	27. b	28. d	29. d	30. d
31. d	32. d	33. d	34.d	35. a	36. b	37. a	38. b	39. a	40. c
41. c	42. a	43. c	44. a	45. c	46. b	47. a	48. c	49. d	50. a
51. d	52. b	53. a	54. b	55. d	56. d	57. a	58. b	59. d	60. a
61. c	62. c	63. a	64. d	65. b	66. d	67. a	68. b	69. d	70. c
71. a	72. c	73. d							

SHORT ANSWER QUESTIONS

1. Define a systematic investigation.
2. Name the general procedure for investigation.
3. What are the satisfy aspects of concrete structures?
4. What are the testing systems of hardened concrete?
5. Name the some non–destructive testing methods.
6. Name the some partially–destructive testing methods.
7. Differentiate non–destructive, semi-destructive & destructive testing methods.
8. What is the principle of rebound hammer test?
9. What are the factors affect the readings of rebound numbers?
10. What are the applications of rebound hammer test?
11. How to assess the quality of concrete from rebound numbers?
12. What is the object of Ultrasonic Pulse Velocity test?
13. What are the types of waves generated in UPV instrument?
14. Relate compression wave velocity & dynamic modulus of elasticity in elastic medium.
15. How do evaluate the strength of concrete by UPV test?
16. What are the factors influenced the calibration charts?
17. What is the use of pulse velocity measurements?
18. How to measure the pulse velocity?
19. What are the parts in pulse velocity measuring equipments?
20. What are the factors influencing the pulse velocity measurements?
21. Write down the criteria adopted for concrete certification by UPV.
22. What is the principle of pulse attenuation method?
23. Which is dominant property in the pulse attenuation method?
24. Write down the structural application of the radar system by pulse attenuation method.
25. What is the need for pulse echo technique?
26. What is the use of echo system?
27. What is the use of Pachometer?
28. What are the determinations by electrical methods?
29. How to identify the delaminated areas under acoustic emission technique?
30. What is the use of thermography test?
31. What is the principle of thermography test?
32. How to measure the hardness by probing technique?
33. Define "Windsor probe technique".
34. What is meant by pull out test?
35. What are the situations when core test performed?
36. Which is the location of core taken?
37. What is the specified size of core?
38. What is the requirement for number of cores?

39. Write about the core drilling equipment?
40. How is the core prepared for testing?
41. How to obtain trimming & capping?
42. What are the identifications by visual examination of core test?
43. Write about the density determination from core sample?
44. How to assess the voidage of concrete from core sample?
45. How do you conduct compression testing on core samples?
46. What are the factors influencing the core compressive strength?
47. How the reinforcement modify the strength of concrete in core test?
48. How to estimate the cube strength from core strength?
49. What is the codal representation of core test?
50. What are the disadvantages of core method?
51. What is meant by acceleration curing test?
52. What are the methods available for accelerated using test?
53. What is the determination of bond strength?
54. What are the uses of resonant frequency test?
55. Write about dynamic or vibration methods.
56. Derive the mathematical relation between velocity and modulus of elasticity?
57. What are the properties evaluating from cores?
58. How to assess the quality of concrete from the core test?
59. What are the tests available to find concrete corrosion?
60. Phenolphthalein test–explain.
61. How can be determined pH value in concrete?
62. How to conduct the chloride test?
63. Give the qualitative guidelines for identification of corrosion prone based on chloride content & pH value.
64. What are the standard electrodes used in half–cell potential survey?
65. What is a measurement by half–cell potential survey?
66. Give the guidelines for identifying corrosion based on half–cell survey.
67. Define–Resistivity
68. What are the factors govern the resistivity value?
69. Give the guidelines of corrosion risk based on resistivity values.
70. Define load tests.
71. What is LOK test?
72. Define TNS Tester.
73. Explain shortly internal fracture test.
74. How do conduct epoxy grouted bolt test?
75. Explain shortly the determination of cement content in concrete.
76. Discuss shortly the determination of water content in concrete.
77. How do find out the sulphate content in the concrete?
78. Define abrasion resistance test in concrete.
79. Rapid chloride test-explain

THEORY QUESTIONS

1. Explain assessment procedure for evaluating damages in structure.
2. Discuss about the evaluation of different types of cracks.
3. Name the various testing system used in the assessment of distressed concrete structures.
4. Differentiate non–destructive testing, partially destructive testing and destructive testing methods.
5. Explain clearly the rebound hammer test and its limitations.
6. Explain clearly the testing procedure of ultrasonic pulse velocity testing of concrete.
7. How the electrical, magnetic and nuclear menaces are used to assess the property of concrete?
8. Explain the radar system used to assess the damages in concrete structures.
9. Explain the pulse echo system used to assess the quality of concrete.
10. How to assess the fire damaged concrete structures by NDT methods?
11. Discuss about the penetration techniques in concrete structures.
12. How do you conduct pull–out test and pull off test?
13. How do you conduct core sampling & testing?
14. Explain in–situ permeability test on concrete.
15. Explain the testing methods to assess the corrosion damages in concrete structures.
16. How do you conduct chloride penetration test and pH value test?
17. Explain the determination of cement content, water content and water cement ratio in concrete structures.
18. Explain the use of half-cell potential survey in the damage assessment of concrete structures.
19. Explain about the destructive testing methods.
20. Discuss briefly the load test on existing concrete member.

PART III
REPAIR MATERIALS

CHAPTER 7

Selection and Evaluation of Repair Materials

7.1. INTRODUCTION

Deterioration of any concrete structure is the natural phenomenon of gradual degradation of constituent materials brought about by physical, chemical or mechanical processes. Occasionally, natural/man-made calamities and accidents accentuate this process. As a consequence, the ability of a structure to perform the intended function is impaired and there arises a need to repair or upgrade a structure to restore the designed level of serviceability or even enhance it. Experience over the last 40-50 years has shown that the current design specifications, even if properly implemented, cannot ensure adequate protection of reinforced concrete in aggressive environs. It is everybody's experience that premature deterioration of concrete is still unacceptably high and corrosion of reinforcement is one of the most dreaded causes of failure.

Repair and rehabilitation have therefore become established facets of construction industry. They have become the preferred alternative compared to demolition and rebuilding of structures due to social, economical and psychological factors. The cost of materials constitutes a substantial portion of the repair and refurbishment cost. Therefore, there is an ever-increasing surge of new products formulated to provide a variety of properties in order to meet the market demand. Since a great majority of these products are proprietary in nature with very little information on their composition and performance, the user is confounded with the problem of selecting the appropriate repair material for a particular job.

7.1.1 Development in repairs

Repair programme is a job of an expert from the beginning and it should be carried out in a planned way. At present there are no widely accepted guidelines for design, specifications and execution of repair projects. Unfortunately, in India whilst a large number of structures are heading for repairs, there is no co-ordination between diagnosis of repairs, understanding and specifications of consultant and manufacturer's products specifications. To top it all the repair person is not knowledgeable and of different chemicals available. This is resulting in to incorrect repairs and a time has come to start repairs to repairs.

7.2 MATERIALS SELECTION

Selecting repair materials is an important and complex process involving an understanding of requirement of the repair by the owner, engineer, service and exposure conditions, and also installation technique. After requirements are established and material properties are defined, the selection of specific materials can be made. Final selection of material is made based on the relationship between cost, performance and risk. Selecting repair materials requires an understanding of material behaviour in the cured and uncured states in the anticipated service and exposure conditions.

One of the greatest challenges facing the successful performance of repair materials is their dimensional behaviour relative to the substrate. Relative dimensional changes cause internal stresses within the repair material and within the substrate. High internal stresses may result in tension cracks, loss of load carrying capability, delamination or deterioration.
Another challenge is selecting repair materials for structural applications. There are two obstacles to achieving 100 % repair efficiency in structural applications. First, Initial loading on repair material and removal of load from the structure during repair. Next, the dimensional behaviour of repair materials on substrate. These are based on load subjected, temperature and moisture change.

For the selection of material of repairs it is necessary to understand the properties of various materials and their limitations. The selection of material also depends upon the basic nature of the structure to be repaired. **Selection of the repair material thus depends on the following:**

(a) Type of concrete to be repaired and its age
(b) Magnitude and thickness of repairs
(c) Site conditions
(d) Temperature and
(e) Environmental influence

7.2.1 Criteria for selection of repair materials

A careful selection of repair material is necessary for the following reasons:

(a) Almost every repair job has unique conditions and special requirements
(b) The composition and properties of repair materials have a profound effect on the performance and durability of a repair
(c) The repair materials perform adequately only if they are prepared, applied and cured as per the specified procedures, which may necessitate the use of appropriate tools and considerable skill.
(d) Repair materials, being generally proprietary in nature, are very costly.

The following factors have to be considered in selecting a repair material:

(a) Function of repair material
(b) Physical and mechanical properties
- Dimensional stability
- Coefficient of thermal expansion
- Strength in compression, tension and flexure
- Stiffness, elastic modulus and Poisson's ratio

(c) Durability related properties
- Permeability to liquids, vapours and gases
- Diffusion resistance

- Water Absorption, Sorptivity and Porosity
- Weathering resistance
- Resistance to chemicals

(d) Chemical compatibility with other repair materials
(e) Identification of anticipated service conditions
(f) Conditions prevailing at the time of repair

7.2.2 Methodology for the selection of repair materials

The basic emphasis of the selection methodology is to maximize the performance of a repair and ensure durability. The following procedure is adopted for selection of repair materials.

(a) **Definition of service condition**-Identification of major and minor chemicals, traces, spillages, cleaning chemicals, slurries and abrasives, their characteristics and their interaction with the environment under stagnant conditions, operating temperature range, dilute conditions, alternate wetting and drying effect

(b) **Determination of appropriate application condition**-Viscosity, flow characteristics, pot life, curing requirements, coverage, film/layer thickness, and size of repair

(c) **Tools and equipments required**-Pump, sprayer, injecting/grouting machine, dryer, mixer, batching plant, heater etc.,

(d) **Repair and maintenance schedule**-Anticipated durability of repair, maintenance requirements, replacement/renewal/strengthening requirements

(e) **Product performance record/history**-Durability, functionality, environment friendliness

(f) **Material testing and assessment for quality assurance and quality control**

(g) **Selection of applicator/contractor**

(h) **Material and job specifications**

7.3 CLASSIFICATION OF REPAIR MATERIALS

The materials used for concrete repairs on the basis of the type of application into following categories:

(a) **Patch Repair Materials**

- Cementitious mortar/concrete
- Polymer modified cementitious mortar/concrete
- Polymer mortar/concrete
- Quick setting compounds
    - High alumina cement based
    - Calcium sulphate based
    - Magnesium phosphates
    - Sulphur concrete

(b) **Injection Grouts**

- Cementitious grouts (with or with out fibres)

- Gas forming grouts
- Sulpho-aluminate grouts
- Polymer grouts

(c) **Bonding Aids**

- Polymer emulsion type
- Polymer resin type

(d) **Resurfacing Materials**

- Protective coatings and membranes
- Impregnants and hydrophobic sealers
- Toppings/screeds
- Overlays
- Gunite/shotcrete

(e) **Other Repair Materials**

- Corrosion inhibitors
- Rebar protective coatings
- Cathodic protection
- Realkalization

7.4 EVALUATION TESTS FOR REPAIR MATERIALS

Presently the construction industry is flooded with large number of materials with claims which often need to be verified. If not properly evaluated, the repair materials may fail during service life due to severity of exposure or incompatibility with substrate or detrimental interaction between the constituents. The best approach to judge the suitability of a repair material for a particular situation is to evaluate the performance of the material under verifiable conditions, which are identical to the expected service history. Generally, the evaluation tests involve the determination of one or more measurable parameters, which bear a fundamental relationship with the resistance and performance of the repair material under given exposure conditions. The tests may be destructive or non-destructive and many times partly destructive in nature. It may often become necessary to resort to accelerated testing to get a rapid assessment of the suitability of repair materials when several candidate materials are available.

This may require the intensification/magnification of the magnitude of the most potentially destructive parameter. However, such tests may not yield a quantitative measure of the prospective service life of a material under actual service conditions. But, nevertheless, they are useful tools for the selection of materials and deciding on the construction features, such as, curing requirements, techniques for finishing, curing and protection of surface treatment. There are a large number of field/laboratory tests developed for cementitious, polymeric and polymer-cement composites. The choice of a particular set of tests for material selection and quality control depends mainly on the magnitude of work and its importance, besides the availability of test facilities and time constraints. Therefore, no generalization can be made as regard to the choice of a particular set of tests for control purposes.

7.4.1 Test for compressive strength and elastic modulus

This is the single most important property for quick quality control checks for all repair materials based on cementitious composites and is important for polymer bound systems. The usual practice is to specify the compressive strengths at the end of 1, 3, 7 and 28 days of curing period for cementitious systems. This is determined by static destructive tests on 50mm or 70mm size cube specimens in case of mortars and 100mm or 150mm cube or cylinder specimens in case of concretes. For finding the compressive strength of resin systems, BS 2728-1970 and BS 6319-1983, Part-2 recommend the use of 12.5mm cubes or 9.5mm diameter × 9.5mm height cylinder specimens while ASTM D 695 specifies 12.5mm × 25 mm cylinders. For polymer mortars, ASTM C 579-91 and IS 9612-1979 suggests the use of 50mm cubes or 25 mm diameter × 25 mm height cylinder specimens. In case of polymer concrete, cylinder specimens of minimum diameter equal to three times the maximum size of aggregate and height of two times the diameter are to be used.

The resin specimens are to be tested so as to complete the test in about 1 to 2 minutes. ASTM C 579-91 specifies that the specimens have to be tested at a load rate of 42 MPa per minute or at a cross head speed of 2.5-3.2 times h/25, where *h* is the height of the specimen in mm. Before testing, the specimens are to be conditioned at temperature and humidity conditioned at specified temperature conditions (25-27°C) for specific duration (up to 7 days). JIS A 6024-1992 suggests the use of 13.8mm × 13.8mm × 40mm prisms for determines the compressive strength of epoxy prisms cube as injecting grouts/adhesives. In case of cementitious systems and resin mortars/concretes, the procedure recommended by IS 516 or BS 1881 or ASTM C-39/C-109/C-579 may be followed. JIS A 6203-1986 recommends the use of broken test piece obtained from flexural tests on 40mm × 40mm × 160mm prisms for determination of compressive strength of polymer modified mortars. For determining the elastic modulus in compression, cylindrical specimens with aspect ratio of 2 may be preferred. Generally the specimens are loaded to 30-40% of their compressive strength and initial tangent/chord or secant modulus obtained from the stress-strain curve.

7.4.2 Tensile strength and elongation

This is a very important property for all repair materials as patch repairs and overlays are often required on the tension faces of concrete beams. For testing pure resin systems, ASTM D-638 and BS 2782 recommends dumb-bell shape specimens, 150mm long × 20mm wide and 3mm thick. The straight portion of the specimen should be 60mm long × 10mm wide, with a recommended gauge length of 50mm for strain measurement. The type of material and temperature dictates the speed of testing. For thermosetting resins like epoxies, BS 2728, BS 6319, Part 7-1981 recommend test duration of 0.5 to 1.5 minutes. IS 9162 recommends

44.5mm × 76mm × 25 mm briquettes with a central neck of 25mm cross section in case of epoxy resin compositions. For tensile modulus, either a secant or chord modulus obtained from the stress-strain plot is used.

For cementitious systems and resin mortars/concretes, ASTM D-190 recommends, direct tensile strength for briquette type specimens 25mm × 25mm in cross section or split tensile strength recommended by ASTM C – 496, IS 516, BS 1881, determined from cylindrical specimens may be used. Tensile elongation at failure is often specified for pure resin systems as one of the requirements and can be measured as per ASTM D-638. ASTM D-882 specifies a minimum elongation of 1% and 30% respectively for bonding materials used for general and skid resistant overlay applications.

7.4.3 Flexural strength and flexural modulus

This is another set of useful measure of relative strength and stiffness of repair materials. For pure resin systems, BS 2728 and ASTM D-790 recommend prisms of size 12.5mm × 9.5 mm × 112.5mm to be tested over a span of 100mm under three point bending. IS 9162 and ASTM C-580 recommend 25mm × 25mm × 250 mm prisms to be tested for flexural strength of resin mortar under centre-point loading at the rate of 4.2 MPa/minute over a span of 229mm. For polymer concrete, specimens of size 50mm × 50mm × 300-400 mm are recommended by ASTM C-580 for small size aggregates. The flexural modulus is determined from the initial straight portion of a load-deflection curve obtained by three point bending test on specimens loaded at the rate of mm/minute. For cementitious systems and resin bound systems, JIS A6203, JIS A6024 and ASTM C-348 suggest the use of 40mm × 40mm × 160mm specimens. IS 516 and BS 1881 give the procedure for determination of flexural modulus. ASTM C-580 defines two types of flexural modulus, which are obtained from the load deflection characteristics. The initial tangent modulus is obtained from the slope of the curve at zero deflection where as secant modulus is obtained from the slope of the chord connecting the origin to 50% of the maximum deflection.

7.6.4 Properties of raw materials and green mixes

For cementitious systems and polymer modified cementitious systems, general sampling procedures laid down for fresh concrete (ASTM C-172) may be used. The minimum numbers of specimens to be tested per batch are generally laid down in most of the specifications and ranges from three to five in most of the cases. The important tests, which are often required to be conducted on fresh cementitious mixes, are:

(a) Slump or flow test for workability
(b) Unit weight and Air content
(c) Sieve analysis
(d) Specific gravity and water absorption for coarse aggregates and fine aggregates
(e) Organic impurities in fine aggregates

The important properties to be tested fin case of or polymeric latexes and resin systems include:

Specific gravity, Solids content, Drying time, Viscosity, Consistency, Thixotropic index, Gel time, Dry film thickness, pH, Glass transition temperature(T_g), Coverage, Minimum film forming temperature (MFFT), Slump, Sag

Specific gravity-Specific gravity of powders and resins is easily obtained by specific gravity bottle/pycnometer method or density gradient method.

Solids content-The non-volatile solids content of a resin solution is determined by finding the loss of weight on heating the resin solution, taken in a small Petri-dish, in an oven and drying off the volatile compounds. The recommended drying time ranges from one hour for phenolics, two hours for emulsions to three hours for unmodified epoxies. This is an important test for emulsions and superplasticizers. For rubber latex, determination of dry rubber content is often necessary. This requires the coagulation of latexes by adding acetic acid, washing the coagulum with water, drying it at a temperature of about 70±2°C, cooling in a desicator and taking the dry weight. Tests are also specified for epoxy and amine content of epoxy resin compositions.

Drying time-This is an important test for coatings and surface treatments. This is usually determined by the touch dry test in which the coating is applied on a glass plate kept inclined at 45° to the vertical and shielded from air currents and direct sunlight. After the completion of specified time, the coated panel is placed horizontally, sprinkled with about 0.5g of sand from a height of about 50mm-150mm and brushed in an inclined position. After 10 seconds, if all the sand can be brushed without damaging the surface, the coating is surface-dry. It is considered to have dried hard when the coating film does not move when pressed with a finger and no impression is noticeable on the coated surface after wiping with a soft cloth.

Viscosity-Viscosity of resins and latexes is a very important property as it is related to the molecular weight and to its solvent strength. The techniques commonly employed for viscosity measurements are:

- Bubble tubes for clear liquids and curing agents
- Ostwald (U tube) viscometer for liquid resins
- Falling sphere viscometer (for high viscosity resins)

Consistency test-Consistency test is specified by ASTM C-881 for epoxy systems. It is defined as the flow of the lower edge of a semi-cylindrical resin bead along its length measured from its original position to the nearest 3mm. Maximum consistency specified for epoxies is 6.4mm at the temperature corresponding to the service condition.

Thixotropic index-Thixotropy is the typical property of viscous liquids to show an apparent increase in viscosity in stationary condition.

Gel time/working time-Gel time is the interval of time, in connection with the use of synthetic thermosetting resins, which extends from the introduction of a catalyst/hardener into a liquid resin until the start of formation of a soft, gelatinous mass. At this time the resin viscosity increase as to a point where it barely moves when probed with a sharp instrument. A soft iron wire terminating into a loop may be used for agitating the resin taken in a test tube. The end point is indicated by the cessation of the motion of test tube relative to the wire agitator. In this condition, the wire agitator can lift the tube filled with gelatinous resin. The minimum specified gel time is generally 30 minutes.

Dry-film thickness (DFT)-This is another important property for coatings. A coatmeter based on magnetic flux principle or electromagnetic induction can be used for this purpose. For low thickness coatings, adhesive tape method is recommended. In this approach, an adhesive tape is placed on the substrate and coated. It is subsequently removed and DFT is obtained by weighing or thickness measurement. Graduated metal wedge scales, which can measure the wet film thickness, are also available.

pH-The pH of latexes measured using a pH electrode is a convenient test for latex quality.

Glass transition temperature (Tg)-Glass transition temperature corresponds to the temperature at which a polymeric material changes from relatively hard, elastic, glass like substance to a relatively viscous rubbery material. Classical methods for determining the T_g include methods based on dimensional change (thermal expansion), optical methods (based on refractive index), electrical methods (based on dielectric properties like loss factor), differential thermal analysis (DTA) and nuclear magnetic resonance NMR techniques. However, for civil engineering purposes, a more simple measure of temperature susceptibility of resin systems called heat deflection/distortion temperature (HDT) is commonly used.

Coverage or wetting power-A checker board is often conveniently used for the test. The number of square meters of a checker board having alternate black and white squares which can be satisfactorily obliterated to remove contrast between black and white squares by a fixed quantity of the coating is determined as a measure of coverage or wetting power. Alternatively, a calibrated metal wedge called (Pfund cryptometer) can be used to check for wet film thickness, which in turn can be expressed in terms of coverage.

Minimum film forming temperature (MFFT)-For polymer dispersions the procedure specified in ASTM D-2354 is used. The ambient temperature at the site should not be less than MMFT to ensure film formation in case of polymer dispersions.

Sag flow test-Sag test is intended for evaluation of the tendency of a coating to sag when applied and allowed to dry on a vertical non-absorbing substrate at a specified film thickness. An anti-sag meter in conjunction with draw down charts is required for this test.

7.4.5 Coefficient of thermal expansion

This is an important measure of the thermal susceptibility of a repair material. ASTM D-696 and BS: 4618 specify the measurement of coefficient of thermal expansion by a quartz tube dilatometer 500mm long. The coefficient of expansion is obtained by measuring the fractional change in length per degree temperature change when the specimen temperature is raise d from –30°C to +30°C using a dial gauge of least count 0.0025mm. For volume expansion, a Pyrex glass tube dilatometer is used to test a bar specimen of size 6mm × 6mm × 150mm with rounded edges. Frequently, direct measurement techniques using a traveling microscope or optical methods are used. In these cases, the relative movement of a mark ruled on the specimen in the form of a strip is observed against an index mark on a steel reference bar, the specimen being placed in a temperature controlled oven/water bath. ASTM C 531-91 recommends the use of 25 × 25 × 250mm specimens for resin mortars.

7.4.6 Shrinkage

Plastic shrinkage of resin binders is a very important property, which determines the performance of a repair system. ASTM D-955 and BS: 2728 require the measurement of diameter of the cold moulded specimen using 50-125mm disks of 3mm thickness. ACI Committee 548 recommends two methods viz. DuPont method and Ohama-Demura method for measuring the plastic/curing shrinkage of PC overlay materials, which can yield a continuous record of shrinkage strain with time. In the former case, a Teflon mould with inside dimension of 75mm × 75mm × 300mm is used and the strain is measured with a displacement transducer by sensing the displacement of a movable prong with reference to a fixed prong. In the latter method, the material is cast in a similar Teflon mould but the displacement is measured by sensing the movement of 25mm dia. brass expansion plugs placed at the ends of the mould using direct current displacement transducers. A very simple test procedure often used for evaluating the shrinkage of repair mortars is the Coutino Ring Test. In this test, the repair mortar is cast to a uniform thickness of 50 mm around the circumference of a 112.5 mm diameter × 65 mm wide steel pipe and exposing it to 50% RH and recording the time of cracking.

7.4.7 Adhesive/Bond strength

ASTM C-882-91 specifies a slant shear test for determining the bond strength. In this method, the bond strength is obtained by testing a cylinder 75mm dia × 150mm height formed by bonding together two replicate halves of Portland cement mortar, each of which has a diagonally cast bonding area at 30 degree inclination to the vertical. After suitable curing of the bonded specimen, the integrated cylindrical specimen is tested in compression and the failure load divided by the bonded area gives the bond strength.

For Latex Modified Cement Concrete overlays, ACI Committee 548 recommends the use of a pull out test as laid down in ASTM C-900. This test involves embedding an insert with a head of 25-30mm diameter at the time of casting to a depth equal to the diameter. The shaft of the insert is about 15mm in diameter. A pullout apparatus bearing around the inserts and secured firmly to the insert is used to rupture the surface within 2 to 2.5 minutes duration. The pullout load divided by the vertical component of the fractured area gives the pull out strength.

JISA 6203 requires the coring of a 40mm × 40mm area through the polymer mortar overlay so as to reach the concrete substrate. A hexagonal head is fixed to the core surface using a high strength epoxy adhesive. The bolt head is pulled out at the rate of about 2.7 kN/min. The pull out loa. is a measure of the bond strength. ASTM C 327-94 & 1S:9162 suggest a cross brick push off test for resin mortars. The bricks should have a flexural strength of at least 6.9 MPa. An instrument called Elcometer adhesion tester is used for evaluating the adhesion of coatings to concrete substrates. For epoxy injection grouts, JISA 6024 recommends a flexural bond test in which two replicate prisms of dimension 40 × 40 × 80 mm are bonded together and then tested under two pint loading with a shear span of 40 mm at a speed of 50 kN/m.

7.4.8 Durability related tests

Thermal compatibility test:

This test specified in ASTM C884-92 consists of applying the repair material to a 300mm × 300mm × 75mm concrete block to a uniform thickness of 12.5mm. The repair mortar is cured for 7 days at 23 ±1°C. At the end of 7 days, the specimen is placed at 0°C for 24 hours in a deep-freezer followed by 24 hours storage at room temperature. Four such cycles are completed. If any delamination of overlay or horizontal cracking of concrete at the interface is observed, it is interpreted as failure of the epoxy system.

Water absorption:

For resin binders, ASTM D-570 specifies 50mm dia × 3mm thick disk specimens to be used for water absorption tests. The specimens are conditioned at 110°C for 24 hours and then cooled in a desicator and immediately weighed. The specimen is immersed on edge in distilled water at 23±1°C for 24 hours and then weighed after surface drying. The change in mass gives the water absorption of the resin system. This procedure is also recommended ASTM C884 for epoxy resin adhesives. IS:9162 and ASTM C 413-94 suggests that 25 mm dia × 25 mm long cylindrical specimens may be dried at 105-110°C, cooled and then soaked in boiling water for 2 hours to determine water absorption of polymer mortar/concrete compositions.

For polymer modified mortars, JISA 5203-1980 specifies the use of 40 × 40 × 160mm specimens. After casting, the specimens are subjected to a 28-day curing cycle involving 2 days moist curing, five days water curing. The specimen is then dried to attain constant mass at 80+2°C for 48 hours. The change in mass gives the water absorption. For purely cementitious systems, BS:1881-Part 5 1970 requires a 75mm dia specimen with thickness less than 150mm and greater than 32 mm to be oven dried at 105 ± 5°C for 3 days, cooled for 24 hours, weighed and then soaked in water for 25 minutes. The change in mass gives the water absorption. Soaking in boiling water for 5 hours is recommended for determination of void content. For chemically resistant resin mortars/concretes, ASTM C413-94 requires 25 × 25 mm cylinder specimens to be kept in boiling water for 2 hours, cooling the cold water and observing the change in mass for obtaining the water absorption.

Water permeation:

This type of test is suitable for polymer modified system, coated specimens and cementitious systems. It may be also used for polymer concrete specimens. As applied to polymer modified mortars, JISA 6203 specifies cylindrical disks of 150 mm dia × 40 mm height prepared and cured as specified for water absorption test. The specimens are then dried to constant mass at a temperature of 80 ± 2°C, brushed lightly at the centre, cleaned and weighed. The specimen is then subjected to a hydraulic pressure of 1kg/cm^2 for one hour. The change in weight of the specimen gives the amount of water permeation.

Carbonation resistance:
ACl Committee 548 recommends an accelerated carbonation test procedure used for polymer modified concrete. In this test, 40 mm × 40mm × 160mm beam specimens coated with epoxy resin paint on their top and bottom faces are placed in a sealed chamber for 14 days. At the end of 14 days, the specimens are split and the exposed faces sprayed with 1% phenolpthalein alcohol. The depth of colourless portion of the rim measured with a slide calipers gives the depth of carbonation.

Chloride penetration and chloride ion diffusion:
In this test, 40 mm × 40mm × 160mm prisms coated with epoxy resin on the top and bottom faces, are immersed in 2.5% Na Cl solution and 20°C for 7 days. At the end of this period, the specimens are split open and sprayed with 0.1% sodium fluoroscein and 0.1 N silver nitrate solutions. The depth of colourless rim is measured using a slide calipers to obtain the chloride ion penetration depth. By periodic observation of chloride penetration depth, the diffusion coefficient can be computed.

ASTM C1202-94 suggests an accelerated chloride diffusion test. This test consists of monitoring the amount of electrical charge passed through a vacuum saturated cylindrical disk of 100mm diameter × 50mm thickness, over a six hour period under a potential difference of 60V DC. One end of the specimen is immersed in 3% Na CI solution while the other end is immersed 0.3N Na OH solution. The total charge passed in six hours is a measure of chloride diffusion resistance. A total charge exceeding 4000 Coulombs indicates high permeability.

Penetration depth of impregnates:
To asses the depth of penetration of hydrophobic treatments like silane-siloxanes and high molecular weight methacrylates, cube samples treated with them may be split and dipped in water. The impregnated/penetrated rim portion will stay dry making it possible to determine the depth of penetration.

Freeze-thaw resistance:
ASTM C 666-92 suggests a procedure for determining the freeze thaw resistance of samples. Cylindrical specimens/prisms of minimum dimension 75-125mm and length 275 to 400mm are tested after the end of curing period. The specimen is kept in a freeze thaw apparatus and cooled to a temperature of about –1.1°C to + 2.2°C. At this temperature, it is tested for fundamental transition frequency, weighed and its initial length determined with a length comparator. The specimen is then subjected to alternate freezing and thawing cycles consisting of alternatively lowering the temperature from 4.4 to - 17.8°C and raising it from –17.8 to 4.4°C in not less than 2 hours and not more than 5 hours. The specimens are

removed at intervals not exceeding 36 cycles and tested for fundamental frequency, weight and length changes. The test is continued for 300 cycles or until the relative dynamic modulus of elasticity reduces to 60% of the initial value whichever occurs first. The durability factor is expressed as DF = PN/M, where P is the relative dynamic modulus, N = number of cycles at which P reaches the minimum value, M = specified number of cycles at which exposure is terminated.

7.4.9 Miscellaneous tests

Crack bridging action:

This is a special requirement for coatings and is intended to ascertain its ability to accommodate the movements in the substrate and allow the opening and closing of cracks without fracturing. Thin slate or ferrocement plates are coated on one face and then tested under monotonic or cyclic tension to induce specified crack width in the substrate. This enables the maximum crack width that a coating can withstand with out fracture to be obtained for different thicknesses.

Weathering resistance:

The weathering of polymeric repair materials is mainly attributed to the break down of polymer chains under concerted action of atmospheric agents such as ozone and UV radiation. To assess the weathering resistance a thin ferrocement or slate slab precracked in tension may be provided with the given repair material and exposed to ozone concentration of 7500 ppm at an elevated air temperature of about 40°C. To test for UV resistance the coated specimens are exposed to UV radiation for duration of about 2000 hours. To simulate the action of rain, water is sprayed on the specimens at fixed time intervals for specified duration. The coating is periodically examined for cracks, blisters, peeling and breakages. A 2000 hours duration is supposed to be equivalent to about 5 years of service life under severe exposure conditions. Occasionally, the specimens are held at the specified crack width in a tensioning device during the period of exposure to accelerate the tests.

Chemical Resistance:

Most of the polymeric materials are resistant to alkalies, gasoline and oils, dilute acids, ammonia, salts and some solvents. However strong oxidizing acids attack the polymeric repair materials. To evaluate the chemical resistance of resin mortars IS: 9162 recommends that 25 × 25 mm cylinder specimens be immersed in the specified chemical for duration of 56 days. It is taken out periodically, cleaned under cleaned under running water and dried wiped with a blotting paper and dried for about 30 minutes and again reimmersed in the chemical solution. The change in mass and physical signs of deterioration, such as, cracking, spalling, softening etc., are periodically recorded.

7.4.10 Test for surface quality

Inadequate preparation of concrete surface to be repaired is often the source of failure of patch repairs and overlays. The tests generally recommended by the American Concrete Institute for evaluation of surface preparation are discussed below:

Cleanliness of the surface:
The following surface conditions are considered:

- Dusts condition-The surface to be repaired is wiped with a clean dark cloth. If any white powder is noticed on the cloth, the surface is considered to be too dusty and therefore unsatisfactory.
- Oily condition-If water sprinkled on the surface immediately spreads out, without forming droplets, the surface may be assumed to be free from oils or dust.
- Acid condition-Using a pH paper, if the pH of the concrete surface is found to be less than 4, it is considered unacceptable.
- Presence of laitance: This may be detected by scrapping the surface with a putty knife and observing for presence of any loose, powdery material.

Dryness of the surface:
Many polymeric bonding agents, repair materials and barrier systems do not perform satisfactorily if moisture is present at the surface of concrete substrate. A simple test for evaluating the dryness of the surface consists of tapping a 1.2m × 1.2m size polythene sheet to the substrate surface and determining the time required for the moisture to collect underneath the sheet, when the ambient conditions are as near to the likely conditions during repair as possible. If the observed time period is less than the cure period of repair material, it may be concluded that concrete is adequately dry. However, decision should be taken giving due consideration to the nature of repair material. Certain cementitious and water-based repair materials are to be applied to damp concrete surface. In those cases, it is enough if care is taken to prevent any water stagnation on the concrete surface.

Test for surface soundness:
This test is recommended to ascertain the quality of surface preparation at the rate of approximately one test for every 10 sq. m of area to be repaired. The test procedure consists of bonding the flange portion of an Aluminium T-section using a small quantity of epoxy compound to the cleaned, dried, and prepared surface, curing it for 24 hours and then measuring the tensile load required to pull out the T-section, when pulled at a uniform rate. A strength of less than 1.2 MPa indicates that the concrete substrate may be suspiciously weak and there is a need for investigation of the quality of concrete before undertaking full scale repairs.

Patch test for adhesion of repair material:
This is another simple field test to check whether adequate bonding has been obtained on properly repaired substrates. Although the test procedure is recommended for epoxies, it may be used for other polymeric materials and barrier systems also. The test procedure consists of drilling through the repaired patch down to the substrate surface to produce a 50mm diameter cored disk. A standard 37.5mm diameter pipe cap with 50mm diameter bottom flange is bonded to the cored disk using a strong epoxy glue. After the curing of the glue, the pipe stub is subjected to tensile pull out and the failure load and mode of failure are observed. Failure

in concrete substrate at stress levels below 1.25 MPa indicates weakness of the substrate. If a cohesive type of failure occurs in the patch or the adhesive fails, the test is repeated. Repeated failures indicate inadequate surface preparation; defect in the repair material or in its application.

7.4.11 Other tests

There are many other important tests to assess the properties of repair materials and repaired structures, such as, impact resistance, abrasion resistance, skid resistance, low temperature flexibility test, thermal stability and degradation, flammability, fire resistance, chemical resistance, climatic cycling resistance/weathering resistance, ozone and UV ray resistance, salt spray resistance etc., which are especially important for polymeric materials used for overlays and, floorings.

CHAPTER 8
Function of Repair Materials

8.1 PATCHING MATERIALS

Patching materials currently available appear to have been formulated to meet the demands of a repair schedule and conditions. Minimum down time has dictated the development of fast setting, high strength-developing materials; poor accessibility has led to the development of high flow, self-leveling materials. While some materials have unique characteristics which offer significant advantages in particular situations, others, when considered against more conventional materials, appear to offer borderline advantages for their higher costs. The success of any patch will depend largely on overcoming the tendency of the concrete/mortar or polymer patching material to shrink after placement, and on securing a bond to the substrate concrete. A variety of methods have been used to overcome shrinkage and to promote a better bond. **The list of requirements for a patching material is given as follows:**

(a) It should be as durable as the surrounding material
(b) It should require a minimum of site preparation.
(c) It must be tolerant of a wide range of temperature and moisture conditions.
(d) It must be chemically compatible with the substrate.
(e) It should have a similar colour and surface texture to the surrounding material.

A wide range of patching materials is available and may be conveniently categorized in the following groups:

(a) Cementitious materials
(b) Polymer concrete and mortar
(c) Quick setting compounds
(d) Bituminous materials
(e) Grouts

8.1.1 Cementitious materials

Portland cement concrete and mortar offer a number of advantages as patching materials: thermal movement similar to the existing concrete, similarity in an appearance, cost, ready availability, and familiarity. Concrete is most often used for complete replacement of sections and deep cavities extending beyond the reinforcing bars, while mortar can be used for cavities as small. Special proprietary mixtures are available that are based on Portland cement and contain hard mineral metallic aggregate to increase the resistance of floors to heavy traffic.

Most of these products use high early strength cement and contain an expansive agent that causes the concrete and mortar to expand either in the plastic stage or after it has hardened. The expansion produced in the plastic state establishes intimate contact with the substrate before it hardens, thus completely filling the space and promoting a good bond. The expansion produced in the mortar or concrete is usually obtained by the use of an expansive agent such as aluminum powder, coke powder, anhydrous calcium sulfo aluminate, or calcium oxide.

Some of the disadvantages of these materials are that high shrinkage and a tendency to crack may result if great care is not taken to keep the water content low and provide proper curing. Also, in cool or cold weather, the rate of strength is too slow. Some of the disadvantages of these materials are that high shrinkage and a tendency to crack may result if great care is not taken to keep the water content low and provide proper curing. Also, in cool or cold weather, the rate of strength is too slow. High early strength cement with a mixture of super plasticizing and non chloride accelerating admixtures can have a more rapid setting time and rate of strength gain. Commonly used admixtures for this purpose include, naphthalene or melamine formaldehyde sulfonates and nitrates, nitrites, calcium formate, sodium sulfate or carbonate.

8.1.2 Polymer concrete and mortar

There are two broad classifications of polymer concrete and mortar. Polymer concrete are materials containing latex emulsions used with hydrating Portland cement, and the second are systems without Portland cement binder. The first class of materials includes latex modified mortars and epoxy emulsions. The second group includes epoxy, polyester, mortars, and concrete. Polymer concrete is comprised of a blend of coarse and fine aggregates held together in a polymerized matrix. Some of the materials are discussed below.

Latex modified mortars:

Latex-modified mortars include mixtures of Portland cement, sand and latex admixtures such as styrene butadiene, polyvinyl acetate, acrylics, and epoxy emulsions. The latex is usually used at a level less than 20% by weight of cement in the mixture. Addition of the latex improves the permeability, bond, tensile, and flexural strengths. Such mortars can be placed in sections ranging from 12 to 50 mm thick in both horizontal and vertical applications such as column and wall repair. The chief advantage of the system is its good workability and ease of application when compared to other similar systems. When the patched area is shallow and extends across a crack or joint that is subject to movement, a flexibilized epoxy should be used to accommodate the anticipated movement. However, if a patch is more than 9 mm deep, the patching material should not bridge the crack.

Epoxy patching compounds:

Epoxy patching materials are usually epoxy mortars. They are generally prepackaged systems containing resin binders consisting of two components that are added to a selected blend of aggregates, which may be contained in the prepackaged system or purchased separately. A typical mortar mix will consists of 1 part of mixed epoxy resin and hardener to

3 parts concrete sand. Due to their strong adhesive character these materials can be placed in very thin layers, which will resist heavy impact loading. Some of the commonly used epoxy compounds for this purpose include black coal tar epoxies, the cream-coloured, petroleum based epoxies and a variety of polysulfide epoxies which range in colour from clear amber to an opaque grey.

Epoxy patching materials present a major advantage over cementitious materials because of their rapid cure and faster rate of strength development, high resistance to aggressive chemicals, high bond strength, good resistance and long-term durability. Normally, traffic can be resumed over a patched area within 24 hours and with a more rapid curing system, the down time can be reduced to 4 to 5 hours. Due to the high cost of epoxy materials, it is too costly to place an epoxy mortar layer more than 9 mm in thick. If extensive resurfacing and /or large patches are required, a thin layer of epoxy or latex cement slurry can be used to bond new regular Portland cement mortar to the existing concrete substrate.

Polyester reins:

Patching compounds based on unsaturated polyester resins have not been used as widely as epoxies. They generally harden by a reaction between the base resin and small amounts of catalyst. Accurate control of proportions and mixing are therefore more difficult than with the two component epoxies. Typical polyesters include unsaturated polyesters such as methyl methacrylates. Polyester patches cure faster than epoxy materials and are less sensitive to lower temperatures. Vinyl ester patching compositions combine resiliency, impact resistance, and excellent chemical resistance. Typical application areas are floors, trenches, and pickling and plating tanks.

Acrylic concrete and mortar:

The most widely used patching materials of this type are acrylic monomers. Acrylic concrete and mortar contains neither water nor Portland cement. Instead, the aggregates are held together by an acrylic polymer. Two types of monomers are usually used: methyl methacrylate (MMA), which has been used for the last 20 years, or high molecular weight methacyrlate (HMWM), a relatively new material. All components of the product can be mixed and placed, or the monomer may be poured over a patch area filled with preplaced aggregate. One commercial, acrylic concrete system consists of two components; liquid acrylic and a package of premixed fine aggregate, promoters, initiators, and pigments.

Polymer concretes are used for patching sections that are generally larger, or more extensive than those suitable for repair with epoxy mortars. A range of properties such as rapid strength gain, low permeability, and good abrasion resistance is achievable by custom design of the mix. The unique features of rapid strength gain and high ultimate strength in conjunction with its excellent durability characteristics make this material suitable for a wide array of applications. Bridge slabs, parking garage decks, industrial warehouse floors, and tanks are potential areas where this repair material could be used.

8.1.3 Quick setting compounds

These compounds are used as repair materials in areas which are heavily trafficked to be achieved with the least disruption to traffic, or, in the case of industrial floors, to production. Also, faster setting and curing materials are required during cooler periods, or in areas, which are subjected to, continued cold temperatures such as food freezer floors. A number of unique and specialized patching materials have been developed for such special repair applications.

High alumina cement-containing compounds:

High alumina cement patching compounds with accelerated set time and strength gain can be obtained by mixing this cement with the appropriate proportion of Portland cement. High alumina cements are three to four times as expensive as other cement, but it produces a faster rate of strength gain and is more resistant to sulphate attack.

Magnesium Phosphates:

These are fast-setting, rapid strength-developing materials. One of the most widely used commercial products is Set-45. In the two-component package, the dry magnesia is mixed with the liquid phosphate in small quantities and worked very rapidly. It usually produces a high strength, low permeability patch with a good bond to many surfaces. However, water will affect hardening; even small amounts produce severe strength reduction. Some evidence indicates that is permeability may be high enough to jeopardize freeze-thaw durability. It is probable that poor mixing, resulting from an increase in the viscosity of the phosphate binder under low temperature conditions, produces a relatively more porous material than under normal temperature.

Magnesium phosphate products cost about the same as epoxy resin mortar. Due to its ability to cure and harden at temperatures at, or below, freezing, it appears to be best suited for repair of sliver spalls and popouts, or for emergency work. Typical applications include repairs to bridge decks, commercial freezer rooms, airport runways, industrial floors, highway pavements, truck docks, ramps, and stairways. Other features which make it a suitable patching material are the similarity of its coefficient of thermal expansion and linear shrinkage value to that of Portland cement concrete. The use of these materials greatly reduces the duration of interruption of the use of the structure under repair.

Molten Sulfur:

Molten sulfur has been used either on its own or in combination with tar-based additives as a quick hardening patch for concrete. It has also been used to impregnate deteriorated concrete and as a binder for concrete.

Calcium sulphate-based materials:

Many of the proprietary, rapid-setting patching products are calcium sulphate-based. Most contain Portland cement in varying amounts; some also contain chlorides. They gain strength very rapidly and can be used at any temperature above freezing. Many have not proved reliable when exposed to moisture and freezing weather. These materials can also promote sulphate attack to the surrounding concrete.

8.1.4 Bituminous materials

Bituminous materials have the lowest cost of all patching materials, are readily available, and are easy to apply. However, when they are of poor quality, and carelessly applied, there is some evidence that they hold water. Poorly graded aggregate and unconsolidated patches are particularly prone to water pick-up, which accelerates deterioration of surrounding concrete. Bituminous patches probably find their best use for repairing pavement prior to the application of an overlay. Hot-mixed, densely-graded asphalt concrete is a wide use for patching. Asphalt emulsions, both anionic and cationic, have also been used with some success for patching. Other bituminous materials used for patching include cold-mixed cutback asphalts, tars and rock asphalt.

8.1.5 Grouts

A grouts is any flowable, plastic, or packable material that can be used to fill the space between the under side of a machine or column and the foundation on which the unit is to rest, then harden there to support the unit. In addition to this principle role, they also find wide use in the repair of deteriorated structures. Typical applications include void filling, and as a mortar for filling large cracks. The most important requirement of a grout is that it completely fills the space in which it is placed. Therefore, it must have negligible shrinkage and remain stable in place, without cracking, delaminating, or crumbling. **There are a variety of grouts in use today which can be classified as follows:**

(a) Nominal cement/sand mixtures
(b) Gas-forming grouts
(c) Metallic aggregate grouts
(d) Inorganic sulfoaluminate cement or expansive additive-based grouts
(e) Fiber-reinforced grouts
(f) Organic resin grouts

Nominal cement/sand mixtures:

Ordinary cement/sand grouts are the lowest in material cost. However, they require high water and cement content to achieve placeable consistencies that shrinkage and cracking promotes on hardening. Dry packing with a stiff grout mix overcomes some of the problems. The use of cement/sand grouts has declined sharply with the advent of specially formulated non-shrink grouts.

Gas-forming grouts:

These grouts contain expansion-producing ingredients which react with the cement liquor to generate gas bubbles. The gas expands the grout to help compensate for any shrinkage that occurs in the plastic state before the grout has hardened. Vertical and lateral restraint is required to achieve specified strength and volume stability in the hardened stage. Aluminum powder and finely divided carbon are used to produce the gas producing reaction. Some of the disadvantages of these products are that they are quite temperature sensitive.

Sulfoaluminate grouts:
Most of the expansion produced by sulfoaluminate grouts occurs after the grout has set. The expansion can be produced by shrinkage-compensating cement or an anhydrous sulfoaluminate expansive additive used with Portland cement. The expansion producing reaction occurs when the anhydrous calcium sulfoaluminate is converted to the hydrated ettringite. Sulfoaluminate cement-based materials are less prone to strength reduction due to increased w/c ratios than the Portland cement -based materials. The initially achieved expansion could well be lost if there is no moist curing after the grout is placed.

Fiber-reinforced grouts:
These grouts may have polypropylene, steel, or glass fibers dispersed in either a Portland cement or a shrinkage compensating mortar. The fibers impart much improved impact resistance and flexural strengths. The stress required to produce cracking is significantly increased and the integrity of the grout after cracking is preserved, thus preventing a catastrophic type of failure.

Metallic aggregate grouts:
These contain iron fillings, to impart toughness, and an oxidation catalyst, such as ferric chloride, that produces expansion to compensate for settling and drying shrinkage. To develop their full strength and stability such grouts require rigid confinement. They are not recommended for areas that are exposed to wet and dry cycling and stray electrical currents. Most of the cementitious grouts are available as one-component products that require only the addition of water in the field. Storage life, under proper conditions of low humidity and temperature, is 12 to 18 months.

Polymer grouts:
Polymer grouts are generally used for equipment bases or concrete repair, especially in areas subjected to chemical exposure, freeze/thaw, impact stresses, or when a fast turn around time is important. Most grouts are supplied as three-component materials having;

- A liquid resin component
- A fluid-reactive curing agent component, and
- A dry filler component

In some cases, two-component products are package with the curing agent mixed in with the aggregate.

Some specialty system formulated for under water applications are available. The composition of these products may consist of resin and hardener systems which are tolerant to moisture. They are usually pumped or poured. Since such materials are unique specialty products. Most products are sold as prepackaged units which avoid the need for field proportioning and the consequent errors and problems that result from it. Although epoxy grouts generally have good chemical resistance to oils and chemicals, there are some applications where the chemical environment may be too severe for epoxy grout. Under such conditions, it is more advantageous to use a vinyl ester-based grout. To prevent potential penetration of the corrosive agent to the concrete, a glass-reinforced monolithic surfacing is installed prior to application of the polyester, vinyl ester, and epoxy.

Because of the more costly installation procedure and the higher shrinkage, polyester, vinyl ester, and epoxy should only be used in situations where rigorous chemical exposure conditions dictate their use. Properties of these grouts include high compressive strength of 3 to 4 times that of standard concrete, and high tensile bond strengths. One of the most important performance factors which distinguish the service capabilities of organic polymer grouts from those of cementitious grouting materials is their excellent adhesion to variety of construction materials.

Acrylate and urethane grouts:
These grouts are used in subgrade injection to repair sewer leaks, control ground water movements, and stabilize soil. The chemical systems used for such purposes include acrylamides, polyacrylamides, urethanes, and acrylates. These grouts have very low viscosities and can be controlled to set almost instantly after predetermined elapsed times. Some, like urethanes, are even activated or catalyzed by water itself. The grouts are true solutions; their constituent materials dissolve completely or combine with water to provide a very thin mix. Since their particle size is less than one third the size of the void being filled, they penetrate completely and prevent fracture of the void being filled; they penetrate completely and prevent fracture of the void.

8.2 RESURFACING MATERIALS

Resurfacing materials are protected concrete structures under certain exposure conditions from spalls, cracks, and disintegrates. Resurfacing materials referred as coatings and toppings. Protective coatings can greatly reduce the effect of deteriorate conditions, and significantly improve the durability characteristics of the concrete. A variety of coatings are available, and some are tailored for greater chemical resistance, while others are formulated to resist wear and erosion. Unfilled and partly filled coatings less than 30 mil are described simply as coating, while filled system with thickness in excess of 30mil, and ranging up to 10mm are described as toppings.

8.2.1 Protective coatings

When considering a coating for concrete, the specifier or buyer must take into account the properties of the concrete and location of the area to be coated in the structure. **The properties of concrete which affect the successful application and performance of a coating are:**

(a) Porosity
(b) Moisture content and
(c) Presence of contaminants on the surface

Porosity will affect coverage rats, interface adhesion, and "pinholing" of the coating. A porous surface will absorb greater quantities of the resin from the surfacing material in comparison to a surface that is dense and of low absorptivity. Coatings of high viscosity may be absorbed to a smaller extent than those of low viscosity. It is a good practice to use priming or first coat which may be tailored to meet a dense or porous substrate. To ensure proper bonding of the coating to the concrete, it is important that the concrete have low moisture content and be surface dry. When a concrete contains a significant amount of water, the moisture may subsequently move to the surface and blister or disrupt the coating film. It is possible, however, to apply latex coatings to slightly damp surfaces because these coatings are water-based emulsions.

The condition of the concrete surface to be coated is of prime importance. A concrete surface that contains residual form-release agent, curing agent, or oil, grime and laitance may not accept the surfacing or not coat the surface uniformly. **The factors due to location of the concrete in the structure that need to be considered include:**

(a) Continuous or intermittent exposure of the coating to moisture
(b) The degree to wear encountered in service and
(c) The temperature fluctuations to which it will be subjected

In some concrete, there is continued ingress of water from an exterior source. This occurs in tanks, tunnels, flatwork on grade, laundry rooms, and improperly constructed basement walls and other such locations. The concrete may contain water soluble salts which can be carried to the surface of the concrete to produce efflorescence. The crystal pressures of such salts, deposited just below a coated surface, can disrupt the film and pry it away for the wall. Therefore, the application of a coating to the surface on which efflorescence might appear is not an effective preventive measure. The degree of wear and traffic anticipated in a particular location may influence the choice of the coating, *e.g.*, a coating for a hospital floor would be expected to be less wear resistant than one intended for a factory floor. **The following factors are considered on the application of resurfacing material.**

Climatic conditions - Coating is best done under favorable climatic conditions because the viscosity of coatings increases as the temperature decreases causing difficulties in application. Coating should not be done below an ambient temperature of 10°C or above 30°C.

Concrete temperature - Some two-component coatings will perform poorly if the concrete temperature is cold, even though the coating is warmed. Therefore, the air and concrete temperature must be above 10°C.

Moisture content of surface - For most polymer-based paints, the provision of a dry surface on which a coating is to be applied is imperative. Water-based coatings may be safety applied to damp surfaces.

The thickness and number of required coats - This will, depend on the texture of the surface and the intended application. A purely decorative application may need a thickness and number of coats to obtain colour uniformity. For abrasion resistance the thickness and the number of coats are required.

Coverage rate - Rough and porous surfaces often require two coatings. The first one is usually thinned to allow for penetration or a compatible primer coat may be used. In estimating coverage rate for a given film thickness, allowances should be made for losses caused by spillage and overspray.

Resurfacing of a concrete surface requires thorough cleaning or removal of the existing coating. If the old coating is in good condition, then compatibility between the old and the new coating should be determined.

Types of coatings:

Many different wall and floor coatings are available which exhibit varying degrees of chemical resistance, physical durability, and ease of application. **Some of the more widely used products are as follows:**

Bituminous coatings and mastics - Asphalt or coal tar coatings may be applied cold or hot. Two coats are usually applied to surface dry concrete: a thin priming solution is applied first and when the primer has dried to a tacky state, it is ready for a finishing coat. Multiple coats at right angles to each other are usually applied to avoid pinholes. Bituminous mastics may be applied cold or as heated fluids. Cold mastics are cutbacks or emulsions produced as thick, pasty materials which contain finely powdered, siliceous mineral fillers and asbestos fibers. Emulsions are slower drying, more permeable, and less protective.

Polyesters and vinyl esters - These coatings are two-or three-part systems consisting of an unsaturated polyester resin, peroxide catalyst, and possibly a promoter. Catalyst and promoter are mixed separately into the resin, and the amount of catalyst must be carefully controlled because it affects the rate of hardening. Fillers such as silica flour and glass fibers are often used to reduce shrinkage, and the coefficient of expansion, and to compensate for brittleness. Selective modifications to resins have produced new, less rigid types, which retain moderate to good chemical resistance. Vinyl esters are different from general polyesters, because the resins are characterized by an acrylic termination of epoxies. They have excellent resistance to oxidizing chemicals, improved handling characteristics, rapid cure, and low cost, and are widely used in the food, chemical process and paper industries for the repair of floors, trenches, storage tanks, scrubbers, and waste treatment facilities.

Urethanes - These coatings may be one- or two-part systems. Most one part systems are moisture cured and must be used on dry surfaces to prevent blistering during the curing period. The two-part system is available in two-types - catalyzed and polyol-cured. Polyol-cured coatings are the most chemically resistant of the urethane coatings but require the greatest care in application.

Epoxies - Epoxy coatings have excellent chemical resistance to most chemicals. They have a better alkali resistance than acid resistance and bond well to concrete. The coatings are resilient and abrasion resistant. They are generally tow-package systems composed of epoxy resin and a curing agent. Common curing agents suitable for room temperature curing are amines, polyamines, amine adducts, polyamides, polysulfides, and tertiary amines. The coating properties are dependent on the type and amount of curing agent used. Generally, three coats must be applied to eliminate pinholes; fillers like silica flour or glass flake may be used to bridge pinholes. Fiber-reinforced epoxy coatings may be used on concrete that may undergo thermal movement.

Neoprene - These coatings may be one-or tow-part systems. One-part systems are generally used in thinner films. The two-part system has lower chemical resistance, cures slowly at room temperature, and usually requires a holding period between mixing and application. Since the materials are sensitive to moisture in the concrete, coating should not be done until the concrete is at least 28 days old.

Coal tar epoxy - Most proprietary products are two-package systems. A combination of coal tar, filler, solvent, and epoxy resin may be in one package and one of the commonly used curing agents in the other. The coal tar, filler, solvent, and curing agent may also be blended together to make up one package and the epoxy in the other package. To ensure proper cure and chemical resistance, the correct mixing ratios must be used, and stipulated mixing times be adhered. Three main types of coal tar epoxy coatings are recognized, the classification being according to epoxy resin content, namely, high epoxy resin content coatings for dry film thickness, medium -epoxy resin content coatings for concrete pipe linings, and low-epoxy resin content materials for nonsag application. The first type requires a primer and its thickness is achieved in two coats, while the two latter materials require no primer and can be applied in a single coat. Spray application gives the best coverage.

Acrylics - These materials are available as solvent-based products and as the more recent 100% solids resins. Solvent-based acrylic coatings have excellent wetting characteristics, a fair degree of chemical resistance, and good colour retention characteristics on outdoor exposure. Because of their good intercoat adhesion they are easily recoated. Because of their low viscosity, these materials are ideally suited for sealing and impregnation of cracks. They can be used at temperatures as low as 10°C because they cure rather rapidly at this temperature. At normal temperatures, treated surfaces are reportedly ready for use after only 1 to 3 hours of curing. Therefore, they can be used for the repair of plant floors without causing prolonged work stoppages. Floors treated with acrylic monomer type materials are skid resistant, hygienic, seamless, easy to clean, and show good resistance to gasoline, oils, alkalis, dilute mineral acids, and organic acids such as lactic acid. They are reported to develop excellent intercoat adhesion as well as bond to older, previously cured acrylic coatings even after years of service, thus achieving easy and effective repairs.

8.2.2 Intumescent coatings

These coatings are specially formulated fire resistant paints applied on structural elements for the purposes of increasing their fire rating. These products are finding increased use in the repair of fire damaged buildings where they are used for fire protection of steel work. The coatings can be brush applied up to a thickness of 3 to 4 mm for thin coatings and up to 30 mm for thicker coatings. These coatings look like ordinary coatings, but are chemically formulated to expand 25 to 30 times their original volume when exposed to temperatures of 200 'c and above. Most system consist of three components, a primer, intumescent base coat, and an external seal coat usually based on chlorinated rubber to retain colour fastness and protect the base coat from wear and tear.

8.2.3 Resin based toppings

When an industrial floor is at an advanced stage of deterioration, heavy duty, monolithic, resinous toppings are probably the best all round materials for resurfacing eroded and spalled floors. In general, the polymer resins, curing agents, catalysts, and fillers used are the same for both the high-build coatings and seamless flooring compositions. Most of the products in use consist of thermosetting resins and special sand fillers that are generally applied as a single 3 mm to 6mm thick layer. Some of the desired properties provided by these toppings include superior abrasion and chemical resistance in comparison to concrete, rapid cure and a tenacious bond to clean sound concrete.

Nearly all toppings are three-component systems consisting of a resin, curing agent and aggregate fillers. The resins used for heavy duty, monolithic toppings are epoxies, polyurethane, polyesters, polyacrylate, and phenolic materials. In addition to binding the aggregate and providing adhesion to the concrete, the resins provide chemical and impact resistance. Epoxies, polyesters, and polyurethanes account for most resins used in heavy duty monolithic toppings. In addition to lowering the cost and providing an abrasion resistant, textured surface, a well graded aggregated provides other advantages, such as increased compressive strength and reduced curing shrinkage. It also reduces the coefficient of thermal expansion to a value closer to that of concrete, increases toughness, and improves trowelability. Some of the materials used as heavy monolithic toppings follow.

Epoxy floor toppings and high-build floor coatings:

These products have excellent resistance to wear and impact, are completely waterproof and impervious to grease, oils, detergents, and many chemicals. Because the epoxy binder used in usually of 100% solids they show negligible shrinkage. Two types of toppings are currently used: a sand-filled epoxy mortar and a glass fiber-reinforced, multicoated system.

Epoxy mortar:

Such systems contain pure or modified epoxy resins of low viscosity blended with silica sands and fillers with mesh sizes ranging from 40 to 200. The blend produces a trowelable mix which is applied at a thickness in the range of 3 to 6 mm. Epoxy/amines combinations have excellent solvent and alkali resistance and good resistance to acids except for strong oxidizing agents. Epoxy polyamides have less chemical resistance than epoxy polyamines, but their physical properties are better, they are tougher, more flexible, have better adhesion, and are also more tolerant of moisture. The polysulfide epoxies are less widely used because they are far less chemically resistant and are more expensive. Since they are more flexible than the other two, they find use in some floor joint sealant applications.

The aggregate fillers are generally silica, quartz and emery. The ratio of aggregate to epoxy resin binder will vary based on mesh size of aggregate and desired consistency anywhere from 3:1 to as much as 8:1. The finer mesh sizes produce a smooth finished surface but the higher resin demand by the large, specific surface area of the filler makes the system quite expensive. A coarser mesh size of 30 produces a more skid resistant surface, loads the resin matrix more fully and proves better troweling properties. A mixture of these, mesh sizes is more widely used where colour is required, chemically resistant pigments may be ground into the epoxy resin or cuing agent, or they may be dry blended with the aggregate.

Epoxy mortar compounds and coatings are used in many areas other than on floors. They are used to line silos, build and repair airport runways, line cooling tower pans, and provide nonskid areas on roadways, bridges ramps, and docks. **A typical floor is applied in three steps:**

- Priming of the concrete surface with a low viscosity epoxy resin to ensure proper wetting of the surface and sealing of the pores
- Trowelling of a 6 mm sand-filled mortar over the primer
- Sealing of the above topping with a low viscosity epoxy resin to fill in any remaining pores and ensure that a complete barrier system is presented to splash chemicals.

Glass fiber-reinforced multicoated system:

The glass fiber-reinforced multicoated system is a heavy duty chemical resistance system used in locations where the exposure conditions involve higher concentration of chemicals and longer periods of exposure. The thickness of the system can range from 0.5mm to 6mm depending on the severity of the in-service conditions. The thicker coats provide a surface where discontinuities or pinholes are minimized or eliminated so that an effective barrier system, which protects concrete against acids and other aggressive chemicals, is produced. **The system is built up in four steps:**

- The surface is primed with a low viscosity epoxy
- A 50 mil thick coat of the filled epoxy is then placed
- A woven glass fabric is then applied
- After the epoxy has partially cured

Polyurethanes, seamless floor toppings:

Many of the synthetic monolithic flooring compounds - termed seamless flooring - are either epoxy - or polyurethane - based products. Polyurethane products result from the chemical reaction between isocyanate groups on the liquid resin, and hydroxyl groups of the curing agent. They are characterized by excellent resistance to chemical attack, impact, abrasion, good adhesion, and flexibility.

These flooring toppings can be one or tow-part systems. A one part system may be moisture-cured. They are usually supplied at a solids contain of 40 to 50% in a blend of solvents. In flooring applications, a base coat of the higher solids content material is first applied. While the base coat is still wet, silica or emery aggregate is sprinkled or blown into it. The aggregate may be applied in one or two applications and a sufficient amount of time must be allowed for the base coat to develop a gel so that the aggregate particles do not settle to the bottom of the coating. Two types of the two-part system are available; catalyzed and polyol-cured. Catalyzed toppings have a limited pot life after mixing and cure rapidly. Solids content range from 55 to 60% and the solvent is xylol. These materials are the most chemically resistant of the urethane floor toppings, but they require the greatest care in application.

Most of the seamless flooring compositions can be readily applied by brush, spray, or roller. **Limitations of toppings made from thermosetting resins are:**

- High coefficient of thermal expansion
- Ultrahigh cohesion and bond strength
- Measurable curing shrinkage
- Small, but measurable, liquid absorption

8.2.4 Surface Hardeners and Overlays

The surface hardeners and overlays are used for the repair and upgrading of industrial floors, where loaded vehicles, such as forklift truck, traverse the floor surface. A number of materials are available that, in specific instances, will improve wear resistance or chemical resistance, reduce dusting, or improve the appearance of concrete floors.

Surface hardeners:

Generally refers to a material used to upgrade a floor's wear resistance, reduce dusting and increase chemical resistance. Two main categories are in current use: sprinkled shake hardeners and liquid hardeners.

Shake hardeners - Shake hardeners produce a special finish that is obtained through the incorporation of selected natural or metallic aggregate into the surface of a newly poured floor. There are many commercial products sold for this purpose, the required characteristics being high aggregate to cement ratio, good finishing characteristics at low cement factor, and capability of being incorporated heavily enough to make a substantial difference in the composition of the floor surface. Two types of aggregates are used in shake hardeners-mineral and metallic. Metallic floors are generally used in floors where there is continuous steel wheel traffic and point impact. Floors containing mineral aggregate tend to be harder, thicker, and will not rust.

Liquid floor hardeners - The liquid hardeners are usually floor sealing liquids, which by their reaction with certain cement hydrates, $Ca(OH)_2$, adhesive character, and sealing ability, are able to bind the floor surface and reduce the rate of wear. More recently, dilute solutions of both solvent and water based emulsions of resins such as epoxies, urethanes, and methyl methacrylates are finding wider use for this purpose. Solvent and water based epoxies and urethanes consisting of 6 to 20% solids solutions are generally used to improve both wear resistance and resistance to oil and grease penetration. Two coats of polyurethanes, epoxy, or methyl acrylates are often used to prevent dusting on concrete floors. Because of their adhesive character, these polymeric materials bind the surface well, providing better abrasion resistance than that obtained by the use of inorganic hardeners.

8.2.5 Overlays

Concrete overlays may be applied as a second stage of construction on a new floor or deck, as preventive maintenance on a deck that has been open to traffic for a short time but was built without a deck protective system, or in the rehabilitation of existing deteriorated floor slabs or decks. The terms overlay and topping are often used interchangeably in the above mentioned applications. **Materials currently available for this method of repair can be categorized as follows:**

(a) Normal or high early strength Portland cement concretes and mortars
(b) Latex modified concrete
(c) High alumina cement concrete and mortar
(d) Fiber-reinforced concrete
(e) Silica fume concrete
(f) Resin based mortars

The materials mentioned above are limited to the rehabilitation of industrial floors and bridge decks. **The general requirements for the above systems include:**

(a) Low permeability to water and deicing chemicals
(b) Adequate skid resistance
(c) High abrasion resistance
(d) Insensitivity to minor variations in ambient temperatures
(e) Sufficient working period to setting time to 15 year service life
(f) Sufficient flexibility to avoid cracking

Several potential advantages of concrete overlays can be identified:

(a) The overlay can be tailored to provide the required thickness of high strength concrete and maximum durability at the deck surface.
(b) Properly proportioned and consolidate concrete is effective in reducing the penetration of chloride ions.
(c) The overlay is an integral component in the load carrying capacity of the deck.
(d) Use of a concrete overlay permits vapor exchange between the concrete and environment, preventing the buildup of vapor pressure that occurs beneath an impermeable membrane.
(e) The overlay has a compatible thermal coefficient of expansion with the base concrete and absorbs less solar radiation than asphalt overlay.
(f) A smooth riding surface can be provided because minor irregularities in profile can be corrected and dead load deflections are minimal.
(g) High quality aggregate can be incorporated into the concrete mixture to provide good wear and skid resistance at a modest additional cost.
(h) In repair work, the overlays will fill in areas where concrete was removed without the need for a separate placing operation. Work can proceed while part of the deck remains open to traffic.

Normal or high early strength Portland cement, concretes, and mortars overlays: Typical overlay mixtures may consist of 1 part Portland cement, 1 part fine aggregate, 1.5 to

2 parts coarse aggregate by volume. The aggregates for such heavy duty overlays should be selected from clean hard and irregular or rounded material such as trap rock, granite, emery corundum, or synthetic products like silicon carbide or heat treated aluminum oxide. Separate bonded overlays, 50 to 76 mm thick are applied after the base structural slab has hardened. The overlays are placed as stiffly as possible. Their performance depends on their bonded to the base slab. For this purpose, the base slab should be roughened with a coarse wire broom before it fully hardens to leave the coarse aggregate partially exposed. After the slab hardens, it should be blown clean of dust and debris, and dampened. Just prior to placing the topping, a grout consisting of a 1:1 mixture of Portland cement and sufficient sand mixed with sufficient water/latex mixture to give a thick slurry consistency, should be broomed into the base slab. The topping must be placed before the slurry dries out. Curing, of the freshly placed topping, is a very important step in securing dense surfaces which will not craze or dust. The success of the overlay is dependent upon good quality control procedures.

Latex-modified concrete overlays:

Latex modified concrete contains the same ingredients as conventional concrete with the addition of polymer latex. The water of suspension in the emulsion hydrates the cement, and the polymer enters the structure of the concrete, there by providing supplementary binding due to the adhesive and cohesive properties of the polymer. This result in concrete having good durability and tensile and flexural strength properties that is well suited to sue in overlays. The structural properties of this type of concrete can vary depending on the type and amount of latex, the type of aggregate, the cement content, and the w/c ratio. The choice between latex-modified mortar and latex-modified concrete is determined by the intend thickness of the overlay. Thickness of less than 25 mm is usually constructed from mortar and overlay greater than 25 mm from concrete.

High alumina cement mortar and concrete overlays:

High alumina cement mortars and concretes have the ability to provide high early strengths at normal temperature or cure and develop higher strengths at lower temperatures faster than Portland cement concretes. In addition, such concretes are more resistant to heat, acidic conditions, and sulphate attack than Portland cement-based concrete. Consequently, their chief use is in locations which are subjected to chemical attack, such as pulp and paper factor floors, refractory conditions such as steel rolling mills, and when repairs are done under cold weather conditions.

Fiber-reinforced concrete overlays:

Several types of fiber have been investigates, but most of the work has focused on the use of steel and alkali resistant glass fibers. More recently, polypropylene fibers have been used as a crack-arresting measure in flat slab applications. The addition of steel fiber to concrete results in substantial increase in tensile and flexural strengths. Most of the high way applications of steel fiber-reinforced concrete have been in pavement overlays where the materials high flexural strength and resistance to dynamic loads can be used to advantage. For pavement rehabilitation projects, the high tensile and flexural strengths of the material, and its resistance to reflective cracking, offer good potential. For bridge and parking decks it is less appealing. No data are available on the resistance of fibrous concrete to deicing salts and chloride ion penetration. The use of fiber-reinforced concrete in the repair of damage caused by cavitations and erosion has had limited success.

Silica fumes concrete overlays:
Silica fume has been used predominantly as performance enhancing admixture in a wide variety of applications where high strength concrete or very durable concrete is required. Recent projects have included high density overlays for parking garage and bridge decks, abrasion-resistant concretes for dam spillways and stilling basins, and shotcrete. Silica fume is a very efficient pozzolana that forms reaction products with lime from hydrated cement and reduces the volume of large pores and capillaries normally found in cement pastes. The second mechanism by which silica fume improves concrete is through the microfiller effect and is credited with greatly reduced permeability and improved paste to aggregate bond of silica fume concrete compared to conventional concrete.

Polymer concrete overlays:
The term polymer concrete is used to describe three types of concretes, namely

- Polymer-impregnated concrete
- Polymer-modified concrete
- Polymer-based concrete

Polymer-impregnated concrete is produced by filling the voids present in a normal concrete with a liquid resin or monomer and then polymerizing or curing that resin. The product is reported to have very high strength, chemical resistance, impermeable to liquids and gases, and good frost resistance. Polymer-modified concrete is produced by adding a latex emulsion to the ordinary concrete mix. The amount of latex used in the mix can range from 10 to 15% by weight of the cement. The improved flexural strength and adhesion to the substrate is of great value to the overlay. Polymer-based concretes are produced by completely replacing the cement and water of a concrete mix with a liquid resin.

Epoxy concrete overlays:
The type, grade, and class of epoxy should be chosen to satisfy job conditions and requirements. The overlay must be compatible with the underlying concrete and bond compatibility is particularly important. Epoxy overlays are placed 6 to 9.5 mm thick. After workers strike off the overlay, they broadcast additional aggregate onto the surface to improve skid resistance.

MMA concrete overlays-(Methyl MethAcrylates) These overlays consist of aggregate combined with a monomer which is polymerized at ambient temperatures producing compressive strengths of 35 Mpa or more within 2 hours of placement. The aggregate used should be as dry as possible with a preferred maximum moisture content of 0.5%. The nominal maximum size of aggregate should be no more than one third of the depth of the overlay. The MMA is preferred for outdoor application particularly under low temperature conditions. The HMWM (High Molecular Weight MethAcrylates) is the preferred overlay system for indoor applications. When thin overlays are applied, the surface temperature of the area to be repaired should be considered, along with the ambient temperature in selecting the monomer system.

8.3 SEALING MATERIALS

There are various forces that can act alone or together to force water through concrete. These are hydrostatic pressure, capillary action, wind driven rain, a difference in a vapour pressure between the two sides of the concrete or some combination of these. Water finds its way most readily through porous concretes, cracks, structural defects, or through joints that have not been properly designed built or sealed. Some of the materials used to seal concrete surfaces and joints to prevent the ingress of moisture into structure are described in the following pages. **Sealants are flexibilized polymeric materials which are used for two purposes:**

(a) To plug irregular gaps between two rigid surfaces and
(b) To provide a dynamic bridge across the gap between two surfaces

In this capacity they perform three functions:

(a) Accommodate joint movement,
(b) Prevent the infiltration of the joint and
(c) Prevent the ingress of water into the structure

8.3.1 Sealant Types

Some of the common types of sealants used in industrial and domestic buildings and other engineering structures include: acrylic latexes, butyls, solvent release acrylics polysulfide urethanes and some flexibilized epoxies.

Oil-based sealants:

Oil-based sealants are easily applied caulks that adhere well to most surfaces and require no special joint preparation. They are used in nonmoving joints and cracks. The selected oils and fillers that comprise the product slowly harden as the oils dry or oxidize. Some of the better oil based products have costs similar to acrylic latex based sealants.

Mastics and thermoplastics:

Mastics are highly filled, inexpensive materials that are used in joints where the anticipated movement does not exceed 5%. The materials used as mastics are asphalt, rubber asphalts and pitch or coal tar which becomes plastic when heated. Thermoplastic are upgraded mastic type materials which show slight improvement in elasticity and chemical resistance when compared to mastics.

Latex-based sealants:

Latex-based sealants are sealants that cure by water evaporation to a resilient material that can accommodate slight movement. The surface usually dries rapidly enough to be appointed 30 to 60 min after application. Due to their quick curing capability and good adhesion to a variety of surfaces without primary they find wide use for residential construction applications.

Butyl sealants:

Butyl sealants are soft, low modulus, that cure by solvent release to a rubber, capable of sealing joints which experience small movement. They have excellent adhesion to wood, metal glass and concrete without primers. Although they possess good water resistance, they are not suitable for immersion service.

Solvent-based acrylic sealants:

Solvent based acrylic sealants cure by solvent release. They are versatile construction sealants providing a high performance seal for many building construction joints without the need for primers. Acrylics have excellent adhesion.

Polysulfide sealants:

The two-component systems are high performance products used to seal joints certain wall construction and for caulking expansion joints in masonry, concrete and glazing applications. They have a history of long term performance having been in use for over 25 years. Polysulfide elastomers have good elastic qualities, good recovery, fast cure, very good adhesion, good resistance to weathering and aging, and reasonably good chemical resistance. Two-component product cures by a chemical reaction between the liquid polymer base and activators component. The resulting synthetic rubber seals accommodate movement of 5.25% of joint width. The bond well to glass and metal but need a primer for full bond to porous surfaces such as concrete, especially for immersion service conditions. The one-part system cures by absorption of moisture from the air to produce a rubber like seal with properties similar to the two component systems. The material cures slowly and requires water spray or wet cooling to speed initial skin cure and prevent dirt pick-up.

Urethane sealants:

Urethane sealants are based on polyurethane elastomers. The two part systems can be modified to provide a variety of physical properties. Most formulations provide good bond to all construction materials, possess high extensibility and a fair water resistance. These sealants cure to a tough, abrasion resistant, elastomeric rubber that facilitates their use in horizontal surfaces such as joints in industrial floors. Other advantageous properties include high tensile strength, good recovery form elongation and compression cycles, outstanding resistance to aging and thermal shock and non staining character. Although urethanes have a great tolerance to a wide range of chemicals than the other sealants, they have less resistance to concentrated chemicals such as acids, alkalis, and some neutral salts. The one-part sealants cure by absorbing moisture form the air to produce a seal with properties similar to the two part systems. They can be sued for general construction sealing, caulking, and glazing applications in which moderate movement be expected. **Two of the main advantages of the material are:**

- The good adhesion obtained without priming of the substrates under dry conditions and
- They are very slow curing

Silicone sealants:
Silicone sealants are high performance, elastomeric sealants which have the widest service temperature and they are the only sealants which have some resistance to nitric acid. They exhibit excellent resistance to weathering and aging and are suitable for sealing all types of construction joints. The sealants cure by absorbing moisture from the air to form a silicone rubber seal capable of a high degree of movement. The presence of moisture in the concrete can significantly reduce adhesion. Silicones are tougher and harder than polysulfide and posses good resistance to organic acids, salts and solvents. The high strength, low shrinkage, low toxicity, and excellent weathering qualities enable these sealants to be used indoors and outdoors without much concern. They are particularly suitable for a chemical environment that is also exposed to elevated temperature levels. The material is always packaged as a single component. **Some limitations of the product are the following:**

- Generation of acetic acid in some formulations
- High price
- The need for a primer on some substrates for some formulations

Epoxy sealants:
Epoxy sealants are modified, flexibilized materials which posses high strength and adhesion in conjunction with their resistance to chemical attack. They can be formulated to produce a grade that is hard enough to resist wheeled traffic and keep the joint edges from fracturing. Such products will usually meet the requirements for an industrial floor joint sealant. Epoxies and modified epoxy sealants have the best chemical resistance to concentrated mineral acids, alkalis, high molecular weight organic solvents, and acids. The products are available mostly as tow component systems consisting of a resign and hardening agent. Two types, self leveling and nonsag grades are available. Limitations of epoxy materials include the service temperature limit of 60°C.

Kraton-based sealants:
Kraton-based sealants are new styrene-butadiene styrene and ethylene-butylenes styrene thermoplastic elastomers, usually formulated with solids contents of approximately 90%. Therefore, there is little solvent release and low shrinkage of the applied sealant. Despite their ability to cure to a hard surface, they possess good elongation and can be formulated to provide clear cured rubber which finds use in architectural applications.

8.3.2 Sealant Properties

Some of the properties that must be evaluated when selecting sealants are: adhering and cohesion, hardness, modulus, stress relaxation, compression set, and resistance to weather and application characteristics.

Adhesion and cohesion:
These are two of the most important properties of a sealant being considered for use in a working joint. The sealant must adhere to the sides of the joint it is sealing, and withstand internal stress caused by joint movement.

Hardness:
Hardness is measured by the resistance to penetration, and is expressed as a reading on the durometer. The higher in the reading, harder the martial.

Modulus of a sealant:
The lower modulus materials tend to dissipate the stress that results from strain induced by joint movement better than high modulus materials. In general, sealants with a high durometer hardness value tend to have high modulus, although low modulus sealants may have high modulus at lower temperature.

Stress relaxation:
In a sealant results from internal relaxation under conditions of strain. Low modules sealants exhibit the highest degree of relaxation by generating stress tending to pull an extended sealant back to its original shape. Consequently, a lower adhesive strength is required to keep it bonded to the faces of the joint.

Compression set:
Compression set it is the inability of a sealant to expand to its original shape after being compressed. A high degree of set is undesirable in a sealant for a working joint, because the result is a permanently reduced width of the sealant bead. When the sealant is so compressed, high internal stresses are produced as the joint widens to its former shape because of it's reduce capability to accommodate movement. Subsequently, these stresses become large enough to cause adhesive failure at the bond line, or cohesive failure within the sealant mass.

8.3.3 Sealant Consistencies

Sealant products are available in four consistencies:

(a) Putty-grade
(b) Gunnable or nonsag
(c) Pourable and self-leveling
(d) Extruded tapes

The **putty grade** is used for glazing applications, and is of a consistency suitable for application by a putty knife.

Gunnable sealants are soft enough to be applied by a hand-operated or air-powered caulking gun, yet stiff enough not to run or sag in vertical joints.

Pourable sealants are designed for joints and cracks in horizontal surfaces such as plant floors. They are usually poured directly from their container into the joint, but some may be applied by gun.

Extruded tapes are soft tacky ribbons of sealant used in glazing applications.

8.3.4 Sealant Classes

Sealants can be grouped into three classes: low, medium, and high range, according to their movement capability, service life, and cost.

Low range sealants:

These materials are cheap and easy to use products, which have a rather short life due to their very limited movement capability. Oil-based and resin-based caulks, bituminous-based mastics, and polybutene products fail into this classification.

Medium-range sealants:

This group can accommodate more movement than products in the low-range category; they are nevertheless limited in use to nonworking joints. Products included in this category are butyls, neoprene, and the higher quality latex materials. The latex used for sealant manufacture is the acrylic latex, rather than PVA latex. Medium-range products are generally used for interior work. Their main advantage over low-range products is longer service life.

High range sealants:

This group includes the silicones, urethanes, polysulfide, solvent release acrylics, and certain proprietary compounds. The products are designed for use in working or moving joints.

8.3.5 Sealant Selection

Movement, width, and depth of the joint affect sealant selection. **Job specification for repair should follow a three step procedure:**

(a) First the anticipated joint movement is determined
(b) Then the most appropriate sealant is selected to meet expected joint moment and other service requirements.
(c) Finally the depth of the sealant bead and the type of backer rod in the joint are specified.

The class of sealant is selected either by considering the type of joint in it will be placed or the type of substrate. Once the minimum sealant class has been determined, matching of sealant substrates and geometry begins. In most joints, any sealant of the recommended class will probably be acceptable. Final product selection requires careful examination of product literature for information on recommended geometry and nature of primers that may required of various substrates. Joint depth affects the stress in a sealant. The base of the sealant bead should be free to move. Therefore a bond breaker must be used to ensure that the sealant does not bond to the base. Of the many joint back up materials available, only closed cell, foamed, polyethylene rod can be used safely with all types of sealants.

8.3.6 Water Stops

In hydraulic structures, where the surface seal in a joint is subjected to water pressure or high rates of flow, special provisions to ensure water tightness should be provided. To ensure water tightness, a water stop must be incorporated.

Types:

The **labyrinth type** is designed to overcome the need to split the concrete forms to install the water stop and is limited in its use to construction joints. The **flat, corrugated type** has numerous ribs running the length of the water stop. The **dumbell types** rely on the bulbs at the end of the section to provide anchorage in the concrete. The **metal types** can be of many forms a typical type is one with a deep, crimped, V-groove in the center. The steel plate consists of a heavy gauge, flat plate, one half of which is embedded firmly in the concrete on one side of the joint, the other half being coated with asphalt to allow it to slip in the concrete whenever the movement occurs.

Properties:

Water stops installed in concrete structure are not easily replaced. Considerable time is consumed in making the numerous joints and splices during water stop installation. Therefore, it is of paramount importance that a water stop should continue to perform well for the life of the structure. The durability characteristics of the materials used in these products thus plays a significant role in their selection for use. **Properties that should be evaluated prior to selection are as follows:**

(a) Mechanical fatigue
(b) Resistance to alkalis and chemical inertness
(c) Resistance to aging when exposed to oxygen
(d) Loss of plasticizer when exposed to liquid media and under stress
(e) Water tightness under hydrostatic conditions
(f) Tensile strength
(g) Flexibility at low temperatures
(h) Stiffness required to resist being folded over during placing of the concrete
(i) Ease of installation

8.4 WATERPROOFING MATERIALS

Waterproofing materials are applied on concrete surfaces to form impervious coatings that prevent the passage of water in liquid form and may also retard vapor transmission in varying degrees, depending on the type of coating. Since the treatment prevents the ingress of water in liquid form, it also stops the transport of water soluble salts in either direction. Some of the marketed materials are given in the following sections.

8.4.1 Cement–based coatings

Metallic waterproofing and cement/latex coatings provide the required adhesive bond to resist hydrostatic pressure. Metallic waterproofing consists of finely divided metallic aggregate and oxidizing agents, mixed with sand and Portland cement. The mixture is mixed with water to form a plaster coat which may be applied by brush or trowel depending on the consistency of the mix. After the coating is applied to the concrete surface, the metallic particles oxidize and expand to at least three times their original size, wedging themselves into pores and interstices thus creating a relatively dense waterproof barrier. However, any subsequent cracking of the concrete wall will open cracks in the coating which would render it ineffective in preventing moisture entry.

8.4.2 Bentonite – based products

Trowel or spray grades - Finely divided bentonite clay, formulated with fibers, is often used in thickness of about 5mm in below-grade locations. In contact with sufficient water, it swells 10 to 20 times its original volume forming a gel that prevents further penetration by water. The material has some ability to reseal itself over cracks that form subsequently in concrete.

Corrugated bentonite cardboard panels - Bentonite clay contained in biodegradable, corrugated, cardboard panels are suitable for application to exterior surfaces below grade. The panels are attached with tempered nails or roofing cement. Water borne bacterial decomposes the cardboard and water swells the bentonite to form the gel.

8.4.3 Mastic asphalts

The products contain asphalt fillers, fibers, polymers, and solvents, and are usually reinforced with fiberglass mesh. These are usually laid by hand using wooden floats or squeegees. They are placed in two layers each of 10mm thickness, as blow holes frequently form in the first layer and the second layer covers these holes. The main advantage of mastic asphalt is that a protective layer is unnecessary, it is however, prone to the problem of blowing and is labour intensive.

8.4.4 Polymer resin–based coatings

These are generally of two types:

(a) Resins blended with organic solvents and

(b) Solvent free coatings

Solvent based coatings are subdivided into single and two component coatings. The coatings on drying produce a smooth dense continuous film that provides a barrier to moisture and mild chemical attack of the concrete. Because of the resistance to moisture penetration, staining, and ease of cleaning, they are preferred for locations of high humidity and those in which a lot of soiling occurs.

Most products are low solids content materials which require multiple coats to produce a continuous film over concrete, since the materials are thermoplastic, and have a significant degree of extensibility they are capable of bridging minor cracks which may develop in the concrete surface if they are applied in sufficient thickness. The number of coats required depends on the surface texture, porosity, and the targeted dry film thickness. Although some of the newer products have some moisture tolerance, enabling them to be applied over damp surfaces, in normal usage they should be applied over dry surfaces. Due to their relative in permeability to water vapor, they could blister when applied to concrete surfaces with high moisture content or where the opposite surface of the concrete is in constant contact with moisture. Careful control of wet film thickness is therefore necessary during application.

Two component polymer coatings consist of a solution of a compounded polymer with or without solvent and a reactive chemical component called the curing agent hardener or catalyst. The materials are usually mixed just prior to use in accordance with the manufacture's instructions. When using two components polymer based coatings the following items are of importance to the application of the materials.

(a) Most produces are supplied as a kit containing the two components in the required proportions. Therefore, in order to realize the full potential of the product the correct mix ratio of the two components must be used.
(b) To ensure a complete reaction of the two components they must be mixed thoroughly
(c) Some two component materials require an induction period of 15 to 40 min after mixing. Therefore, such products cannot be used immediately after mixing.
(d) Viscosity reduction by the use of thinners should be resorted to only after the manufacturers are consulted.
(e) The storage temperature of solvent based coatings is not critical. They should be stored at a temperature 16 to 32°C just prior to use.

Some of the generic resin types used includes thin adhesive coatings based on epoxy, polyurethane, chlorinated rubber, polyvinyl chloride, neoprene, and chlorosulfonated polyethylene. To overcome some of the drawbacks of solvent-based coatings a significant number of solvent-free coatings are on the market. These coatings are often specified for their high build, reduced risk of toxicity and fire hazard during applications where adequate ventilation of solvent-based materials would be uneconomic. Their major disadvantage is that they are relatively high viscosity materials that are difficult to apply in comparison to their solvent-based counterparts. Suitable methods of application are rollers but in some cases it may be preferable to use squeegees. Their favored application is to large areas such as floors.

8.4.5 Membranes

Membrane waterproofing is applied to prevent water intrusion thorough openings resulting from construction defects, shrinkage cracking, and structural movement. Vertical surfaces are waterproofed from the potentially wet side- the earth side of a foundation wall or the liquid side of a tank wall. Slab construction is treated similarly, except the membrane is often encased between two courses of concrete to protect if from mechanical damage. More recently, traffic bearing membrane systems which combine an abrasion resistant top coat with an electrometric under layer have been used to protect bridge and parking decks. **Membrane waterproofing systems can be grouped according to fabrication and application methods as follows:**

(a) Built up sheet membranes

(b) Prefabricated sheet, and
(c) Liquid applied materials.

Built-up membranes:
Hot-applied membranes, built up from multiple plies of saturated felt fabric or coated sheets, set in interplay applications of hot coal tar pitch or asphalt are the oldest type of membrane waterproofing. **Application of built up membranes involves the following:**
(a) Priming the concrete surface
(b) Reinforcing the corners with two or more plies of felt and adhesives
(c) Sealing the openings
(d) Applying a coating of hot waterproofing bitumen
(e) Brushing or rolling the sheet into hot bitumen
(f) Applying additional courses and
(g) Applying protection board over the membrane.

The advantages generally associated with hot applied built up waterproofing system are:
(a) Workmen are familiar with materials and methods
(b) Membrane is continuously bonded to concrete
(c) Allows the correction of filed errors
(d) Bridges existing joints and cracks in concrete

The disadvantages of the system are:
(a) Hot kettles of bitumen must be kept at correct temperature
(b) Hot bitumen causes burns and the fumes cause worker discomfort
(c) The method is labor intensive
(d) Organic felt and fabric deteriorate
(e) Conforming membrane to substrate contours and projections through concrete is difficult

Prefabricated sheet membranes:
Sheet materials are typically 1.25 to 2.4 mm thick, and are applied by three methods: loosely laid, mechanically fastened and fully bonded. Loosely laid systems are usually used in the waterproofing of roof and require an overlay of smooth stones or earth for ballast. Mechanical fastening is usually required at perimeters and openings. Fully bonded systems use contact adhesives. **Sheet membrane system includes:**
(a) Vulcanized rubbers
(b) Thermoplastics
(c) Ethylene interalloy, and
(d) Impregnated asphalt composites

An advantage normally associated with prefabricates sheet waterproofing include the following:
(a) Workmen are familiar with application techniques

(b) The membrane is more elastic than built-up membrane
(c) Materials bridge existing joints and cracks in concrete
(d) The membrane has uniform thickness controlled at the factory and not on site.

The disadvantages are:

(a) The seal at sheet laps and joints in some systems can be the weak link in membrane
(b) Conforming the membrane to irregular surfaces and projections through concrete is difficult

Liquid applied membrane:

A number of liquid applied membranes are available:

(a) Hot-applied rubberized asphalts
(b) Liquid applied neoprene, and
(c) Cold applied polyurethane

Many of these materials can be used en vertical surfaces as well as horizontal slabs. Most must be covered with protection board or a traffic bearing slab or coating. A few formulations with sand sprinkled on the laid coating can accept pedestrian traffic. The performance of liquid applied waterproofing systems is related to membrane thickness, which should be checked during application using a wet film thickness gauge.

Advantages provided by the system are:

(a) Application is easy by several methods-spray, roller, squeegee etc.
(b) The continuous bond formed prevents lateral movement of water between membrane and concrete
(c) No hot material is used
(d) The membrane is more elastic than built-up membranes

The disadvantages associated with the system include the following:

(a) The thickness depends on the skill of the applicator and may therefore vary
(b) The material may require accurate mixing at job site
(c) Compatible sheet material or fiberglass reinforcement material must be added to bridge larger existing joints and cracks in concrete.

Multilayer membrane systems:

Multilayer membrane systems are effective in reducing the ingress of moisture. They provide a watertight barrier that has elasticity to maintain integrity under the influence of temperature-induced structural movement. Membranes have proven to be especially effective at preventing the onset of corrosion in new structure by preventing contamination of reinforced concrete by salt. **There are tow basic types of membrane systems:**

(a) Liquid applied, traffic bearing elastomeric membrane system and
(b) Asphalted systems.

Traffic bearing elastomeric membrane systems should have the following basic properties:

(a) Impermeability

(b) Adhesion
(c) Crack bridging characteristics
(d) Durability
(e) Ease of applications

The asphaltic systems are much thicker than liquid applied elastomeric membranes and use asphalt cement based waterproofing membrane, which is either hot or cold applied. An asphalt paving wearing course 32mm thick or more, is applied on top of the membrane. The asphalt paving may be latex modified for added flexibility, or may be applied as a mastic system which does not require compaction. Due to the softness of the bituminous membranes, the thicker the membrane layer the thicker the asphalt paving wearing course required to obtain a stable surface.

Final selection of a membrane system is based on material properties or field experience and should involve the following items for consideration:

(a) Anticipated degree of structural movement
(b) Anticipated traffic loading, exterior or interior application
(c) Consideration of live load capacity and height limitations
(d) Nature and extent of deterioration of the concrete deck
(e) Installation procedure recommended by the manufacture
(f) Compatibility with expansion and contraction joint seals
(g) Cost
(h) Long standing example of successful application available for inspection
(i) Good standing of the manufacturer and the length and conditions of the guarantee
(j) The case of repair of the system and the effectiveness of the bond between old and new membranes

8.5 BONDING MATERIALS

Bonding materials are natural, compounded or synthetic materials used to joint individual members of a structure without mechanical fasteners. These products are often used in different repair applications, such as bonding of new concrete to old concrete. Two main types of bonding materials namely, latex emulsions and epoxies, are frequently used in the building industry. **Critical factors which govern bonding between new and old concretes, provided that sound concrete technology is followed, include:**

(a) The strength and integrity of the old surface and
(b) The cleanliness of the old surface.

Adequate bonding can be achieved by placing repair material directly against properly prepared substrate. The flow chart figure 8.1 has shown the development procedure of bonding material. There are special conditions when bonding agents are used. The surface condition plays an important role in bond development, although the magnitude of the bond varies depending upon other factors. **Some of these factors are as follows:**

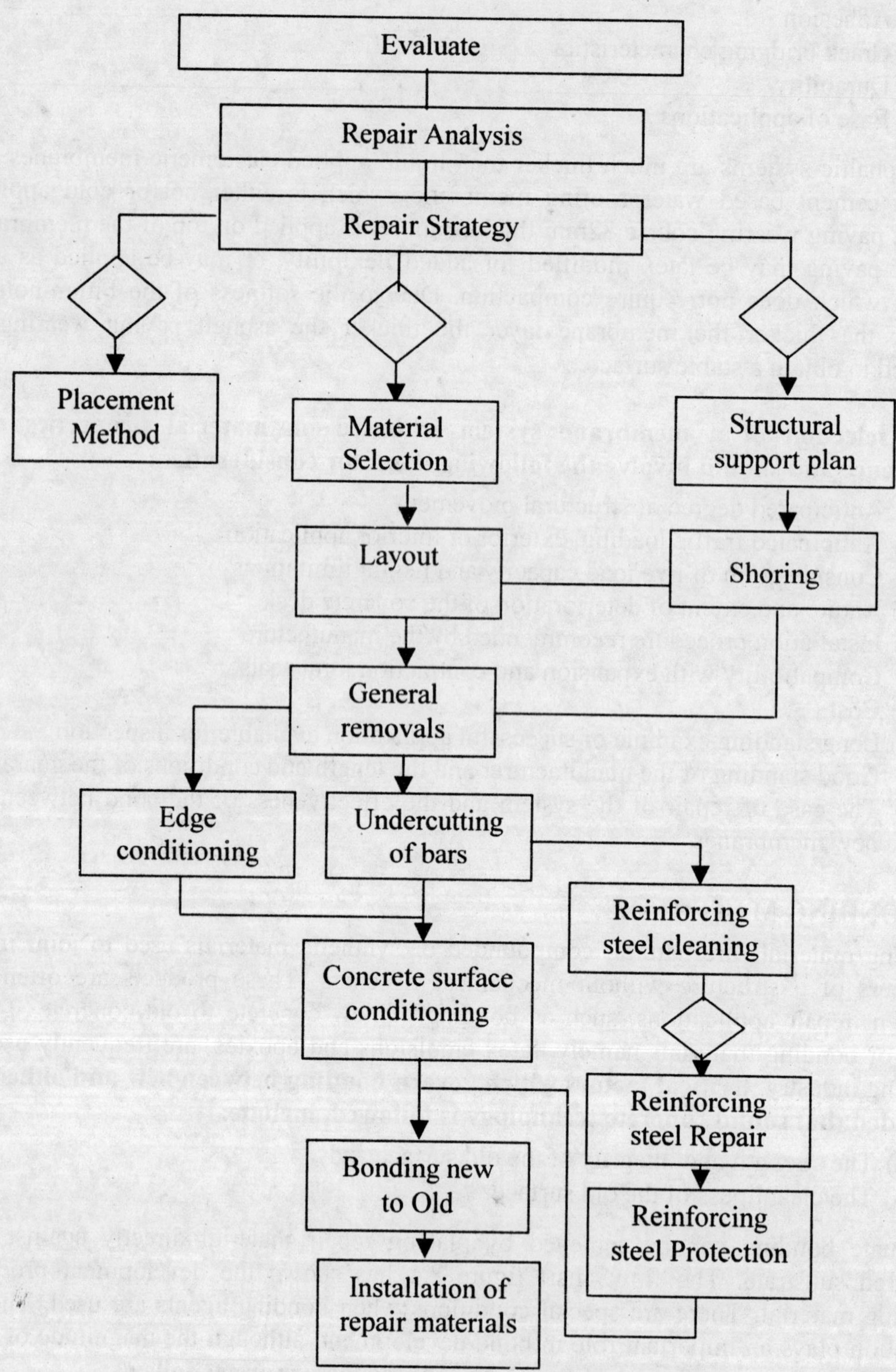

Figure 8.1. Flow chart diagram for the process of bonding materials

(a) Although a good bond may be obtained with the without a grout bonding layer, generally a grout layer consisting of cement and sand slurry or cement/latex slurry increases bond strength.
(b) Proper compaction of the new concrete improves bond.
(c) The method of surface cleaning that produces a good bonding is determined by the density of the base concrete, *e.g.*, for a sound base, concrete acid etching would suffice.
(d) Mechanical cleaning is essential if a weal or deteriorated surface layer existed on the old concrete.

8.5.1 Bonding Agents

Three main types of bonding agents are frequently used:

(a) Cement-based slurries,
(b) Epoxies, and
(c) Latex emulsions.

For **Portland cement based** repairs and overlays, cement or sand cement slurry is used. After the substrate has been prepared and immediately before placing the repair materials, a thin coating of creamy grout must be vigorously and thoroughly broomed or brushed into the prepared surface.

In the case of **latex-modified** repair materials or overlays, the bonding grout can be broomed in from the mix itself. For latex-modified materials, it can also be mixed separately. There are, now available, factory blended latex-modified cement slurry bonding agents which are mixed with water on site and applied to the prepared concrete surface. **The following latex products are used as bonding agents:**

(a) Styrene butadiene (SBR)
(b) Acrylic
(c) Polyvinylacetate (PVA)

Latex emulsions:
There are a variety of different applications for latex emulsions as bonding agents. Some types have a greater degree of water resistance than others. The emulsions generality used in cementitious compositions and environments are of the oil-in-water type, and sometimes contain more than 50% of water. They are generally stable in the cement/water system. The following description of the advantages and limitations of the various types of emulsions used as bonding agents is intended to serve as a preliminary guide for the user and specifier. **A latex may be sued as a bonding agent in three ways:**

(a) Preparing a neat cement slurry, utilizing the latex as part of the gauging water
(b) The use of a 1:1 water-latex diluted materials
(c) The use of a re-emulsifiable latex which can be softened and retackified upon contact with water

Styrene butadiene rubber (SBR):
SBR lattices which are compatible with cementitious compounds are copolymers. They show good stability in the presence of multivalent cations such as calcium and aluminium and are unaffected by the addition of relatively large amounts of electrolytes. The SBR lattices may coagulate if subjected to high temperatures, freezing temperatures, or severe mechanical action for prolonged periods of time.

Polyvinyl acetate (PVA) lattices:
Commercial materials are copolymers manufactured by the emulsion polymerization process. Two main types of polyvinyl acetate lattices are used in repair: nonremulsifiable PVA, and emulsifiable PVA.

Non emulsifiable PVA - This PVA forms a film which offers good water resistance, good light stability, and good aging characteristics. Because of its compatibility with cement, it is widely used as bonding agent and a binder for cementitious water-based paint and waterproofing coatings.

Emulsifiable PVA – These lattices produce a film that can be softened and retackified with water. Such a latex permits application of a film to a surface long before the subsequent application of a water-based overlay. They are the most widely used as a bonding agent for plaster, and to bond finish or base coat gypsum, or Portland cement plaster to interior surfaces of cured cast-in-place concrete.

Acrylic lattices:
Acrylic lattices are produced from acrylic resins which are polymers and copolymers of acrylic and methacrylic acids. These resinous materials range in physical properties from soft elastomers to hard plastics. The emulsion is used in cementitious compounds in much the same manner as SBR lattices.

Epoxy lattices:
Epoxy emulsions are produced from liquid epoxy resin when mixed with the curing agent. The curing agent serves the additional functions of an emulsifying and wetting agent and as a surfactant. Emulsions are stable and water dilutable from the time of mixing until gellation occurs. Pot life may vary from 1 to 6h, depending upon the curing agent selected, by adding high amounts of water.

100% - Solids epoxy bonding agents:
Various epoxy products are available for bonding of freshly placed concrete to cured concrete as well as concrete to steel. Most products contain 100% solids resins which may or may not contain thixotropic fillers and other additives to enhance a particular property or reduce cost. Current products are available in a verity of consistencies, ranging from a highly filled paste to that of liquids with viscosities of 100 centipoises similar to that of water. It is a performance specification based on end use and there are no specific limits on chemical

composition. Instead the materials must meet requirements related to physical properties such as viscosity, bond strength, shrinkage, and thermal compatibility.

The specification classifies the epoxy resin bonding system by type, grade, and class. The properties stipulated in the specifications are of great importance: namely, bond strength, shrinkage, and thermal compatibility. **The materials are placed in one of four categories:**

(a) General repairs – these materials do not fit the other three categories, but are suitable for general repair work.
(b) High early strength – these materials have a compressive strength of 8 Mpa at 4 hrs.
(c) Self-leveling – these materials are very flowable and will spread out to fill a cavity. They are used on horizontal surfaces and are useful for hard to reach areas.
(d) Slurry coats – these are used for very thin repairs or for coatings.

8.5.2 Keys to developing bond

(a) Clean, sound substrate
(b) Roughened profile of substrate for mechanical interlock.
(c) Open pore structure in substrate
(d) Repair materials/bonding agent with sufficient paste for absorption into substrate pores.
(e) Repair materials are applied with sufficient pressure to facilitate contact between the repair materials and the substrate at the bond line.

Achieving an adequate bond between repair materials and existing concrete is a critical requirement for durable surface respires. Various techniques are available to achieve the required bond. The bond at the interface between the repair materials and concrete substrate is likely to be subject to considerable stress from volume changes, freeze thaw, force of gravity, and sometimes, impact and vibration. The stress states that develop at the bond lines will vary considerably, depending on the type and use of the structure. For example, the bond on the bridge deck overlay may be subject to shear stress in conjunction with tensile or compressive stress induced by shrinkage or thermal effects, and to compression and shear from service loads.

It is essential that the repair materials achieve a string bond to the substrate and the subsequent stresses are not sufficiently great as to cause depending. Repair which have bond lines in direct tension have the greatest dependency on bonding. Repairs that are subject to shear stresses at the bond line are capable of stress resistance not only by bonding mechanisms, but also by aggregate interlock mechanism, which add greatly to shear bond capacity. It should be remembered that high initial bond strength is generally not as important as bond durability.

CHAPTER 9
Special Repair Materials

9.1. INTRODUCTION

Concrete is no longer a simple mixture of cement, sand and aggregates, but it has matured into an engineered material containing a mixture of admixtures. Admixtures are materials other than water, aggregates and hydraulic cements. They are used as ingredients of concrete or mortar and added to the batch immediately before or during mixing. Almost every property of concrete can be modified to some extent. It can be used to improve concrete properties either in the handling process or consolidation of fresh concrete, in the performance of concrete both in fresh and hardened stages and even for economy in the cost of construction. However, the effectiveness of an admixture in a concrete mix depends on several factors such as type and amount of cement, temperature of concrete, etc. It is important to note that no admixture of any type or amount should be considered as substitute for good concreting practices. Admixtures should be used judiciously and not for the purpose of covering up deficiencies in the mix.

In spite of several significant beneficial effects and extended applications provided by admixtures, it is important to note the serious problems presented by the user of a mixture of admixtures in concrete. Compatibility of the ingredients becomes an important issue and verification by trials with the actual materials to be used on a job site under the actual field conditions is therefore crucial.

Admixtures can generally be classified as chemical and mineral admixtures. Chemical admixtures can be further be classified into:

(a) Surface active chemicals, which are used for air entrainment or allowing reduction of water in concrete mixtures and
(b) Set-controlling chemicals

These may be either powders or in liquid form which are added in small quantities. Mineral admixtures are finely divided materials having pozzolanic characteristics, which can be added, to concrete in relatively large quantities (20% to 100%) by weight of cement. Power plants using coal or rice husk as fuel and metallurgical furnaces producing pig iron, silicon and ferrosilicon alloys are main sources of artificial mineral admixtures. Naturally available mineral admixtures are generally of volcanic origin. Improvements in impermeability, resistance to thermal cracking and chemicals, ultimate strength, durability, better finish etc. are some benefits of using mineral admixtures.

9.2 CHEMICAL ADMIXTURES

Chemical admixtures such as retarder and accelerators are helpful in avoiding the undesirable effects on rate of cement hydration due to ambient temperature conditions. Plasticisers and superplasticisers can be used to produce concretes of any required workability to suit the actual construction practices and any type of structural members such as for pumping, members with highly congested reinforcement, self-leveling purposes, massive structures.

9.2.1 Superplasticisers (SP)

These are the most recognised category of water reducing admixtures having a remarkable plasticising action on concrete and produce substantial enhancements in workability. Water reduction as high as 30% may be achieved. They may not affect the setting and hardening behaviour of the system provided a proper compatibility with cements is established for the mix under consultation. This admixture was first introduced in the Japanese market in the 1960s followed by Germany. **ASTM C494 classifies super plasticisers into two types:**

(a) High range water reducer and
(b) High range water reducer with retarder.

Chemically, superplasticisers are organic polyelectrolytes, which belong to the category of polymeric dispersants. **Some SPs are synthetic while others are derived from natural products and may be classified into the following four categories:**

(a) Sulphonated melamine formaldehyde condensate (SMF)
(b) Sulphonated naphthalene formaldehyde condensate (SNF)
(c) Modified lignosulphonates (LS)
(d) Hydroxy-carboxylic acid such as polyacrylate ester (PAE)

However, the performance of a superplasticiser in a cementitious system depends on its chemical composition, dispersion ability, thermal stability and molecular weight, and also on cement fineness and its composition, Table 9.1 gives typical results of the elemental analysis of superplasticisers.

Table 9.1. Elemental chemical composition of some superplasticisers

	Carbon	*Hydrogen*	*Oxygen*	*Nitrogen*	*Sulphur*	*Sodium*
LS	42.7	4.5	35.4	0.0	7.8	7.1
SNF	48.3	2.6	22.6	0.0	13.8	12.7
SMF	22.0	3.5	26.7	25.5	12.4	9.3
PAE	47.8	8.0	37.8	0.0	0.9	4.3

Sulphonated melamine formaldehyde condensate:

Though this product was originally developed in the 1950s as a dispersant for a variety of industries, it was not until some 10 years later that the possibilities for its use in concrete were recognized. The molecular weight is generally about 30,000. It has minimal effect on air-entrainment and setting time.

Sulphonated naphthalene formaldehyde condensate:

This was one of the first materials referred to in the literature as a water-reducing agent but only since 1970, it has found extensive application as admixture. The solid content of commercial products varies from 25 to 45% .The molecular weight is of the order 2000.

Modified lignosulphonates:

The crude lignosulphonates derived from wood pulp are commonly used as plasticisers, and in order to improve their .effectiveness they can be refined and modified. The processing involves the removal of sugars and other impurities, selection of a higher molecular weight fraction and optimally, further sulphonation or partial polymerisation.

Polyacrylate ester:

Carboxylated synthetic Polymers act as highly effective dispersant

Mode of action of superplasticisers:

The superplasticisers are adsorbed on to the initial hydrates of C_3 A, C_3S and C_2S to form a complete monolayer around them. In case of SNF and SMF, the ionised sulphonate groups on the adsorbed superplasticiser molecules are strongly negatively charged and repel the cement particles thus breaking the weaker Van der Waal forces of attraction, resulting in a well-dispersed system. The polyacrylate materials are similarly adsorbed and cause dispersion of the cement particles by the same electrostatic mechanism through the ionised carboxyl groups. The long flexible side chains attached to these polymers act as physical barriers, preventing the cement particles from coming within the range of the Van der Waal forces. In doing so, water trapped within the original flocs is released and can contribute to the mobility of the cement paste and hence to the workability. Superplasticiser does not cause much reduction in the surface tension of water; hence there is little tendency to entrain excessive air even at high dosages. There is some retardation in cement hydration but at 28 days, the products of C_3S hydration are the same as an unsuperplasticised concrete.

Problems associated with the use of superplasticisers:

Some of the problems associated with the use of superplasticisers are discussed below:

Rapid loss of slump with time - This is especially true when a high early strength superplasticiser is used, along with the higher-grade cement. It is therefore recommended that the superplasticiser be added only at the point of discharge of the concrete. One of the ways to deal the issue of slump loss is to use repeated dosages of superpiasticisers. By the addition of second and third dosages, large increases in workability of superplasticised concrete can be maintained for several hours, the amount of second and third dosage being the same as the initial dosage. Another method of extending the slump in superplasticized concrete is to use blended admixtures such as a blend of naphthalene based SP and lingo-sulphonate or a gluconate retarder.

Loss of slump with increase in ambient temperature - Loss of slump in concrete mix is accelerated when the concrete mixing temperature is high. When superpiasticisers are mixed in such a situation, bleeding of the concrete mix might take place. This problem is true in the case of superplasticiser based on naphthalene sulphonate when the mixing temperature is around 40°C. The use of melamine-based superplasticiser may be tried in such a case.

Prolonged retardation of concrete setting time - This is a typical problem when a retarding superplasticiser is used. This is mostly due to excessive dosage of the superplasticiser. It is always advisable to check the dosage of the superplasticiser to be used for given cement for its slump, loss of slump with time and air entrainment. When there is a change in the brand or type of cement, it pays to check the fresh concrete properties once again. Superplasticisers, because of their tendency to slightly increase air entrainment, may enhance the cohesion of the mix. Some category of SPs has little effect on air, while others reduce the natural air entrainment, thereby slightly reducing the cohesion. These factors should be borne in mind when selecting the superplasticiser category especially if changes in mix constituents are not feasible for producing a good cohesive mix.

Strength characteristics of superplasticised concrete:

The dispersing capability of superplasticiser is also used to reduce the water requirements in comparison with control mix. In effect, the same slump as that of the control mix can be obtained with a lesser water cement ratio, leading to high strength of concrete. Increasing the cement content to achieve early high strength may result in excessive heat output, causing cracking as well as undesirable shrinkage. Increasing the quantity of cement alone is not desirable for a reduction in water/cement ratio to achieve high strength. It is clear, therefore, that any means, consistent with good concrete practice, of reduction of water-cement ratio closer to the theoretical level is desirable.

Superplasticisers are claimed to go a significant way towards satisfying this requirement, as water reductions in the region of 20% to 35% are obtainable by their proper use, although limited by grade of cement, type of aggregate and temperature. By comparison, water reductions of about 15% are obtainable with lignosulphonate based plasticisers. Concretes with low water content and containing superplasticisers are noted for their high early and ultimate strengths, excellent durability and water-proofing characteristics. The technique of producing low water/cement ratio concrete for the purpose of obtaining high strength or retaining workability at reduced water content is of considerable interest to concrete technologists.

The higher strength of the admixtured concrete is achieved by a reduction in w/c ratio, the influence of which is more pronounced at early ages, but less effective at later ages. Superplasticisers strongly bond onto cement particles and makes them disperse preventing flocculation. In India, with the advent of higher grades of cement like 43 & 53, there is a tremendous potential for SPs to produce high strength concrete. Superplasticisers, both high early strength type and retarding type, based on naphthalene and melamine are available. In comparison to other methods like a polymer impregnation, etc., water reduced concrete using superplasticiser offer the easiest and most economical way of producing high strength concrete.

9.2.2 Accelerators

Cement accelerators can be classified into two distinct types:

(a) Set accelerators, which reduce the period over which the cement remains workable, leading to early stiffening of the mix.

(b) Strength accelerators, which give an increase in the early age strength of the cement.

An admixture, which falls into one of these categories, does not necessarily also meet the requirements of the other and may even have the reverse effect. Temperature is also an important factor and accelerators, which are very effective at 20°C, may be of little benefit at 5°C. The converse is also true, some accelerators only being effective at low temperatures. Accelerators are used for cold weather concreting, in repair applications where the concrete must be put into service at an early age and also in seawater concreting. **They fall under three categories:**

(a) Chloride based,

(b) Chloride free, and

(c) Very rapid set accelerators

Chloride based accelerators:

Calcium chloride has been used as an accelerator for concrete since at least 1885. Both setting time and early strength are accelerated while at later ages strengths remain similar to or above the control mix. Calcium chloride, aluminium chloride, sodium chloride and ferric chloride are a few accelerators that can be used for cement concrete.

Chloride free accelerators:

These can be divided into three categories.

(a) Materials containing active anions

(b) Materials containing active cations

(c) Organic chemicals

Substances in the first category include calcium or sodium salts of formates, nitrates, nitrites, thiocyanates, thiosulphates, permanganates and chromates. Triethanolamine, 8-hydroxyquinolonet and acetinst come under the category of organic accelerators Triethanolamine though most popular, is generally used in combination with other formulation as its effect is very sensitive to dosage and the chemical composition of cement.

Very rapid set accelerators:

This type of accelerator can be used in applications where the desirable initial set is less than 3 minutes and the final set less than 12 minutes. These accelerators are an essential ingredient of mixes for certain specialised applications, particularly in mining, tunneling and sea defence works. The most active ingredients of these accelerators are alkali metal silicates, aluminates, hydroxides, sulphates, or chlorides.

Mode of action:

Calcium chloride added to Portland cement affects the rate of hydration from an early age. When a 2% addition of anhydrous calcium chloride is made, the rate of stiffening may commence after 30 minutes compared with 180 minutes for the control mix. The increase in the degree of hydration is assessed by heat evolution using calorimeter. Calcium chloride reacts relatively rapidly with tricalcium aluminate paste to form chloro-aluminate but in Portland cement, this reaction is inhibited by the presence of gypsum which reacts with tricalcium aluminate to form ettringite. Only when all the gypsum has reacted does the chloride again start to react with C3A. Thus, the chloride ion is left free to accelerate the silicate hydration reaction.

Performance:

Dosage of anhydrous calcium chloride is approximately 1.5 to 3% on cement weight although up to 6% is used for some applications where rapid set is required. Typical results show that both initial and final setting times are accelerated progressively with increase in dosage and that the time between initial and final set is also reduced. Concrete containing calcium chloride develops from a very early age a higher compressive strength than a corresponding mix without the admixture. The effectiveness of accelerators is temperature dependent.

9.2.3 Retarders

Retarders slow down the rate of setting of cement and assist in hot weather concreting by extending the workable time, and in the casting and consolidating of large area concreting preventing the formation of cold joints. However, once the reaction, starts, the hydration of cement continues at the normal rate and sometimes greater. Retarders are also used in massive structures to slow down the heat built-up, reducing the thermal stresses. These are also called as set retarding admixtures. They prolong the setting and hardening times. They can be used whenever it is desirable to offset the effects of high temperatures which decrease setting time, or to avoid complications when unavoidable delays may occur between mixing and placing. Prolonging the plasticity of fresh concrete is useful in placing mass concrete. Successive lifts can be blended together by vibration, with the elimination of cold joints that would occur if the first lift were to harden before the next is placed. Basic research on the effect of retarders has shown that they slow down the rate of early hydration of tri-calcium silicate. Gypsum is commonly used as a retarder. Plastic shrinkage from evaporation is likely to be more in retarder added concrete. Curing therefore is very important. Fly ash, which is a waste product from thermal power stations, may also be used as an economical admixture to delay the setting time of cement

9.3 MINERAL ADMIXTURES

Mineral admixtures are finely divided siliceous materials, which can be added to concrete in relatively large amounts, generally in the range of 20 to 100 % by weight of cement. They may be pozzolanic, or cementitious or both. A pozzolan is a siliceous or siliceous and aluminous material. Mineral admixtures can be divided into two groups-natural and artificial. Naturally available mineral admixtures are of volcanic origin. However, they are not available in India. Power plants using coal or rice husk as fuel, and metallurgical furnaces producing pig iron, steel, copper, nickel, lead, silicon, and also ferrosilicon alloys are main sources of artificial mineral admixtures. Benefits of using mineral admixtures in concrete include improvements in impermeability, resistance to thermal cracking and chemicals, higher ultimate strength, better durability, etc. Fly ash (FA), Ground granulated blast furnace slag (GGBS) and condensed silica fume (CSF) are the major industrial by products suitable as mineral admixtures.

9.3.1 Fly ash

Fly ash is the finely divided residue resulting from the combustion of ground or powdered coal which is transported from the fire box by the flue gases and is subsequently removed from the gas by electrostatic precipitators. During combustion of coal in thermal power plants, the volatile matter and carbon are burnt off, whereas mineral impurities in the coal, such as clays, quartz, feldspar etc., melt and are transported to lower temperature zones where it solidifies as spherical particles of glass. Some of this mineral matter agglomerates to form bottom ash, but most of it flies out with the flue gases called fly ash.

Classification:
According to ASTM C618, mineral admixtures are classified into three:

(a) Class N: Low or calcined natural pozzolanas such as diatomaceous earth, opaline cherts, clays, and shale's, tuffs and volcanic ashes or pumices.
(b) Class F: Fly ash from anthracite or bituminous coal.
(c) Class C: Fly ash from lignite or sub-bituminous coal.

Influence of physical characteristics on properties of concrete:
The shape, fineness, particle size distribution and density of fly ash particles influence the properties of fresh and hardened concrete. This is primarily due to the particle influence on the water demand of the concrete. The majority of fly ash is glassy, solid or hollow and spherical in shape. Clean, glassy, spherical particles of fly ash act as good void fillers in concretes and reduce water requirement. Rough textured and porous glass spheres in fly ash possess more reactivity than smooth spheres and hence particles size distribution alone may not reflect the potential activity of a fly ash. Fineness of fly ash has an influence on the strength of concrete.

Strength Development:

When fly ash is added as replacement of fine aggregates and not that of cement, the strength of concrete at given ages is enhanced. When used as a replacement of cement by fly ash, concretes exhibit lower gain in strengths at all ages. However at ages beyond 28 days. they attain strengths equal to or greater than that of concretes without fly ash. Test results indicate that silica fume can be used to increase the early strength of fly ash concrete.

Increases in impermeability and strength of fly ash concrete are due to:

(a) Presence of hydrated phases with low-density phase products and smaller voids in the hydrated paste
(b) Refinement of pore system
(c) Improvement in structure and strength of transition zone between the aggregate and hydrated cement

9.3.2 Ground granulated blast furnace slag (GGBS)

Slags are by products of the metallurgical industry. The most important slag from the standpoint of the quantity used as building material is iron blast furnace slag. Slags that have crystalline, stable and solid structure are used as aggregates while glassy structured material is used as a hydraulic binder. GGBS exhibits cementitious as well as pozzolanic characteristics. Hydrated GGBS gives the same hydrates as that of Portland cement. It reacts more slowly with water than Portland cement but can be activated with lime and sulphates either physically, by grinding or thermally. Typically silicon, calcium, aluminium, magnesium and oxygen constitute 95% or more of the blast-furnace slag.

Specifications:

ASTM C989 provides for three strength grades of GGBS, depending on their respective mortar strengths when blended with an equal mass of Portland cement. The classifications are made based on the slag-activity index and are classified into three grades as 80, 100 and 120. The slag activity index gives an indication for evaluating the relative cementitious potential of slags. The specification includes two chemical requirements: one limiting the sulfide sulphur to a maximum of 2.5 % and the other limiting the sulphate content to a maximum of 4 %.

Effect on properties of hardened concrete:

The extent to which GGBS affects strength is dependent on the slag activity index of the particular GGBS and the replacement level used in the mix. Other factors that affect the performance of GGBS in concrete are w/b ratio, physical and chemical characteristics of cement, curing conditions etc. Concrete containing GGBS is more susceptible to poor curing conditions than concrete without GGBS. The rate of strength gain of Grade 120 slag is greater in concrete mixtures, which have high w/b ratio than in mixes with a low w/b ratio. Highest strengths at 28 days are obtained with 40 to 50 % level of replacement. Atleast 70% GGBS may be needed to achieve significantly low heat of hydration. Comparatively, lower heat of hydration can be always expected when GGBS is used to replace equal mass of cement. Permeability is considerably reduced due to the pore filling with C-S-H and reduction in pore size.

9.3.3 Condensed silica fume (CSF)

Condensed silica fume is a by-product from electric arc furnaces used in the manufacture of silicon metal or silicon alloys. This product is excellent for use as a supplement to Portland cement. Being composed of essentially noncrystalline SiO_2 in a finely divided state, CSF exhibits excellent pozzolanic characteristics. Unlike fly ash, the lime-silica reaction involving CSF is rapid and, therefore a long curing period is not necessary before the desired high strength and impermeability are realised. Due to very fine particle size, bulk density of CSF is generally between 200 and 250 kg/m^3. This low bulk density causes great difficulties in handling and disposal of the materials. Silica fume is available in slurry, compacted and pellet forms.

9.4 ADMIXTURES FOR REHABILITATION

Since the beginning of the use of concrete, strength has been regarded as the most significant and important property of concrete. However it is now well recognized that this strength of concrete may deteriorated with time, is a result of a combination of various factors that may include moisture ingress leading to carbonation of concrete and/or reinforcement corrosion, chemical attack on concrete, poor quality of materials and workmanship. Concrete admixtures are vital in repair and rehabilitation works where rapid gains in strength, nor shrink characteristics, adequate bonding to parent concrete are a necessity. Although the use of admixtures for rehabilitation can encompass a fairly wide range of application in concrete from patch work of concrete in relatively small area to injection grouts for cracks, a few of the most common admixtures that are useful in rehabilitation are dealt with here.

9.4.1 Polymer dispersions or lattices

Various polymer systems may be used as an emulsion or dispersion of polymer droplets in water. One of the most common of these is styrene butadiene rubber latex solution, properly known as SBR latex. These are commonly used as an effective bonding agent when made in to a slurry with cement and water. Screeds, toppings, mortars and concrete containing such admixtures are often described as latex modified concrete. SBR latex is a milky white liquid which contains microscopic particles of synthetic rubber dispersed in an aqueous surfactant solution produced from styrene and butadiene. **When SBR latex is added to a standard concrete or a mortar mix the following improvement in properties can be obtained:**

(a) The w/c can be reduced from 0.5 to 0.3 due to the plasticizing effect of the latex.
(b) The adhesion of mixes to the apparent material can be increased by a factor.
(c) The tensile strength is normally doubled.
(d) The shrinkage during drying/curing is drastically reduced
(e) The water penetration of latex modified mixes is greatly reduced
(f) A slurry of SBR latex and Portland cement gives prolonged corrosion resistance to steel, because of adhesion, alkalinity, and low permeability
(g) The greatest advantage of the SBR latex lies in the fact that it has similar properties like modulus of elasticity and thermal expansion.

9.4.2 Latex modified system

Cement products are weak in tension because they contain numerous voids. As an latex modified mortar dries out, the rubber particles coalesce together to form a continuous film and strands which stitch the opposite sides of the voids together. These films and strands also block up the voids thus increasing resistance to water penetration. They also adhere excellently well with old concrete, brick, steel asphalt, wood, etc. **Recommendations for the use of SBR latex cementitious mixes:**

(a) Always use fresh cement – air set or warm cement leads to a rapid setting.
(b) Use washes sand complying with zone II – not more than 0.5% of the sand should be below 75microns.
(c) Use the recommended dosage of the latex – normally 10literes for every 50kg of cement.
(d) Avoid excess mixing water – for example when used as a floor screed, the reasonable w/c is of the order of 0.3.
(e) Always apply a bond coat before applying the latex modified mortar/concrete to the old concrete.
(f) Allow a little time for the cement – latex slurry bond coat to be tacky.

9.4.3 Epoxy resins

Amongst the synthetic resins such as epoxy, polyester, acrylic, polyurethane, etc. epoxy resins possess very high mechanical and adhesive strength properties most desirable for civil engineering applications. **Epoxy resins when cured with different hardness offered a wide range of properties:**

(a) High adhesive strength to almost all materials
(b) Low shrinkage during curing
(c) Exceptional dimensional stability
(d) Natural gap filling properties
(e) Thermosetting
(f) Resistance to most chemical and environments

Epoxies for crack repair:

The most appropriate method of crack repair depends on whether the crack is still actively moving or not. Active cracks may be due to inadequate provision of movement joint in the structure. The repair process involves converting the crack in to a movement joint, by the use of a suitable sealant. The injection of a low–viscosity epoxy is a possible repair method for cracks between 0.02mm to 6mm in width. There are wide variations in the properties of different low–viscosity epoxy systems and it is therefore necessary to choose carefully to match the individual job requirements in relation to such matters as temperature of application, capability of bonding to moist concrete, shrinkage, thermal and elastic properties of the hardened resin. For epoxy injection to be effective the cracks must be free of dirt, grease, etc. Before injection can begin, the crack, where it appears on the surface of the member, must be sealed, so that the liquid resin does not leak and flow out of the crack before it begins to gel and cure.

Epoxy mortars/concretes:
Epoxy resin systems are superior in almost all properties to concrete. The adhesive strength of the epoxy resin system is 50 times than that of cement concrete. How ever, the use of epoxy in mortar is recommended only in emergency repair, since they have a low modulus of elasticity and high creep than most Portland cement products. Fillers are a must when workability is to be improved and to minimize the use of resin. They are Portland cement, fly ash or silica flour.

Epoxy coatings for reinforcement:

Extensive experimental investigations reveal that epoxy coatings when properly applied could be expected to adequately: product, reinforcing steel bars from corrosion. However, powder based epoxies performed better than the liquid epoxies.

Epoxy resins find use in grouting, bonding of old to new concrete, repairs of eroded concrete, coatings for reinforcements, etc. A judicious use of epoxy is called for in the light of their very high cost. Epoxies are also used for bonding of steel plates to the external faces of structural members like beams, etc. which have suffered damage revealed by cracks in the flexural, shear modes. The technique consists of in-situ strengthening of failed concrete elements by boundary steel plates to concrete surface with epoxy resins. A large amount of experimental work done all over the world including in India show that repaired and externally reinforced beam exhibit a 24 % and 20 % increase in strength over the original RCC beams in flexure and shear.

Other important admixtures include shrinkage: compensating admixture that can be used for filling of grouts, cracks, etc. and accelerators for guniting/shotcreting works. A number of admixtures are now available for rehabilitation of concrete, but there are no cheap solutions to the problem of repair. Proper assessment of the problems and selection of the right combinations of repair materials are imperative in ensuring the success of the rehabilitation work.

9.5 POLYMERIC MATERIALS

Recent years have seen an increased emphasis on the repair and rehabilitation of all types of structure in preference to demolition and rebuilding. As a result, there is an increasing demand for repair materials. Durable repair can be obtained only by matching the properties of the base concrete with those of the repair materials intended for the use. The selection of repair materials is a predictive effort to maximise future performance of durability. Variety of repair materials has been identified and used for repair, rehabilitation and retrofitting of reinforced concrete structural elements. Polymeric materials, concrete polymer composites, fibre reinforced concrete/mortar, fibre reinforced polymer concrete/mortar-composites, fibre reinforced polymer composites, etc, are identified from extensive research work as high-performance materials for repair, rehabilitation and retrofitting of damaged structural elements.

All matter in this world is composed of extremely small units called molecules. They are too small to be seen even under the most powerful microscope and are a complex association atom. Molecules come in different sizes and shapes. Molecules of plastics are much larger than the ordinary molecules. They are giant molecules in the form of long chain which are called polymers.

9.5.1 Types of Polymers

The distinction between types of polymers is based on their reaction to heating and cooling.

Thermoplastic polymers:

Thermoplastic polymers soften upon heating, and can be made to flow when a stress is applied. When cooled again, they reversibly regain their solid or rubbery nature. Continued heating of thermoplastics will lead ultimately to degradation, but they will generally soften at temperatures below their degradation points.

Thermosetting polymers:

Thermosetting polymers are materials which can be heated to the point where they would soften and made to flow under stress. However, they do not revert to the original solid state as the heating causes the material to undergo a curing reaction. Often, these polymers emerge from their synthesis reaction in a cured state. Further heating ultimately leads only to degradation and not softening and flow.

9.5.2 Applications of Polymers

In building construction the application of polymers can be classified in various ways, for example:

(a) Nonstructural polymers
(b) Structural and semi-structural polymers
(c) Auxiliaries to other materials

The first group constitutes, by far, the greatest volume and number of different uses. The second group includes patch repair, overlays, linings to concrete/ steel products, injection to structural cracks, strengthening of structural elements, etc. Auxiliaries include adhesives, bonding agents, sealants, and decorative and protective coatings.

9.6 ORGANIC POLYMERS

Organic polymers are complex chemical compounds derived mainly from the petrochemical industry. These materials are often referred to as resins and the principal resins used in the construction industry are epoxide, polyurethane, polyester, acrylic, polyvinyl acetate and styrene- butadiene.

9.6.1 Epoxide resins

The formulators market epoxide resins and have the special properties required for the specific use to which they will be put. **The basic characteristics of epoxide resins include the following:**

(a) Outstanding adhesive qualities to such materials as concrete and steel
(b) Resistance to wide range of acids, alkalies and other chemicals
(c) Rather vulnerable to organic solvents
(d) Low shrinkage
(e) High coefficient of thermal movement compared with concrete.
(f) High compressive, tensile and flexural strength
(g) Appreciable loss of strength at high temperatures
(h) High rate of gain of strength, which can be varied to suit the particular application
(i) Rather poor resistance to fire compared with concrete and clay bricks
(j) To obtain satisfactory results, the site conditions under which the resin will be applied must be stated to the formulator.

9.6.2 Polyurethanes

Polyurethanes, like epoxide resins, are products of the petrochemical industry, They can be obtained as elastomers, solid and rigid materials and as flexible coatings. They are very durable in external conditions and retain their gloss well. Polyurethanes can be specially formulated to meet specific site requirements. They can be combined with epoxies and will withstand relatively high temperatures as well as sudden changes in temperature, *i.e.* thermal shock.

9.6.3 Polyester resins

Polyester resin can be divided into two classes:

(a) The so-called saturated polyester resins such as polyethylene terepthalate or terylene
(b) The unsaturated polyester resins which can be cross-linked

Commercially available unsaturated polyester can be divided into eight different classes or groups as follows:

(a) Gel coat/topcoat resin
(b) General purpose resins
(c) Chemical resistant resins
(d) Reduced flammability resins
(e) Low styrene emission resins
(f) Low profile/low shrinkage resins
(g) Special purpose resins
(h) Casting resins

All the strength properties decrease under load at elevated temperatures. The resins had good electrical properties such as loss factor, dielectric strength and are resistance chemicals. Polyester resins are used with Portland cement and selected aggregates to form a polymer-cement-aggregate mortar. Such mortars possess a number of desirable qualities such as good resistance to wide range of chemicals, high resistance to abrasion, waterproofness under the range of heads likely to be met in liquid-retaining structures, and high bond strength with most common building materials.

9.6.4 Polyvinyl acetate (PVA)

This material is used as a bonding agent and as an admixture in mortar to improve pertain properties of the mortar. Manufacturers of proprietary compounds based on PVA claim increased tensile and flexural strength reduced drying shrinkage and reduced permeability.

9.6.5 Styrene-butadiene and acrylic resins

These materials are also known as latexes and polymer emulsions. Some of the properties of a proprietary styrene-butadiene copolymer emulsion used with Portland cementing grout, mortar and concrete are:

- pH = 11.0
- Total solid content = 47% (nominal)
- Specific gravity = 1.1
- Butadiene content = 40%

9.7 TYPES OF POLYMER CONCRETE COMPOSITES

Depending on the manner in the polymeric materials are incorporated, polymer concrete composites can be classified under the following three major types :

9.7.1 Polymer-Modified Cement Composites (PMCC)

In PMCCs, polymeric materials are incorporated into cement composites during the mixing stage. The composite is then cast to the required shape in the conventional manner and is cured in a manner similar to the curing of cement concrete. The hydrated cement and the polymer film, formed due to the curing of the polymeric material, form an inter-penetrating network that binds the aggregates as shown in figure 9.1.

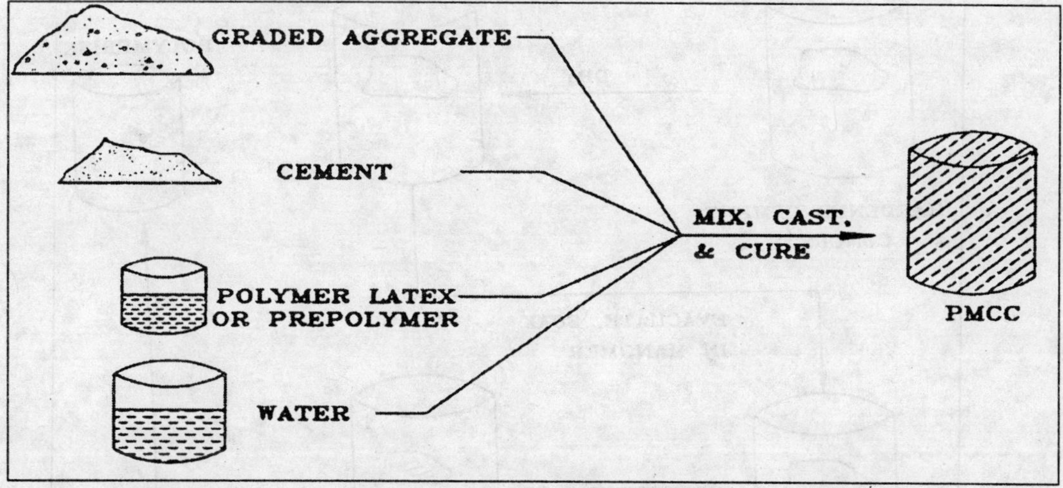

Figure 9.1. Process technology for polymer modified cement composite

9.7.2 Resin Concrete (RC)-also called polymer Concrete (PC)

In these, polymers are used as the binders of the aggregates, in lieu of the cement -water binder system adopted in cement composites. Monomers or pre-polymers are mixed with the aggregates and the mix is cast to the required shape or form. The mix is then polymerized either at the room temperature. The polymer phase binds the aggregates to give a strong composite as shown in figure 9.2.

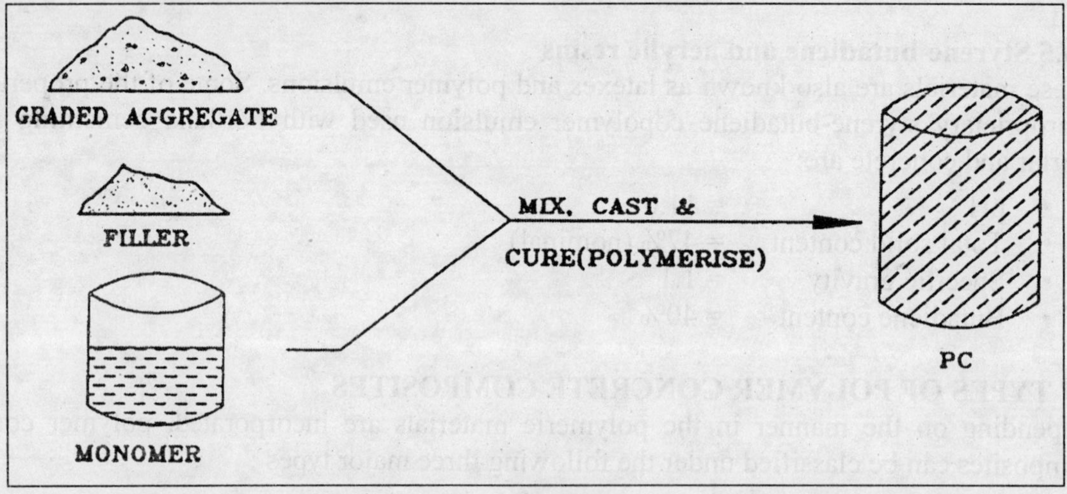

Figure 9.2. Process technology for polymer concrete or resin concrete

9.7.3 Polymer Impregnated concrete (PIC)

In PIC, monomers or pre-polymers are impregnated into the pore system of hardened cement composites and are then polymerized. A very strong composite viz., PIC, results, in which cured polymer fills almost all the pores as shown in figure 9.3.

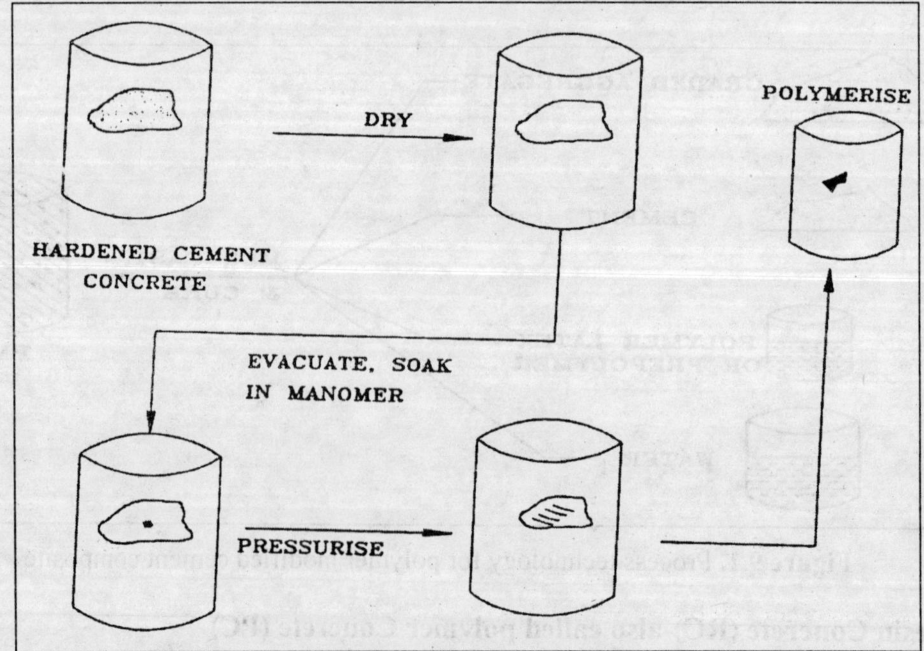

Figure 9.3. Process technology for polymer impregnated concrete

9.8 POLYMERIC REPAIR MATERIALS

Polymeric resins are complex chemical compounds derived from the petrochemical industry. The principal resins used for repair works are epoxides, polyesters, polyurethanes, acrylics, and vinyl esters. They may be used in the form of latexes such as styrene butadiene rubber or acrylics. While the range use of these materials is now very wide, it is convenient and practical to divide into two categories; namely coatings and bonding aids in which the formulated compound is used on its own, and polymer mortar/concrete composites in which the polymer is mixed with aggregate and sometimes cement.

9.8.1 Polymeric coatings

Surface coatings and sealers relay upon the formation of a pin-hole-free film of finite thickness over the concrete surface that acts as a protective barrier for concrete. They include epoxies, acrylics, polyurethanes, vinyl esters etc. The hardened film may form by solvent evaporation, reaction with moisture or atmosphere, chemical reaction with a catalyst or hardener, or oxidation.

Acrylics:

These form the base for a large proportion of coatings available on the market viz., methacrylates, styrene methacrylate, styrene acrylics, and synthetic rubber latex. They have good wetting characteristics, satisfactory chemical resistance, and good colour retention characteristics. They also show good inter-coat adhesion.

Urethanes:

These can form an ideal coating for concrete. They can be easily applied by brush, spray or roller. Their main disadvantage is that they require careful surface preparation to ensure proper adhesion to the substrate and possess less inter-coat adhesion.

Chlorinated Rubber:

Coatings which were traditionally used for concrete were based on chlorinated rubber. While they have excellent diffusion resistance to carbon dioxide, they also do not permit the water vapour from inside the concrete to move out. This vapour barrier effect has resulted in many failures of this type of coating. Additionally, they are not permanently elastomeric and, subsequently, crack and debond.

Epoxides:

These can provide excellent protection to concrete in extremely aggressive environs where high chemical resistance is the main parameter. They have excellent alkali resistance also and bond well to concrete. The coatings are resilient and abrasion resistant but inter-coat adhesion is poor. However, they may not be cost effective under other conditions. Coal tar epoxy systems are widely used as protective coatings for pipe lines and sewer appurtenances.

Silanes/Siloxanes:

These are gaining popularity under the category of surface impregnants or pore liners.

Hydrophobic materials like silicones line the pores of concrete and rapel water. Silicone resins are slow drying and pick-up dirt. They should be applied on a perfectly dry concrete surface only. Silanes have high penetrating capabilities because of their smaller particle size. Their hydrophobic nature expels water from any pore or capillary into which they penetrate. This opens the surface of the substrate and allows gases to pass in more easily.

Latex coatings

These are most widely used as interior coatings. Because they possess the desirable property of breathability, they are finding use in exterior applications, also. Some of their advantages include ease of application, clean-up, and wide range of available colours. Typical emulsions used are based on vinyl acetate, acrylic and styrene, and acrylic resins. Some of these resins have limitations such as sensitivity to moisture and UV exposure.

9.8.2 Polymer concrete/mortar composites

Polymer concrete/mortar (PC/PM) composites are being increasingly employed in repair and rehabilitation jobs owing to their very high strength and durability characteristics. **There are essentially three basic types of PC composites:**

(a) Polymer concretes (PC) formed by polymerizing a monomer binder and an aggregate mixture,

(b) Polymer modified concretes (PMC) in which a monomer or polymer is added to a fresh ordinary concrete mix in a liquid, powder or dispersed phase, and subsequently allowed to cure, and

(c) Polymer impregnated concretes (PIC), in which ordinary hydrated Portland cement concrete (PCC) is impregnated with a monomer and is subsequently polymerized in-situ.

Polymer modified concrete:

PMC has been of considerable interest to engineers because of its similarity in process technology to conventional concrete. Polymer latexes are usually copolymer systems such as vinyl acetate, vinyl chloride, and butadiene, besides elastomeric systems like acrylonitrile-butadiene (NBR), neoprene, and styrene-butadiene (SBR). PMC and PMM are increasingly used for rehabilitation because they are cement-based and therefore, give homogeneity to the system and the repair materials, and are more compatible with concrete compared to all other PC composites. An advantage of solvent-free PMC/PMM is the ease of adjusting the working rhythm as against the pot life and film -forming resistance of polymer solutions. They are economical while maintaining the technical value.

Polymer concrete:

Initial use of PC was almost exclusively for repair of ordinary PCC. The excellent bond of PC to PCC and very rapid cure time make PC an ideal repair material, especially in urban areas where fast, permanent repairs are essential. Added to these advantages are the possibility of tailoring its properties to suit any particular situation, excellent chemical resistance, and high bond strength. The most common types of monomers used to produce

PC are methyl methacrylate, unsaturated-polyester resin, and epoxies, besides furan, urethane, furfuryl alcohol, and vinyl ester which are also occasionally used. The materials used for patch repair range from unfilled resin binder to thick mortar sections. **The PC repair can be carried out in two ways:**

(a) Dry pack system in which the aggregates are prepacked and vibrated into the crater location and then infiltrated by a low-viscosity monomer like MMA

(b) Premixed PC in which the aggregates and monomer are mixed together in a wheel barrow or a conventional concrete mixer and then directly applied to the surface and leveled.

PC toppings consist of thermosetting resins with special sand fillers and are applied as a single 3-6mm layer by troweling. They have the advantage of superior abrasion and chemical resistance, rapid cure, and tenacious bond to concrete base. Epoxies, polyesters, and polyurethanes, are more widely used while polyacrylics and phenolics are occasionally used for specific applications. PC overlays constitute the major field of PC application in civil engineering, MMA, UP resin, and epoxy-resin based PCs have been used for bridge deck overlays. Furan resins have been used for floor overlays where chemical resistance is important. PC overlays have an important role to play in situations requiring thin overlays which are impermeable, possess high bond strength with the substrate, besides high corrosion resistance, and low shrinkage. However, the most enticing aspect of PC overlays is the rapid turn-around-time from rehabilitation to reopening of bridges/ pavements.

Polymer impregnated concrete:

PIC was developed in the 1950s and received wide publicity in the 60s and 70s. Recently, concrete sealing compounds like alkoxy silanes, alkoxy siloxanes, and metallic stearates have entered the market with claims of providing surface protection like surface impregnation. However, they do not provide the same extent of surface penetration and abrasion resistance, and their long-term durability and performance is suspect, due to possibility of removal from surface due to shallow depth of penetration. The process technology of PIC particularly for in-situ applications.

9.9 FIBRE REINFORCED CONCRETE

Fibre reinforced concrete can be defined as a composite material consisting of mixtures of cement, mortar or concrete and discontinuous, discrete, uniformly dispersed suitable fibres. Fibres used in concrete can be classified in to two groups: natural fibre materials-asbestoes, cellulose, sisal, coir, jute, coconut and cane splits, artificial fibre materials-steel, glass, carbon, polymer etc. Fibre reinforced concrete is one of the most promising composites used in the construction. Natural fibres are not suitable for rehabilitation work but artificial fibres are suitable for rehabilitation work. The mechanical properties of fibre reinforced concrete (FRC) are influenced by: the type of fibre; fibre length-to diameter ratio (aspect ratio); the amount of fibre; strength of matrix; the size, shape and method of preparation of the specimen; and the size of the aggregate. Fibres influence the mechanical properties of concrete and mortar in all failure modes. The strengthening mechanism of the fibres involves transfer of stress from the matrix to the fibre by interfacial shear or by interlock between the fibre and matrix if the fibre surface is deformed.

Besides the matrix itself, the most important variables governing the properties of FRC are the efficiency factor and the fibre content. Fibre efficiency is controlled by the resistance of the fibres to pullout which intern depends on the bond strength at the fibre matrix interface. Also, since pullout resistance is proportional to interfacial area, non-round fibres offer more pullout resistance per unit volume than larger diameter fibres. Therefore, for a given fibre length, higher aspect ratio is more beneficial. Most mixes used in practice employ fibres with an aspect ratio less than 100, and failure of composites, therefore is, due primarily to fibre pullout. However, increased resistance to pullout without increasing the aspect ratio is achieved in fibres with deformed surface or end anchorage; failure may involve fracture of some of the fibres, but it is still usually governed by pullout. An understanding of the properties of FRC is an important aspect of successful design.

9.10 BEHAVIOUR OF STEEL FIBRE REINFORCED CONCRETE

9.10.1 Compression

The effect of steel fibres on the compressive strength of steel fibre reinforced concrete (SFRC) concrete varies with fibre content. It is interesting to note that both increase and decrease of compressive strength with different fibre types have been experimentally observed. Even for the same material, there is mounting evidence to show that compressive strength may first rise, then drop, with increasing fibre volume fraction. These observations suggest that the addition of fibres in a cement composite lead to a likely manifestation of increased resistance to micro crack sliding and extension, where as strength degradation is a likely manifestation of increase in either pore or micro crack density, as a result of fibre addition.

9.10.2 Direct tension

Because of the brittle nature of concrete, valid direct tensile testing of concrete and FRC is always difficult to carry out. Due to the importance of the tensile behaviour of steel fibre reinforced concrete and concrete, many direct tensile tests of these materials have been attempted, using different designs of loading grips. Indirect methods of measuring the stress-strain curves have been attempted. SFRC has superior tensile properties, particularly ductility, over plain concrete. Studies have indicated that the tensile stress-crack separation curve is the best alternative to characterise the tensile behaviour of SFRC.

9.10.3 Flexure

The influence of steel fibres on the flexural strength of concrete and mortar is much greater than for direct tension and compression. Two flexural strength values are commonly reported. One corresponds to the first crack and the other corresponds to the maximum load. For large amounts of fibres, the two loads are quite distinct, but for very small fibre volumes, the first crack load may be the maximum load as well. Ultimate flexural strength generally increases in relation to the product of fibre volume concentration and the aspect ratio l/d.

Concentrations less than 0.5 volume percent of low aspect ratio fibres have negligible effect on the static flexural strength properties.

9.10.4 Flexural toughness

Toughness is an important characteristic for which SFRC is noted. Under static loading flexural toughness may be defined as the area under the load-deflection curve in flexure, which is the total energy absorbed prior to complete separation of the specimen.

9.10.5 Fatigue strength

The behaviour of SFRC in cyclic fatigue, despite its importance, has reckoned relatively little attention. FRC improves the dynamic properties like energy absorption, behaviour under fatigue loading over the plain concrete.

9.10.6 Behaviour under cyclic loading

The objective of subjecting the plain and SFRC specimens to cyclic loading is to investigate whether the specimens after subjecting them to cyclic loading would continual possess their original integrity (*i.e.* without suffering damage). Since the peak strain for plain concrete is around 0.002, and these specimens fail suddenly, it is possible to subject them to cyclic loading only at very low strain levels. The performance of SFRC is fond to be far superior to plain concrete even with 0.75% fibre volume fraction. The SFRC specimens, when loaded monotonically after cyclic loading at a strain of 0.003, reached almost the same peak load as was obtained under monotonic loading. All the SFRC specimens were able to sustain higher strain even after being loaded cyclically for fifteen cycles at a high strain of 0.007. This particular characteristic of SFRC could be beneficially used in the design of seismic resistant structures.

9.10.7 Shear and torsion

Studies in the last few decades indicate that use of steel fibres as shear reinforcement in reinforced concrete beams helps in enhancing the tensile strength, resulting in increase in shear strength and possible prevention of shear failure. Studies carried out so far have shown that steel fibres up to about 1.5% by volume are effective as shear reinforcement either by themselves or in combination with vertical stirrups. The studies carried out on the torsional behaviour of SFRC shown that there is an improvement in the torsional strength of concrete on addition of steel fibres of various types in varying volume fractions. Various researchers also studied the effect of reduction in longitudinal and transverse steel due to fibre addition.

9.10.8 Impact

Impact is a complex dynamic phenomenon involving crushing, shear failure and tensile fracturing. It is also associated with penetration perforation and fragmentation end scaling of the target being hit. The addition of fibres improved the impact resistance of the plain concrete to a great extent. The improvement in the strength is dependent on the fibre types and fibre volume fractions. As there is no acceptable standard methods for determining impact resistance of SFRC, several types of tests have been used, namely, weighted pendulum charpy-type impact test, drop-weight test, rotating impact test, blast impact test, projectile impact test and instrumented impact test. The simplest of the impact tests is the drop-weight test. This test yields the number of blows necessary to cause prescribed levels of distress in the test specimen. The test can be used to compare the relative merits of different fibre concrete mixes and to demonstrate the improved performance of FRC compared to conventional concrete.

9.10.9 Abrasion/Cavitation/Erosion

Both laboratory tests and full-scale trials have shown that SFRC has high resistance to cavitation force resulting from high-velocity water flow and the damage caused by the impact of large water borne debris at high velocity. Several tests indicate that steel fibre addition do not improve the abrasion/erosion resistance of concrete caused by small particles at low water velocities. This is because adjustments in the mixture proportions to accommodate the fibre requirements reduce coarse aggregate content and increase paste content.

9.10.10 Creep and shrinkage

There has been little work on the creep of steel fibre reinforced concrete. Fibres generally reduce the compressive and tensile creep. Tests have shown that steel fibres have little effect on free shrinkage of SFRC. However, when shrinkage is restrained, steel fibres can substantially reduce the amount of cracking and mean crack width.

9.10.11 Freeze-thaw resistance

Steel fibres do not significantly affect the freeze- thaw resistance of concrete, although they may reduce the sensitivity of the visible cracking and spalling as a result of freezing in concrete with inadequate air-void system. The freeze-thaw resistance of non-air-entrained concrete is similar for FRC and control concrete, where SFRC was found to better in the case of air entrained concrete.

9.11 APPLICATION OF SFRC TO REPAIR OF DISTRESS STRUCTURES

The applications of SFRC fall in two categories-repairs and new construction. Repairs are invariably required to tackle problems of abrasion, cavitations or impact damage in various components of hydraulic structures, such as, spillways, stilling basins, baffle blocks, outlet conduits, etc. Other area of applications of SFRC for repair purposes is in the field of shotcreting and overlays. Both steel and polypropylene fibres are used in shotcreting applications. Shotcreting with fibres improved the properties of the repaired concrete very much when compared to shotcreting with plain concrete. Similarly the behaviour of overlays using SFRC is far superior to the conventional concrete.

9.12 BEHAVIOUR OF FRC WITH OTHER FIBRES

9.12.1 Glass fibre

The glass fibers are divided into three classes—E-glass, S-glass and C-glass. The E-glass is designated for electrical use and the S-glass for high strength. The C-glass is for high corrosion resistance, and it is uncommon for civil engineering application. Of the three fibers, the E-glass is the most common reinforcement material used in civil structures. Some typical properties of the glasses are shown in table 9.1. It is produced from lime-alumina-borosilicate which can be easily obtained from abundance of raw materials like sand. The fibers are drawn into very fine filaments with diameters ranging from 2 to 13 $\times 10^{-6}$ m. The glass fiber strength and modulus can degrade with increasing temperature. Although the glass material creeps under a sustained load, it can be designed to perform satisfactorily. The fiber itself is regarded as an isotropic material and has a lower thermal expansion coefficient than that of steel.

Table 9.1. Typical properties of E & S glass

Typical Properties	*E-Glass*	*S-Glass*
Density (g/cm^3)	2.60	2.50
Young's Modulus (GPa)	72	87
Tensile Strength (Gpa)	1.72	2.53
Tensile Elongation (%)	2.4	2.9

Glass fibres possess high tensile strength and modulus of elasticity, but serious concern is expressed regarding their durability in an alkaline environment. For low w/c pastes, compressive strength is reduced by about 20% and for higher w/c ratio the decrease can be as high as 30%. Uniaxial tensile strength increases with age and amount of fibre. Aggregate grading does not influence the strength. In spite of the enhanced mechanical properties, question of the durability of alkaline resistant glass fibre concrete composite in alkaline environment remain unresolved.

Glass fibre reinforced concrete:

For the precast concrete industrial sector, the use of light weight composite components (*e.g.* 60-80 kg/m² for main slab elements and 30-50 kg/m² for secondary slab elements), able to withstand to structural loads, would represent a significative improvement with respect to the current production. In fact the possibility to manufacture precast elements weighting about one third of the conventional concrete products (*e.g.* 250-300 kg/m² for main and 100-150 kg/m² for secondary slab elements), would allow to achieve some important objectives:

(a) A substantial reduction of overall costs;
(b) A major improvement in the safety of the structure with respect to the seismic risk;
(c) The possibility to construct buildings also on low bearing capacity soils.

The reduction in weight and of carriage costs also allows to think about exportation of structural elements produced in an European factory to a foreign Country. An opportunity has been recognized by the industrial proposes to realize these light weight structural elements using Glass-fibre Reinforced Concrete (GFRC). Until now its main application has been in architectural cladding, in the form of panels 10-20 mm thick. Due to the tendency of GRC to *age*, or loose ductility with time, it never has been considered for *structural* applications. Although the estimated cost of such innovative, light-weight GRC components (*i.e.* 100-150 ECU/m² for main slab elements) is comparable or lightly higher than conventional ones, a reduction of overall costs of about 30% could be foreseen due to the following reasons:

(a) A slender carrying structure of the building (foundations, pillars, beams, etc.) is required;
(b) A complete automation of the whole production cycle is allowed;
(c) Carriage and assembly costs are minimal;
(d) The surface finishing (*e.g.* thermal insulation, waterproofing, painting, etc.) of precast elements can be carried out directly at the factory and no further operations are required on site.

Applications:
Potential applications of structural GRC components are copious also limiting our consideration to the precast sector; some of these have already been identified by the proposers and represent an interesting market opportunity:

(a) Main and secondary elements for slabs (*e.g.* in industrial sheds or in building retrofitting);
(b) Facade panels, wall panels and roof elements (*e.g.* in stand-alone technical shelters for mobile phone repeaters and in prefabricated houses).

The time-to-market for the above-mentioned or similar applications is estimated to require at least one year after the end of the project, because it will be necessary to modify the current manufacturing plants for GRC production or to install new ones.
The lack of specific standards for designing and testing of these components could represent an obstacle to a wide acceptance of new products. To avoid this risk, one of the actions carried out by the consortium members in the exploitation phase will consist of a collaboration with national and international Standardisation Bodies in order to ensure the issuing of a specific normative in the shortest possible time.

9.12.2 Polypropylene

Polypropylene fibres have high tensile strength and low modulus of elasticity and they are dependent on the isotacticity molecular weight and other structural features of PP. The elastic properties would be influenced by the extent to which calcium hydroxide interacts with fibre at the interface and therefore dependent on the fibre type. The brittleness of the composite is probably also affected by the amount and size of the calcium hydroxide crystals present. Further crystallisation of the calcium hydroxide in the contact zone may actually result in a weakening of the bond between the fibre and matrix. Surface modification of PP can result in improving interfacial bond. Reinforcement of cement matrices with continuous fibres, fibrillated filaments, fibrillated films; tape or woven fabric generally results in increased flexural and tensile strength.

9.12.3 Natural fibres

Some investigation have already been carried out on the use of natural fibres from coconut husk, sisal, sugar cane bagasse, bamboo, akara, plantain and musamba in cement paste, mortar, and concrete. These investigations have shown encouraging results. Flexural strength increases with fibre addition to a maximum and then decreases. The decrease at higher fibre content is due to incomplete compaction and increased porosity. The decrease in the strength for longer fibres was mainly due to bundling effects of the fibres. Impact strength depends on curing period and fibre volume fraction for both jute and their fibre reinforced concretes. Durability is of major importance in evaluating the suitability of natural fibres for inclusion in cement matrix. Coir fibre exhibits ductile failure characteristics while most of the other fibres exhibit brittle failure. Coir also shows the greater resistance to alkali attacks. Resin coatings provide a reasonable measure of protection against alkali attack.

9.12.4 Aramid fibers

These are synthetic organic fibers consisting of aromatic polyamides. The aramid fibers have excellent fatigue and creep resistance. Although there are several commercial grades of aramid fibers available, the two most common ones used in structural applications are Kevlar® 29 and Kevlar® 49. Typical properties of aramid fibres are shown in table 9.2.The Young's Modulus curve for Kevlar® 29 is linear to a value of 83 GPa but then becomes slightly concave upward to a value of 100 GPa at rupture; whereas, for Kevlar® 49 the curve is linear to a value of 124 GPa at rupture. As an anisotropic material, it's transverse and shear modulus are an order of magnitude less than those in the longitudinal direction. The fibers can have difficulty achieving a chemical or mechanical bond with the resin.

Table 9.2. Typical properties of aramid fibres

Typical Properties	*Kevlar 29*	*Kevlar 49*
Density (g/cm^3)	1.44	1.44
Young's Modulus (GPa)	83/100	124
Tensile Strength (GPa)	2.27	2.27
Tensile Elongation (%)	2.8	1.8

Experimentation was conducted on the use of a new aramid fiber for the reinforcement of Portland cement-based concrete and slurry. The study showed that synthetic fibers manufactured by cutting an epoxy-impregnated, braided bundle made of aramid filaments act similarly to steel fibers in reinforcing both portland cement-based concrete and slurry matrices. The advantage of epoxy-coated aramid over steel is in the lack of corrosion problems, and over polypropylene in the higher performance.

9.12.5 Carbon fibres

The graphite or carbon fiber is made from three types of polymer precursors—polyacrylonitrile (PAN) fiber, rayon fiber, and pitch. The tensile stress-strain curve is linear to the point of rupture. Although there are many carbon fibers available on the open market, they can be arbitrarily divided into three grades as shown in Table 9.3. They have lower thermal expansion coefficients than both the glass and aramid fibers. The carbon fiber is an anisotropic material, and its transverse modulus is an order of magnitude less than its longitudinal modulus. The material has a very high fatigue and creep resistance.

Table 9.3. Typical properties of carbon fibres

Typical Properties	*High Strength*	*High Modulus*	*Ultra-High Modulus*
Density (g/cm^3)	1.8	1.9	2.0-2.1
Young's Modulus (GPa)	230	370	520-620
Tensile Strength (GPa)	2.48	1.79	1.03-1.31
Tensile Elongation (%)	1.1	0.5	0.2

Since its tensile strength decreases with increasing modulus, its strain at rupture will also be much lower. Because of the material brittleness at higher modulus, it becomes critical in joint and connection details, which can have high stress concentrations. As a result of this phenomenon, carbon composite laminates are more effective with adhesive bonding that eliminates mechanical fasteners. Wide spread use of carbon fibre in cement has been limited due to cost consideration. Initially it was used in the pipe manufacturing only. Alternative uses are now being exploited as a result of the development of less expensive discontinuous fibre. Although discontinuous randomly distributed carbon fibres are less efficient than continuous aligned fibres, the properties of composites containing these carbon fibres are significantly improved. Tensile and flexural strengths increase with fibre content and they are generally less than for those with continuous fibres. At low water-cement ratios, the strengths are similar. Compressive strength of carbon fibre reinforced cements generally decreases with fibre addition.

9.13 FIBRE REINFORCED POLYMER COMPOSITES

Although the concept of fiber reinforced materials can be traced back to the use of straw as reinforcement in bricks manufactured by the Isrealites in 800 B.C., and in more recent times to the use of short glass fiber reinforcement in cement in the United States in the early 1930's, fiber reinforced resin matrix materials (or fiber reinforced composites as we know them today) were not developed until the early 1940's. Here the term composites is used in the context of advanced fiber reinforced polymers (FRP's). Because of the highly desirable light weight, corrosion resistance, and high strength characteristics in composites; research emphasis went into improving the material science and manufacturing process. That effort led to the development of two new manufacturing techniques known as filament winding and pultrusion, which helped advance the composite technology into new markets. With the increasing demand for composites, new and improved manufacturing processes such as pultrusion, resin transfer molding, and filament winding were developed and implemented in the early 1990s. With these enhancements in place, the current focus is to rebuild the infrastructure using FRP composites for maintenance and rehabilitation of existing bridges as well as new construction.

There are three basic manufacturing techniques in producing composite structural products, with many variations and patented processes: 1) The pultrusion process involves a continuous pulling of the fiber rovings and mats through a resin bath and then into a heated die. The elevated temperature inside the die cures the composite matrix into a constant cross-section structural shape. 2) The filament winding process can be automated to wrap resin-wetted fibers around a mandrel to produce circular or polygonal shapes. 3) The layup process engages a hand or machine buildup of mats of fibers that are held together permanently by a resin system. This method enables numerous layers of different fiber orientations to be built up to a desired sheet thickness and product shape.

Characteristics of Composites:

The mechanical properties of composites depend on many variables such as fiber types, orientations, and architecture. The fiber architecture refers to the preformed textile configurations by braiding, knitting, or weaving. Composites are anisotropic materials with their strength being different in any direction. Their stress-strain curves are linearly elastic to the point of failure by rupture. The polymeric resin in a composite material, which consists of viscous fluid and elastic solids, responds viscoelastically to applied loads. Although the viscoelastic material will creep and relax under a sustained load, it can be designed to perform satisfactorily. Composites have many excellent structural qualities and some examples are high strength, material toughness, fatigue endurance, and light weight. Other highly desirable qualities are high resistance to elevated temperature, abrasion, corrosion, and chemical attack.

Some of the advantages in the use of composite structural members include the ease of manufacturing, fabrication, handling, and erection. Project delivery time can be short. It took the Russell county engineer one day to install the deck panels in the first vehicular composite bridge. Composites can be formulated and designed for high performance, durability and extended service life. They have excellent strength-to-weight ratios. If durability can be proven to last 75 years, composites can be economically justified using the life-cycle cost method.

Some of the disadvantages in the use of composites in bridges are high first cost, creep, and shrinkage. The design and construction require highly trained specialists from many engineering and material science disciplines. The composites have a potential for environmental degradation, for examples, alkalis' attack and ultraviolet radiation exposure. There are very little or nonexistant design guidance and/or standards. There is a lack of joining and/or fastening technology. Because of the use of thin sections, there are concerns in global and local buckling. Although the light weight feature may be an advantage in the response to earthquake loading, it could render the structure aerodynamically unstable. In manufacturing with the hand layup process, there is a concern about the consistency of the material properties.

As the name implies, advance fiber reinforced polymer composites is made of fiber reinforcements, resin, fillers, and additives. The fibers provide increased stiffness and tensile capacity. The resin offers high compressive strength and binds the fibers into a firm matrix. The fillers serve to reduce cost and shrinkage. The additives help to improve not only the mechanical and physical properties of the composites but also workability. The discussions that follow immediately will explain the basic functions and behaviors of the constituents.

9.13.1 Fiber reinforcements

The fiber is an important constituent in composites. A great deal of research and development has been done with the fibers on the effects in the types, volume fraction, architecture, and orientations. The fiber generally occupies 30%-70% of the matrix volume in the composites. The fibers can be chopped, woven, stitched, and/or braided. They are usually treated with sizings such as starch, gelatin, oil or wax to improve the bond as well as binders to improve the handling. The most common types of fibers used in advanced composites for structural applications are the fiberglass, aramid, and carbon. The fiberglass is the least expensive and carbon being the most expensive. The cost of aramid fibers is about the same as the lower grades of the carbon fiber. Other high-strength high-modulus fibers such as boron are at the present time considered to be economically prohibitive.

9.13.2 Resin Systems

The resin is another important constituent in composites. The two classes of resins are the thermoplastics and thermosets. A thermoplastic resin remains a solid at room temperature. It melts when heated and solidifies when cooled. The long-chain polymers do not chemically cross link. Because they do not cure permanently, they are undesirable for structural application. Conversely, a thermosetting resin will cure permanently by irreversible cross linking at elevated temperatures. This characteristic makes the thermoset resin composites very desirable for structural applications. The most common resins used in composites are the unsaturated polyesters, epoxies, and vinyl esters; the least common ones are the polyurethanes and phenolics.

9.13.3 Fillers

Since resins are very expensive, it will not be cost effective to fill up the voids in a composite matrix purely with resins. Fillers are added to the resin matrix for controlling material cost and improving its mechanical and chemical properties. Some composites that are rich in resins can be subject to high shrinkage and creep and low tensile strength. Although these properties may be undesirable for structural applications, there may be a place for their use.

9.13.4 Additives

A variety of additives are used in the composites to improve the material properties, aesthetics, manufacturing process, and performance. The additives can be divided into three groups-catalysts, promoters, and inhibitors; coloring dyes; and, releasing agents.

9.14 FIBRE REINFORCED POLYMER (FRP) COMPOSITE LAMINATES

Fibre reinforced polymer {FRP) composite materials are produced from three main fibre types, namely carbon, glass and aramid. Each fibre has different engineering properties and therefore selection must be made to suit the requirements of a particular application. Carbon fibres have great strength. Stresses at failure can be in excess of 3000 N/mm^2. However, they are very expensive. Conversely, glass fibres are relatively inexpensive but have less strength and greater elasticity. Aramid fibres, renowned for their high impact resistance, tread the middle ground and may be yet to realize their true potential in structural engineering applications.

9.14.1 Fibre reinforced polymer composites for repair and strengthening

Perhaps the most commercially successful use of FRP composite construction materials has been in the strengthening existing structures. In the life time of building modifications are common. Changes in use often involve increases in loading or necessitate the introduction of alternative load paths. The solution in many cases has been to install FRP laminates or wraps as externally bonded reinforcement. The use of FRP materials for structural repair presents several advantages and has been recently investigated all over the world. An extensive experimental work has been carried out on strengthening of RC elements/beams by external plate bonding technique using carbon fibre reinforced polymer laminates and glass fibre reinforced polymer laminates.

9.14.2 FRP Laminates technology and benefits

The pultruded FRP laminate reinforcing consists of bonding the strip with the concrete structure using a high-strength epoxy resin as the adhesive. The strips are manufactured using a pultrusion process. The pultrusion principle is comparable with a continuous press. Normally 24,000 parallel filaments are pulled through the impregnated bath, formed into strips under heat, and hardened. These strips are uni-directional; the fibers are oriented only in the longitudinal direction. Correspondingly, the strip strength in this direction is proportional to the fiber strength and, thus, very high. Strips are produced with strengths of approximately 3,000 MPa in the longitudinal direction, and with a thickness of up to 1.5 mm and widths of up to 150 mm.

In order to achieve an optimum composite action, the preparation of the bonding surfaces of the strip and concrete is critical. The strips must have the outermost layer of their bonding face, normally matrix-rich, removed to expose the fibers. Just before the bonding, the bonding surface is carefully cleaned with acetone. The concrete surface is treated by sand blasting, high-pressure water jets, stoking, or grinding. Shortly before the bonding, it is cleaned with a vacuum cleaner. Concrete must be at least 6 weeks old, and have a minimum tensile strength of 1.5 MPa. Highly filled epoxy resin adhesive is used for the bonding.

FRP laminate reinforcing technology provides a solution for strengthening problems of concrete structures. It provides great strength, high modulus of elasticity, and outstanding fatigue resistance It is a very lightweight non-corrosive material, that requires minimal preparation of laminates, and it's alkali resistant. It is an economic method that requires very short contract times.

9.15 FERROCEMENT

As per ferrocement modal code, ferrocement is a type of reinforced concrete commonly constructed of hydraulic cement mortar reinforced with closely spaced layers of relatively small wire diameter mesh. The mesh may be made of metallic or other suitable materials. The fineness of the morter matrix and its composition should be compatible with the opening and tightness of the reinforcing system it is meant to encapsulate. The matrix may contain discontinuous fibers.

9.15.1 Materials used

Cement:

The cement should comply with ASTMC 150-859, ASTMC 595-85, or an equivalent-standard. The cement should be fresh, of uniform consistency, and free of lumps and foreign matter. It should be stored under dry conditions for as short duration as possible.

Fine aggregates:

Normal weight fine aggregate used in ferrocement, should comply with ASTMC 33-86 requirements or an equivalent standard. It should be clear, inert, free of organic matter and deleterious substances, and relatively free of silt and clay.

Mixing water:

The mixing water should be fresh, clean and potable. The water should be relatively free from organic matter, silt, oil, sugar, chloride, and acidic material. Salt water is not acceptable, but chlorinated drinking water can be used.

Admixtures:

Conventional and high–range water reducing admixtures should be used.

Reinforcement:

Wire mesh with closely spaced wires is the most commonly used reinforcement in ferrocement. Expanded metal mesh, welded wire fabrics, square steel wire mesh, hexagonal or chicken wire mesh are used. The various types of reinforceent is shown in figure 9.4. Properties and types of constituent materials used in ferrocement construction as shown in table 9.4.

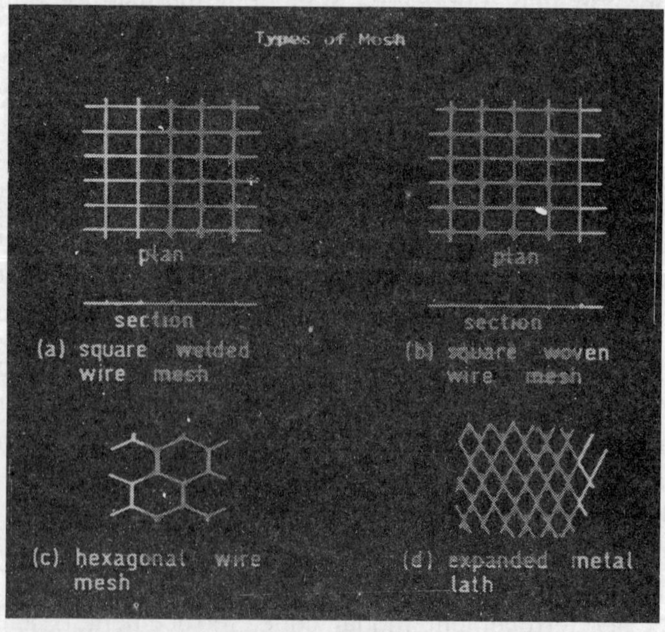

Figure 9.4. Typical sections of wire meshes

Table 9.4. Guidelines for material selection

Materials	*Range*
Wire mesh:	
Diameter of wire φ	$0.5 \leq \varphi \leq 1.5$ mm
Types of meshes	Chicken wire or square woven or welded galvanized mesh or expanded metal
Size of mesh opening (S)	$6 \leq S \leq 25$mm
Volume fraction (Vr) of reinforcement in both direction	$2\% \leq Vr \leq 8\%$
Specific surface (Sr) of reinforcement in both direction	$0.1 \leq Sr \leq 0.4 mm^2/mm^3$
Elastic modulus (Er)	140 – 200 kN/ mm^2
Yield strength (Fy)	250 – 460 N/ mm^2
Ultimate tensile strength (Fu)	400 – 600 N/ mm^2
Skeletal steel:	
Type	Welded mesh, steel bars, strands
Diameter (d)	$3mm \leq d \leq 10mm$
Grid size (G)	$50mm \leq G \leq 200mm$
Yield strength	250 – 460 N/ mm^2
Ultimate tensile strength	400 – 600 N/ mm^2
Mortar composition:	
Cement	Any type of Portland cement (depending on Application)
Sand/cement ratio (s/c)	$1 \leq s/c \leq 3$ by weight
Water/cement ratio (w/c)	$0.35 \leq w/c \leq 0.65$ by weight
Gradation of sand	5mm to dust with not more than 10%* passing 150μm BS test sieve
Compressive strength (cube)	30 – 50 N/ mm^2

* For crushed stone, sand may be increased to 20%

9.15.2 Ferrocement as a repair material

The major objective of repair or renovation work for concrete is to return the concrete to a satisfactory level of performance at a reasonable cost. For many repairs and renovation programs of civil Engineering structures cite the suitability of ferrocement because of,

(a) Better cracking behaviour
(b) Capability of improving some of the mechanical properties of the treated structures.
(c) Further modification or repair of ferrocement treatment is not difficult.
(d) Imposition of little additional dead load requiring no adjustment of the supporting structures.
(e) Ability to withstand thermal changes very efficiently.
(f) Ability of achieving water proofing property with out providing any surface treatment.

(g) Readily available constituent materials.
(h) No need for special equipment.
(i) Ability to be used in repair program with no distortion or downgrading of the architectural concept of the structures.
(j) Flexibility of further modification.

Ferrocement has been used extensively in the area of homing application for the rehabilitation of distressed elements due to overloading, fire damaged beams and columns and general repair in the deteriorating structure. The ferrocement material is ideally suitable due to its ability to arrest crack and high tensile strength-to-weight ratio.

9.16 SIFCON AND SIMCON MATERIALS

Many infrastructure deterioration problems and failures can be traced back to the cracking and brittle nature of concrete. It is no wonder that significant research efforts have gone into attempts at enhancing the ductility of concrete materials. To date, the most effective means of imparting ductility into concrete is by means of fiber reinforcement. Most fibre reinforced concrete results in enhanced post peak tension-softening behavior under tensile load. This improved toughness is useful in resisting the propagation of cracks into major fractures by energy absorption in the bridging actions of fibers in the fracture process. While the fracture toughness of concrete can increase by an order of magnitude by fiber reinforcement, the tensile strain capacity usually remains little changed. In recent years, efforts to convert this quasi brittle behavior of fiber reinforced concrete (FRC) to ductile strain-hardening behavior resembling ductile metal have met with varying degrees of success. In most instances, the approach is to increase the fiber content as much as possible and processing techniques are adapted to overcome the resulting workability problem. Cement-based composites such as slurry infiltrated fiber concrete (SIFCON) and slurry infiltrated mat concrete (SIMCON) are good examples of this approach, which utilizes from 5 to 20% volume content of steel fibers.

9.16.1 Sifcon

Slurry infiltrated fibre concrete is a combination of cement, aggregate, fibre and water with some admixture. SIFCON is a recently developed construction materials using steel fibers with cement matrix. The matrix consists of cement mortar slurry. The fibre bed is infiltrated with cement based slurry by gravity flow. Hence the optimization of the flow of matrix of cement slurry is, as important step in the design of its composition. Slurry-infiltrated fibrous concrete (SIFCON) can be considered as a special type of fiber concrete with high fiber content. The matrix usually consists of cement slurry or flowing mortar. SIFCON has excellent potential for application in areas where high ductility and resistance to impact are needed. Only very limited information is available about its behavior under different types of loading. Tests on 22-mm-thick SIFCON specimens were carried out to study their behavior in flexure and under subjection to abrasion and impact loads. Toughness characteristics of SIFCON were also evaluated by testing another set of specimens 100 × 100 × 500 mm per ASTM C1018. Both strength and deformation characteristics of the specimens were studied. The results obtained from these tests were compared with those carried out on companion plain mortar and conventional fiber-reinforced mortar (FRM) specimens.

The investigations confirm the superior characteristics of SIFCON as compared with plain mortar and normal FRM. SIFCON (slurry infiltrated fiber concrete), which have number of outstanding properties such as strength (compression, tension, bending, and shear), ductility, toughness, durability, stiffness, and energy absorption capacity under monotonic and cyclic loads. The properties of SIFCON are achieved through an optimized combination of matrix properties, fiber reinforcing parameters, fiber content, and interface characteristics between fiber and matrix. It is observed that the presence of SIFCON in over reinforced concrete beams leads to ductility indexes exceeding three times those obtained without it. Crack widths and spacing are more than an order of magnitude smaller than in conventional reinforced concrete. Experimental results also suggest that there is no need for stirrups in flexural members with a SIFCON matrix.

9.16.2 Simcon

The cost of civil infrastructure constitutes a major portion of the national wealth. Its rapid deterioration has thus created an urgent need for the development of novel, long-lasting and cost-effective methods for repair, retrofit and new construction. A promising new way of resolving this problem is to selectively use advanced composites, such as High-Performance Fiber Reinforced Cementitious Composites (HPFRCCs). With such materials, novel repair, retrofit and new-construction approaches can be developed that would lead to substantially higher strengths, seismic resistance, ductility, durability, while also being faster and more cost-effective to construct than conventional methods. SIMCON which stands for Slurry Infiltrated Mat Concrete is well suited for the development of novel repair, retrofit and new-construction solutions that lead to economical and improved structural performance.

SIMCON (slurry infiltrated mat concrete) is a non-woven steel fiber mat that is infiltrated with concrete slurry. It is shown that the SIMCON concept of reinforcement represents a significant improvement in the reinforcement efficiency in high density, high strength fiber reinforced concrete. Comparable levels of flexural strength and flexural energy absorption capacity can be achieved at greatly reduced fibre loadings relative to the short, discrete fiber reinforcement approach. SIMCON, uses a manufactured continuous mat of interlocking discontinuous steel fibers, placed in a form, and then infiltrated with a flow able cement-based slurry. The use of continuous mats, typically made with stainless steel to control corrosion in very thin members, permits development of high flexural strengths and very high ductility with a reduced volume of fibers.

9.17 MISCELLANEOUS MATERIALS

It is imperative that materials be tested and evaluated, prior to use in the repair of abrasion-erosion damaged hydraulic structures. Abrasion-resistant concrete should include the maximum amount of the hardest coarse aggregate that is available and the lowest practical w/c. In some cases where hard aggregate was not available, high-range water-reducing admixtures (HRWRA) and condensed silica fume have been used to develop high compressive strength concrete 97 MPa to overcome problems of unsatisfactory aggregate. Apparently, at these high compressive strengths the hardened cement paste assumes a greater role in resisting abrasion-erosion damage, and as such, the aggregate quality becomes correspondingly less important. The abrasion-erosion resistance of vacuum-treated concrete, polymer concrete, polymer-impregnated concrete, and polymer Portland cement concrete is significantly superior to that of comparable conventional concrete that can also be attributed to a stronger cement matrix.

The increased costs associated with materials, production, and placing of these and any other special concretes in comparison with conventional concrete should be considered during the evaluation process. While the addition of steel fibers would be expected to increase the impact resistance of concrete, fiber-reinforced concrete is consistently less resistant to abrasion-erosion than conventional concrete. Therefore, fiber-reinforced concrete should not be used for repair of stilling basins or other hydraulic structures where abrasion-erosion is of major concern. Several types of surface coatings have exhibited good abrasion erosion resistance during laboratory tests. These include polyurethanes, epoxy-resin mortar, furan-resin mortar, acrylic mortar, and iron aggregate toppings. However, some difficulties have been reported in field applications of surface coatings, primarily the result of improper surface preparation and thermal incompatibility between coatings and concrete.

Conventional materials – While proper material selection can increase the cavitations resistance of concrete, the only totally effective solution is to reduce or eliminate the causes of cavitations. However, it is recognized that in the case of existing structures in need of repair, the reduction or elimination of cavitations may be difficult and costly. The next best solution is to replace the damaged concrete with more cavitations-resistant materials. Cavitations resistance of concrete can be increased by use of a properly designed low w/c, high-strength concrete. Vital to increased cavitations resistance are the use of hard, dense aggregate particles and a good aggregate to- mortar bond. Typically, cement-based materials exhibit significantly lower resistance to cavitations compared to polymer-based materials.

Other cavitations-resistant materials – Cavitations damaged areas have been successfully repaired with steel fiber concrete and polymer concrete. Some coatings, such as neoprene and polyurethane, have reduced cavitations damage to concrete, but since near-perfect adhesion to the concrete is critical, the use of the coatings is not common. Once a tear or a chip in the coating occurs, the entire coating is likely to be peeled off.

9.18 FULFILS FOR THE REPAIR MATERIALS

The repair materials should fulfill the following requirements:

(a) It must be thoroughly bonded to concrete
(b) Shrinkage of the material should be small enough not to jeopardise the bond.
(c) The material and old concrete should respond to changes in temperature, moisture and load similarly enough to avoid gross differences in movement.
(d) The permeability of the material should be sufficiently low.
(e) The material should be resistant to weather action.
(f) In addition to them requirements, it is some times necessary to match the colour and texture of the old concrete.

OBJECTIVE TYPE QUESTIONS

1. Polymer modified cementitious materials are used for a repair of concrete structures up to a depth of
   (a) More than 12 mm
   (b) Less than 12 mm
   (c) Between 25-30 mm
   (d) None of the above
2. The polymer latex acts in a concrete as a
   (a) Water – reducer
   (b) Strength – reducer
   (c) Binding – reducer
   (d) Permeability – reducer
3. Polyvinylidene dichloride latexes are not recommended for repairs because
   (a) Free sulphates released in the long term
   (b) Free chlorides released in the long term
   (c) Free carbonation occurs in the short term
   (d) All the above
4. Where the cover is less than 12mm and the areas to be repairs
   (a) By cementitious mortar
   (b) By super plasticizer
   (c) By resin mortar
   (d) All the above
5. Acrylic resin in classed as
   (a) Thermoset material
   (b) Thermoplast material
   (c) Organic material
   (d) None of the above
6. The curing of epoxy resin system is an
   (a) Endothermic reaction
   (b) Electrochemical reaction
   (c) Exothermic reaction
   (d) All the above
7. ---------------- are generally used for the structural repair of cracks
   (a) Low viscosity epoxy resin systems
   (b) High flowable material systems
   (c) Non-destructive material systems
   (d) All the above systems

8. Modified resins which flow readily in to fine cracks under low pressures but stop flowing immediately the pressure is released, are termed
   (a) Exothermic
   (b) Endothermic
   (c) Thixotropic
   (d) Diotropic

9. Epoxy resin systems has excellent
   (a) Chemical instability
   (b) Volume stability
   (c) Electrochemical ability
   (d) All the above

10. Improving the abrasion resistance of concrete floors by
    (a) Polymer impregnation
    (b) Vacuum impregnation
    (c) Chloride impregnation
    (d) None of the above

11. The name of the material has be used effectively to stop leakage in existing home basement is
    (a) Bituminous watering
    (b) Bentonite
    (c) Grouts
    (d) All the above

12. Latex modified concrete are used to improve
    (a) Shear strength
    (b) Bond strength
    (c) Durability
    (d) All the above

13. Materials used for jacketing purposes
    (a) Ferrocement
    (b) SBR
    (c) Linseed oil
    (d) All the above

14. Materials used as a coating compound
    (a) Fibrous material
    (b) Styrene butadiene rubber
    (c) Ferrocement
    (d) None of the above

15. The material used to prevent penetration of water through masonry structures
    (a) Pumice

(b) Blasting sand
(c) Bentonite
(d) None of the above

16. For preventing differential shrinkage between hardened concrete and fresh concrete
(a) Shrincomp grouting
(b) Polymer grouting
(c) Cement grouting
(d) None of the above

17. The injection of a low viscosity epoxy is a possible repair method for cracks between
(a) 1 mm to 2 mm in width
(b) Greater than 3 mm in width
(c) 0.02 mm to 6 mm in width
(d) All the above

18. A judicious use of epoxies is called for in the light of their
(a) Very low cost
(b) Very high cost
(c) Medium cost
(d) None of the above

19. Impact is a complex dynamic phenomenon involving
(a) Crushing
(b) Shear failure
(c) Tensile fracturing
(d) All the above

20. Fibres generally reduces
(a) Compressive creep
(b) Shear creep
(c) Shrinkage creep
(d) Dynamic creep

21. Glass fibre possess high
(a) Compressive strength
(b) Tensile strength
(c) Bond strength
(d) Shear strength

22. Most of the natural fibres exhibit
(a) Brittle failure
(b) Ductile failure
(c) Impact failure
(d) None of the above

23. Polyethylene terepthalate is called
   (a) Polyester
   (b) Polyester resin
   (c) Terylene
   (d) Saturated polyester

24. The polyester resin has good electrical properties such as
   (a) Trielectric strength
   (b) Monoelectric strength
   (c) Chemical strength
   (d) Dielectric strength

25 Polyvinyl acetate claims increased
   (a) Tensile strength
   (b) Shrinkage
   (c) Permeability
   (d) None of the above

26. PVA claims reduced
   (a) Flexural strength
   (b) Tensile strength
   (c) Permeability
   (d) None of the above

27. SBR & AR is also known as
   (a) Latexes and polymer emulsions
   (b) Styrene butadiene rubber latex
   (c) Acrylonitrle
   (d) None of the above

28. Polymeric resins are complex chemical compounds derived from the -------- industr
   (a) Chemical
   (b) Petrochemical
   (c) Physical
   (d) None of the above

29. Repairs to areas which are heavily trafficked need to be achieved with the lea disruption to traffic by
   (a) Cementitious mortar
   (b) Polymer mortar
   (c) Quick setting compounds
   (d) Bituminous mortar

30. One of the quick setting compounds
   (a) Molten sulfur
   (b) Sulphur phosphate

(c) Ammonium sulphate
(d) None of the above

31. Rapid setting patch products are -------------- based
(a) Calcium chloride
(b) Calcium sulphate
(c) Magnesium sulphate
(d) Magnesium chloride

32. ------------------ have the lowest cost of all patching materials
(a) Quick setting compounds
(b) Grouts
(c) Bituminous material
(d) None of the above

33. A grout is a ------------------ material
(a) Non flowable
(b) Packable
(c) Non plastic
(d) None of the above

34. These are natural or synthetic materials used to join individual members of a without mechanical fastners
(a) Quick setting compounds
(b) Bituminous materials
(c) Grouts
(d) Bonding agents

35. The properties of concrete which affect the successful application of coating are
(a) Porosity
(b) Moisture content
(c) Pressure of contaminants
(d) All the above

36. Fire resistant coatings applied on structural elements are called
(a) Intumescent coatings
(b) Bituminous coatings
(c) Epoxy coatings
(d) Neoprene coatings

37. Filled system with thickness in excess of 30 ml, and ranging up to 10 mm are described as
(a) Coatings
(b) Toppings
(c) Screedings
(d) Varnishing

38. A heavy duty chemical resistance system is called
   (a) Glass fiber – Reinforced multicoat system
   (b) Seamless floor toppings
   (c) Epoxy floor toppings
   (d) High build floor coatings

39. The term --------------- refers to a material used to upgrade a floor's wear resistance
   (a) Shake hardeners
   (b) Sprinkled
   (c) Surface hardeners
   (d) None of the above

40. These are high performance, elastomeric sealants which have the widest service temperatures
   (a) Urethane sealants
   (b) Silicone sealants
   (c) Butyl sealants
   (d) Polysulfide sealants

41. ---------------- sealants are styrene – butadiene styrene and ethylene butylenes styrene thermoplastic elastomers
   (a) Latex based
   (b) Solvent based
   (c) Epoxy based
   (d) Kraton – based

42. The treatment of a surface to prevent the passage of liquid water under hydrostatic pressure is called
   (a) Water proofing
   (b) Damp proofing
   (c) Surface proofing
   (d) All the above

43. Damp proofing can be achieved by the use of a
   (a) Air entrained cement
   (b) Coloured cement
   (c) Hydrophobic cement
   (d) High alumina cement

44. --------------- coatings contain a copolymer, a plasticizer, and a solvent
   (a) Polyurethanes
   (b) Polyvinyl chloride
   (c) Neoprene
   (d) Polyethylene

45. Ferrocement consists of
    (a) Cement mortar & meshes
    (b) Cement concrete & reinforcement
    (c) Cement slurry & fibres
    (d) None of the above
46. Closely spaced fine meshes with cementitious mortar are called
    (a) GFRP
    (b) CFRP
    (c) Ferrocement
    (d) None of the above
47. SIFCON stands for
    (a) Slurry infiltrated fibre concrete
    (b) Steel infiltrated concrete
    (c) Steel in fibre concrete
    (d) None of the above
48. SIMCON stands for
    (a) Steel in mat concrete
    (b) Slurry in mat concrete
    (c) Slurry infiltrated mat concrete
    (d) None of the above
49. Repair material need for Jacketing techniques of repair
    (a) Fibre reinforced concrete
    (b) Ferrocement
    (c) Grouts
    (d) None of the above
50. To resist high tensile strength, to need
    (a) Fibre reinforced plastics
    (b) FRC
    (c) Ferrocement
    (d) None of the above

ANSWERS

1. a	2. d	3. b	4. c	5. a	6. c	7. a	8. c	9. b	10. a
11. b	12. d	13. a	14. b	15. c	16. a	17. c	18. b	19. d	20. a
21. b	22. a	23. c	24. d	25. a	26. c	27. a	28. b	29. c	30. a
31. b	32. c	33. b	34.d	35. d	36. a	37. b	38. a	39. c	40. b
41. d	42. a	43. c	44. b	45. a	46. c	47. a	48. c	49. b	50. a

SHORT ANSWER QUESTIONS

1. What is meant by binder?
2. What is the benefit of binder?
3. Define fine aggregate as a repair material.
4. Write the benefit of fine aggregate.
5. Define coarse aggregate as a repair material.
6. What is the benefit of coarse aggregate?
7. What is meant by special fillers?
8. What is the benefit of special filler?
9. Define polymer modifiers.
10. Write the benefits of polymer modifier.
11. What are the uses of chemical modifier?
12. Write the benefit of chemical modifier.
13. What is meant by Repair material?
14. What are the basic needs for selection of repair materials?
15. What is the material properties considered when selecting a repair material?
16. What is PMCC?
17. What is PC &RC?
18. What is PIC?
19. How do the polymers used in repair system?
20. What is the action of polymer latex?
21. What are the properties of polymer based repair materials?
22. Explain about polymer bonding aids.
23. What are the points fulfills for the repair materials?
24. Write short notes on bentonite.
25. What is the use of bituminous coat as a repair material?
26. Write short notes on epoxies.
27. Write about polymer.
28. Name the materials available for repair techniques.
29. Define SBR latex.
30. What are the properties obtained from SBR Latex with cementetious mixer?
31. How is the latex works in cementitious materials?
32. Recommendation for the use of SBR Latex cementitious mixes
33. Name the epoxy resins for structural repairs and Rehabilitation
34. What is the necessity of careful selection of repair material?
35. What are the factors considered in selecting a repair material?
36. Define the methodology for the selection of repair materials?
37. Define the classification of repair materials?
38. How do evaluate the repair materials?
39. What is meant by thixotrophic index?
40. What are the properties involved in green mixes?

41. What is the limitation of compressive strength and elastic modules of repair materials?
42. In which type of repair tensile strength and its elongation is required for repair materials?
43. Write about flexural strength and flexural modules of repair material?
44. What are the properties to be tested in case of polymeric latexes and resin systems?
45. What are the tests under durability related?
46. Name the tests for surface of repair materials.
47. Write about patch test for adhesion of repair material.
48. What are the mechanical properties of fibre resin forced concrete as a repair material?
49. What is meant by aspect ratio in FRC?
50. Name some of the natural fibres used as a repair material?
51. Expand the terms FRC, GFRP, CFRP & SFRC
52. What are the types of polymers used as repair materials?
53. Write shortly the application of polymers.
54. What are the characteristics of epoxide resins as a repair material?
55. What is the use of polyurethanes as repair materials?
56. Classify polyester resins as a repair material.
57. What is the function of polyvinyl acetate as a repair material?
58. What are the properties SBR & AR with Portland cement mortar?
59. Define PMCC, RC/PC & PSC
60. Differentiate polymer coatings & latex coatings.
61. Compare Silanes/Siloxanes.
62. Define admixtures and classify it.
63. Write short notes on chemical admixture.
64. Write short notes on mineral admission.
65. Define SIFCON.
66. Define SIMCON.
67. What is meant by Ferro cement?
68. What are the materials used in ferrocement?
69. Write about the reinforcement used in ferrocement.
70. What are the applications of ferrocement?

THEORY QUESTIONS

1. Explain the classification of repair materials with selection criteria.
2. Discus the methodology used in the selection of repair materials.
3. Explain about the various tests for surface quality of repair materials.
4. Discuss about various miscellaneous test/chemical test for repair materials.
5. Explain about the durability related tests for repair materials.
6. Write about test for physical and mechanical properties of repair materials.
7. Explain the behaviour of steel fibre reinforced concrete as a repair material.
8. Explain the behaviour of glass fibre as a repair material.
9. Explain the application of carbon fibres as a repair material.
10. Write about organic polymer as repair materials.
11. Explain the application of fibre reinforced polymer composite for repair & strengthening of concrete structures.
12. Explain the types of polymer concrete composites with neat sketches.
13. Discuss the process technology of polymer composites.
14. Explain about the fibre reinforced polymer composite laminates as a repair & rehabilitations material.
15. Explain about the polymeric repair materials.
16. Discuss about chemical admixtures as a repair material.
17. Discuss about mineral admixtures as a repair material.
18. Explain about ferrocement as rehabilitation material.
19. Define SIFCON and explain the characters and application in construction industry.
20. Explain briefly about SIMCON and its application in retrofitting work.
21. Write about the role of epoxies and latexes as repair materials.
22. Write about the general considerations in the repair materials.

PART IV
REPAIR AND REHABILITATION

CHAPTER 10

Repair of Cracks

10.1 INTRODUCTION

Cracks are almost common in all buildings resulting in ugly appearance and a feeling of instability to the occupants. But all cracks are not harmful to the life of a building. Non-structural cracks do not affect the main structure if they are treated appropriately in time whereas structural cracks can be eliminated by proper design and through supervision.

Cracks may be differentiated into following kinds:

(a) With regard to depth – surface, shallow, deep, through cracks

(b) With regard to direction – map cracks or pattern cracks roughly hexagonal and single continuous cracks

(c) With regard to width – thin (less than 1 mm), medium (1-2 mm), and wide (more than 2 mm) cracks

(d) With regard to load transfer – structural and non- structural

10.1.1 Factors that affect cracking

Before going into details it is necessary to know the factors that affect cracking. The various factors are:

Water – the more is the water the greater is cracking tendency. Water both increases shrinkage and reduces strength.

Cement – richer concretes crack more. Increasing the volume of aggregates by 10% can be expected to reduce shrinkage by about 50%.

Aggregate – presence of clay binders in aggregates cause high shrinkage and cracking. The smaller the maximum size of well-graded aggregates the greater the shrinkage of concrete at the same strength.

Bleeding – excessive bleeding induce cracks.

Improper curing – when rate of evaporation is very high so that fresh concrete dries rapidly early curing and appropriate screening is essential.

Exposure – the southern and western exposures of s structure are usually more severe than northern and eastern exposures.

Cover – clear cover less than the specified value expedites corrosion of reinforcement.

10.1.2 Measures of cracking

Following measures may be adopted to rectify the cracking:

(a) The fresh concrete should have low water/cement ratio to reduce shrinkage and well compacted to prevent settlement due to shrinkage.

(b) The aggregates used should be well graded to give high quality concrete.

(c) In order to avoid formation of hair cracks, dry cement should not be allowed to be sprinkled over the wet surface to absorb bleeding water, but finishing operation should be delayed till the evaporation of waters is complete.

(d) The curing of concrete should be allowed for a sufficient time.

(e) In order to avoid absorption of moisture from the concrete by the surface over which concrete is to be placed, should be made sufficiently moist. This will avoid drying shrinkage.

(f) Formwork used should be sufficient strong.

The repair of concrete structures may vary between a cosmetic treatment and a total replacement. By a proper investigation and by using well designed equipment, tools, and materials can be reinstated economically. An appropriate repair method can be selected depending upon the cause and extent of damage, importance of the structural element, and its location. The choice of the method will determine its success. **Repair of concrete structures is decided upon the factors such as:**

(a) The cause of damage

(b) Type, shape and function of the structure

(c) The type and extent of damage

(d) The capabilities and facilities available with builders

(e) The availability of repair materials

10.1.3 Stages of concrete repair

Repair of concrete structures is carried out in the following stages:

(a) Removal of damaged concrete

(b) Pre treatment of surfaces and reinforcement

(c) Application of repair material

(d) Restoring the integrity of individual sections and strengthening of structure as a whole

Removal of defective concrete:

Prior to the execution of any repair, one essential and common requirement is that all deteriorated or damaged concrete be removed. Removal of defective concrete can be carried out using tools and equipment, the types of which depend on the damage. Normally, removal of concrete can be accomplished by hand tools, or when that is impractical because of the extent of repair, it can be done with a light or medium weight air hammer fitted with a spade-shaped bit. Care should be taken not to damage the unaffected concrete portions. For cracks and other narrow defects, a saw-toothed bit will help achieve sharp edges and a suitable under cut.

Pre treatment of surfaces and reinforcement:

The preparation of a surface/pretreatment for repair involved the following steps:

(a) Complete removal of unsound material.

(b) Undercutting along with the formation of smooth edges.
(c) Removal of the cracks from the surface.
(d) Formation of a well defined cavity geometry with rounded inside corners.
(e) Providing, rough but uniform surface for repair.

The cleaning of all loose particles and oil and dirt out of the cavity should be carried out shortly before the repair. This cleaning can be achieved by blowing with compressed air, hosing with water, acid etching, wire brushing, scarifying, or a combination. Brooms or brushes will also help to remove loose material.

Application of repair material:
After the concrete surface has been prepared, a bonding coat should be applied to all the cleaned exposed surfaces. It should be done with minimum delay. The bonding coat may consist of bonding agents such cement slurry, cement sand mortar, epoxy, epoxy mortar, resin materials etc., Adequate preparation of surface and good workmanship are the ingredients of efficient and economical repairs.

Repair procedure:
The repair of cracked or damaged structures is discussed under two distinct categories, namely, ordinary or conventional procedures; and special procedures using the latest techniques and newer materials such as polymers, epoxy resins, etc. **A repair procedure may be selected to accomplish on or more of the following objectives:**
(a) To increase strength or restore load carrying capacity.
(b) To restore or increase stiffens.
(c) To improve functional performance.
(d) To provide water tightness.
(e) To improve appearance of concrete surface.
(f) To improve durability.
(g) To prevent access of corrosive materials to reinforcement.

10.2 DURABILITY OF CONCRETE REPAIR

The objective of any repair should be to produce repair – which means a repair carried out at a relatively low cost, with a limited and predictable degree of change with time, and without premature deterioration and/or distress throughout its intended life and purpose. To achieve this goal, it is necessary to consider the factors affecting the durability of a repaired structural system as part of a whole, or a component of composite system. Summarized some of the findings and recommendations may be grouped in to three categories:
(a) Durable repair design
(b) Durable repair application and
(c) Evaluation of the repairs

10.2.1 Durable repair design

Modulus of elasticity and strength of repair material:
The modulus of elasticity of the repair material not only affects the resultant flexural stiffness

of the repaired members, but also the tensile stress present with in the repair material and the debonding stress at the interface when differential movement occurs between the repair material and substrate. A higher difference in modulus of elasticity between the repair material and substrate may adversely affect the stress distribution with in the repaired composite material's cross section and may lead to considerable stress concentrations. Therefore, the repair material selected should have as similar a modulus of elasticity as the substrate as possible. Thus considering the strength of repair material alone seems less important. In fact, an overemphasis on strength may cause repairs to experience cracking arising from excessive drying shrinkage, creep, and heat of hydration.

Coefficient of thermal expansion of repair material:
Tensile stresses in the repair material caused by changes in the temperature of the surrounding environment are proportional to the difference in the coefficient of thermal expansion and the change in temperature. Therefore, the repair material selected should have as similar a coefficient of thermal expansion as the substrate as possible.

Thickness of repair:
The internal stresses with in the repair material and substrate are affected not only by the differential movements, but also by the relative thickness. A thinner repair layer is more easily cracked or debonded by the higher tensile stress which occurs in the repair material. For most available repair materials, there seemed to be an optimum thickness of repair material which results in the lowest tensile stress occurring within the repair material for a given amount of differential movement between the repair material and substrate. This optimum value is affected by the ratio of the modulus of elasticity of the repair material to substrate and thickness of repair material to substrate.

Shrinkage and creep of repair materials:
Differential shrinkage of the repair material and substrate is another important consideration for a durable repair. The most common damages in concrete between substrate and repair materials and it is proportional to the differential shrinkage. Therefore, the repair material selected should have free shrinkage properties that are as low as possible. For the repair material under tension, creep may mitigate against the tensile stresses caused by differential shrinkage. However, for the repair material under compression, creep may decrease the compressive stress within the repair material and aggravate the compressive stress in the substrate caused by the differential shrinkage. The creep of the repair material should be controlled based on the state of stress that the repair material will be subjected to in service.

10.2.2 Durable Repair Application

Preparation of the repaired surface:
The most important surface characteristics of the receiving substrate are its roughness, soundness, cleanliness and moisture condition prior to application of the repair materials. The first step in a repair to be carried out is the removal of the damaged concrete. It is very important to select a, method most appropriate for the specific in-situ condition. Any method

that weakness the sound concrete and creates micro cracking should be avoided. Otherwise, the durability and bond will be decreased by these defects. Commonly used methods in – situ include: sand blasting, chipping with jack hammers, and hydro demolition among which, the last is highly recommended. A sound surface of adequate roughness can be created by this method.

Higher plastic shrinkage of the repair material near the interface should be avoided. This requires that the substrate be pre – wetted for atleast 7 hours prior to the application of the repair material in order to decrease the absorption by and expansion of substrate caused by the up take of moisture from the repair material. Otherwise, the higher uptake of moisture by the substrate after the repair material is cast may lead to higher plastic shrinkage of the repair material near the interface and higher expansion of the substrate, and their by resulting in the possible debonding of the repair material at an early age.

Repair method:

The application method and surface preparation are equally important considerations with regards to the performance of the repaired structures. The repair method adopted not only affects the resultant quality of the repair material, but also the quality of the interfacial transition zone. Shotcrete seems to be an ideal method because it has good compatibility with the substrate concrete. Further more, good compaction with a relatively lower water/cement ratio of the repair material can be achieved using the shotcreting process. This ensures good/high mechanical properties of the component parts and durability of the repair structure.

Bonding agents:

Use of polymer bonding agents is not recommended as their modulus of elasticity is substantially different from that of the substrate. However, use of a cementitious bonding agent with a low water/cement ratio may be considered. This type of bonding agents not only has good compatibility with the substrate and repair material, but can also alleviate the effects of differential shrinkage and thermal movement between the repair material and concrete substrate thus enhancing the bond strength and durability.

Curing of repair materials:

Excessive loss of water may result in higher shrinkage (plastic and drying) and cause debonding failure of the repair material at an earlier age. Therefore, specification of proper curing after completion of the repair is very important. Curing time should be atleast the same as that adopted for usual concrete practice or in accordance to manufacturers recommendation if a commercially available material is used due to the restraint afforded by the substrate.

Workmanship:

It is customary to specify that the specialized contractor should provide satisfactory proof of experience in concrete repair works. Carelessness in any of the above aspects may affect the performance of the repaired structures and lead to premature failure of the repair. Experienced contractors may be more equipped to deal with the unexpected problem occurring during execution of the repair, thus ensuring the quality of the repair work carried out.

10.2.3 Evaluation of the Repairs

Behaviour of the interfacial transition phase:

The formation of the interfacial transition phase is affected by many factors. Defects such as micro-cracks and pores may be formed within this phase caused by the differential movements between the substrate and repair material and a lack of aggregate interlock action between the two materials. Its mechanical behaviour and durability affects directly the performance of the repaired members in service. Thus the mechanical properties and durability of this phase should be evaluated after the completion of the repair work.

Behaviour of the repaired structure:

Differential movements between the repair material and the substrate may result in the cracking of the repair material and thereby decrease the flexural stiffness and durability of the repaired members. The degradation in flexural stiffness of the repaired beams under static and cyclic loading was related to the appearance and development of cracks. However, the presence of steel fibres within the repair material may improve resistance against cracking and fatigue resistance. Therefore, the flexural stiffness can be enhanced and the deflection of the repaired members reduced. At the same time, the fatigue resistance of the interface between the repair material and substrate may also be evaluated by cyclic loading test of representative samples. The test results can also form a database for the formulation of guidelines for use in practice.

10.3 TYPES AND CLASSIFICATION OF REPAIR

Types of repair:

(a) Cosmetic treatments on surfaces
(b) Partial replacement of surface and subsurface material
(c) Addition of reinforcements and bonding materials to strengthen the element
(d) Total replacement of the structural element

The classification of repair and its requirements are given in table 10.1.

Table 10.1. Classification of repair and requirements

Class of damage	*Classification of repair*	*Repair requirements*
1.	Superficial	Cement mortar or cement slurry bonding by trowelling.
2.	General	Non structural or minor structural; restoring cover to rebars; use polymer modified cement mortar or slurry and light fabric reinforcement; or use epoxy mortar over epoxy primer coat.

Class of damage	*Classification of repair*	*Repair requirements*
3.	Principal	Significant loss of concrete strength; shotcreting for slabs and beams, jacketing for columns, using epoxy formulations as bonding materials; post repair strength assessment required.
4.	Major	Demolition and recasting required.

Bonding cracks of width less than 0.075 mm with epoxy compounds by pressure injection make a grove 12 mm by 12 mm by chiseling; wash crack with water under pressure to remove any debris and drying by hot blown air; drilling 30 mm to 40 mm deep, 15 mm diameter holes at 100 mm spacing along the length of the crack; clean holes by compressed air, inserting 12 mm diameter aluminium nipples to the holes, sealing the crack lengthwise with epoxy putty and pressure grouting the holes using epoxy grout mix until the crack fully filled; after curing projecting part of nipples are cut and surface finished smooth.

10.4 METHODS OF REPAIR

In dealing with cracks, the following considerations are to be taken care of and observed:

(a) Determination of extent, location and width of cracks
(b) Classification of cracks as structural and non-structural

Dormant cracks – are caused by some event in the part (eg. drying shrinkage), which is not expected to recur. They remain constant in width, and may be repaired by filling then with a rigid material.

Active cracks – do not remain constant in width, but open and close as the structure in loaded, or due to thermal and hydral changes in the concrete.

Growing cracks – increase in width become the original reason for their occurrence persists, *e.g.* Continuing foundation settlement or reinforcement corrosion.

The repair of cracks can be achieved with the following techniques:

(a) Resin injection
(b) Routing and Sealing
(c) Stitching
(d) External stressing
(e) Bonding
(f) Blanketing
(g) Overlays
(h) Dry pack
(i) Vacuum impregnation
(j) Polymer impregnation
(k) Autogenous healing

(l) Flexible sealing
(m) Drilling and Plugging
(n) Bandaging
(o) Coating
(p) Grinding
(q) Sand blasting
(r) Resurfacing
(s) Acid etching
(t) Caulking

10.4.1 Resin injection

Epoxy resins are usually selected for crack injection because of their high mechanical strength and resistance to most chemical environments encountered by concrete. However, epoxies are relatively rigid and not suitable for active cracks. The method is used to restore structural soundness of members where cracks are dormant or can be prevented from further movements.

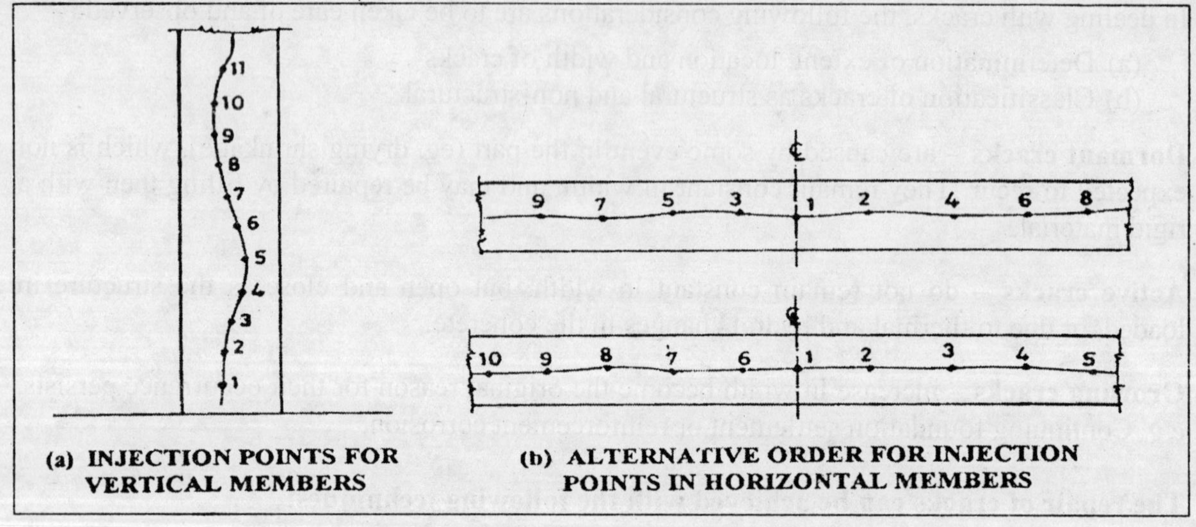

Figure 10.1. Diagram of Crack Injection

The polymer/resin injection under pressure will ensure that the sealing material or sealant penetrates to the full depth of the crack. The injection holes are drilled at close intervals along the length of the crack and the epoxy is injected under pressure in each hole in turn till the injection material starts to flow out of the next hole. The hole which is injected is then sealed off and the next hole is treated. Before injecting the epoxy, it should be ensured that the crack at surface is sealed between the holes with rapid curing resin. Usually pressure injection is resorted to in sealing the cracks. Figures 10.1, 10.2 and 10.3 shows schematically crack injection procedures generally adopted for concrete structures.

CRACK IN CONCRETE OPEN AND ACCESSIBLE
PIPE
TUBE
EPOXY MORTAR
IF CONTACT SURFACES DAMP, USE WATER RESISTANT EPOXY GEL TO SEAL CONCRETE SURFACE
RESIN SPILLOVER
T-INJECTION PORTS PLACED DIRECTLY ON CLEAN CONCRETE SURFACE AND GROUTED IN PLACE USING EPOXY MORTAR
INJECT EPOXY UNTIL GROUT FLOWS OUT VENT IMMEDIATELY ADJACENT TO INJECTION INJET
BINDING WIRE (OR OTHER MEANS) FOR SEALING
TO INJECTION EQUIPMENT

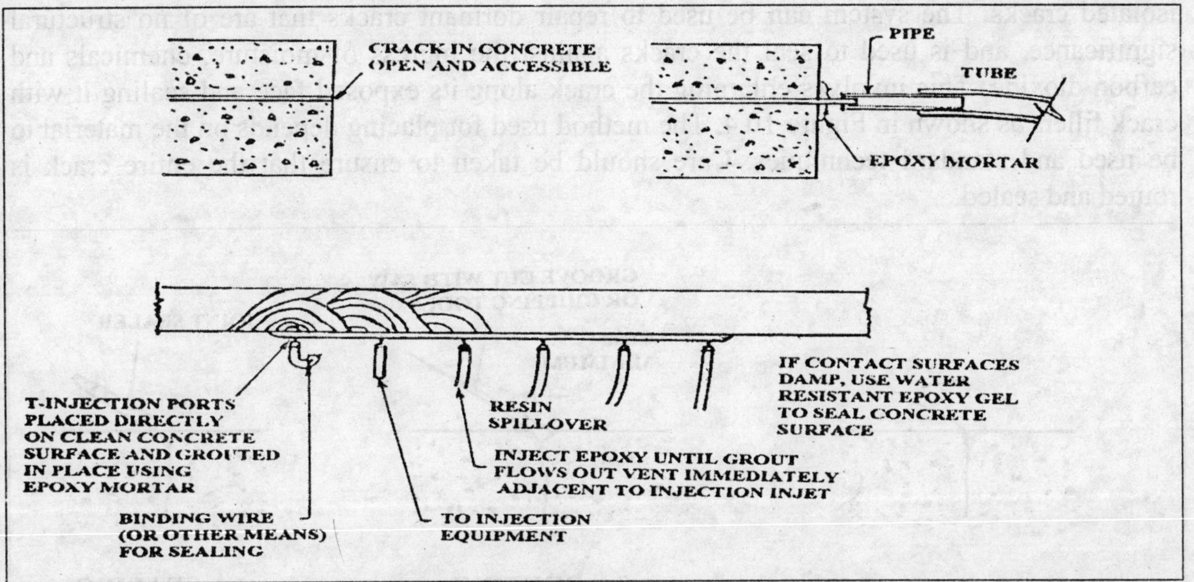

Figure 10.2. Use of injection ports in drilled holes

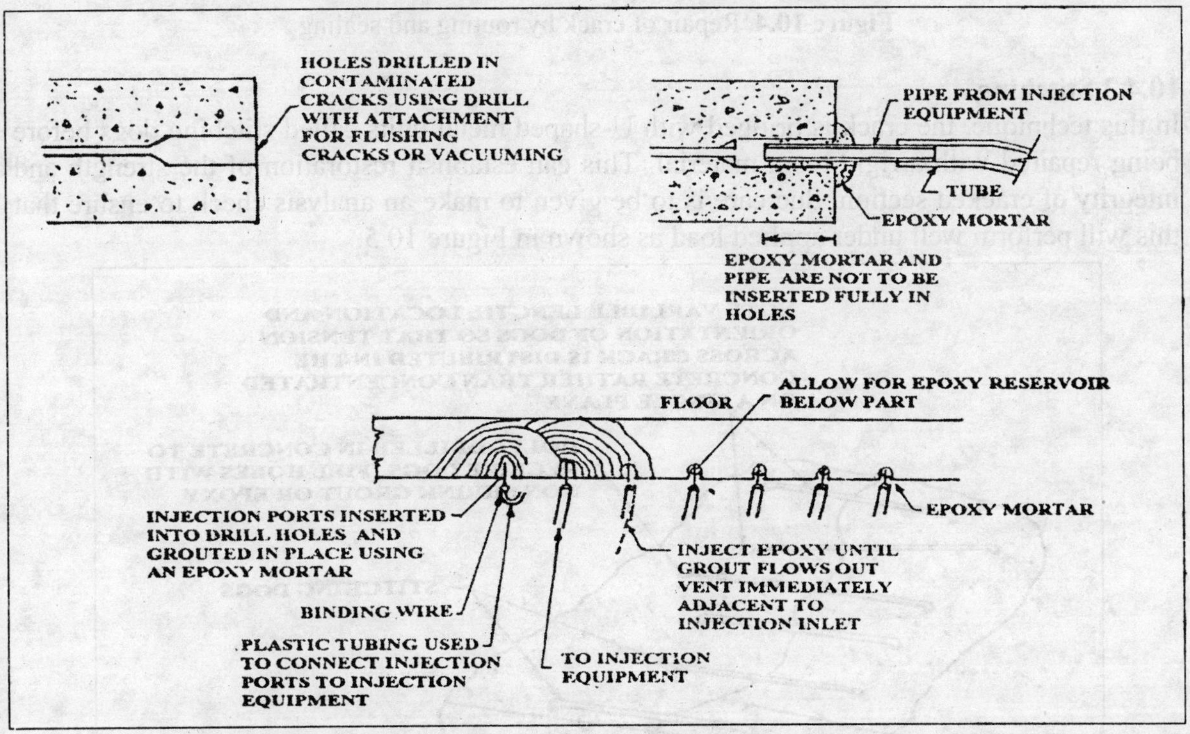

HOLES DRILLED IN CONTAMINATED CRACKS USING DRILL WITH ATTACHMENT FOR FLUSHING CRACKS OR VACUUMING
PIPE FROM INJECTION EQUIPMENT
TUBE
EPOXY MORTAR
EPOXY MORTAR AND PIPE ARE NOT TO BE INSERTED FULLY IN HOLES
FLOOR
ALLOW FOR EPOXY RESERVOIR BELOW PART
EPOXY MORTAR
INJECTION PORTS INSERTED INTO DRILL HOLES AND GROUTED IN PLACE USING AN EPOXY MORTAR
INJECT EPOXY UNTIL GROUT FLOWS OUT VENT IMMEDIATELY ADJACENT TO INJECTION INLET
BINDING WIRE
PLASTIC TUBING USED TO CONNECT INJECTION PORTS TO INJECTION EQUIPMENT
TO INJECTION EQUIPMENT

Figure 10.3. Use of T-injection ports placed flush on concrete surface

10.4.2 Routing and sealing

This is the simplest and most common method of crack repair. It can be executed with relatively unskilled personnel and can be used to seal both fine pattern cracks and larger isolated cracks. The system can be used to repair dormant cracks that are of no structural significance, and is used to seal the cracks against the ingress of moisture, chemicals and carbon dioxide. This involves enlarging the crack along its exposed face and sealing it with crack fillers as shown in Figure 10.4. The method used for placing depends on the material to be used and standard techniques. Care should be taken to ensure that the entire crack is routed and sealed.

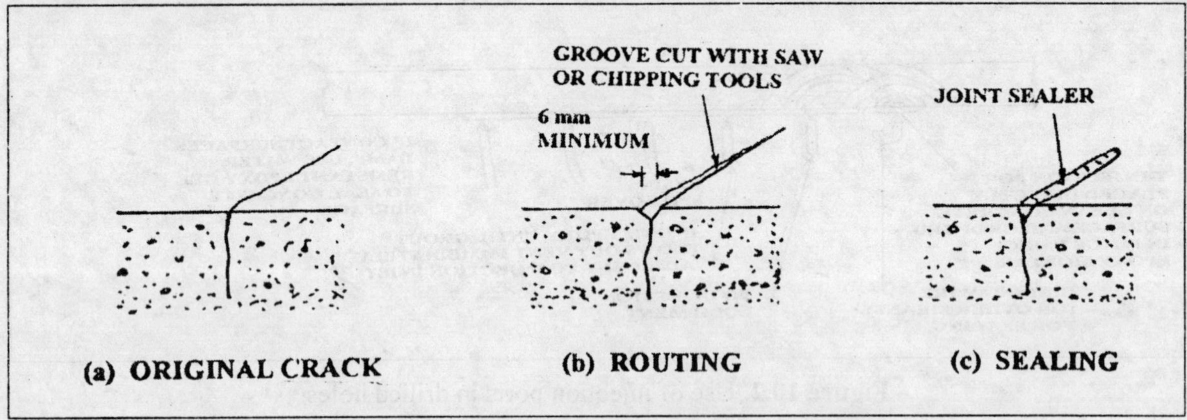

Figure 10.4. Repair of crack by routing and sealing

10.4.3 Stitching

In this technique, the crack is bridged with U-shaped metal units called stitching dogs before being repaired with a rigid resin material. This can establish restoration of the strength and integrity of cracked section; due care is to be given to make an analysis check to ensure that this will perform well under applied load as shown in Figure 10.5.

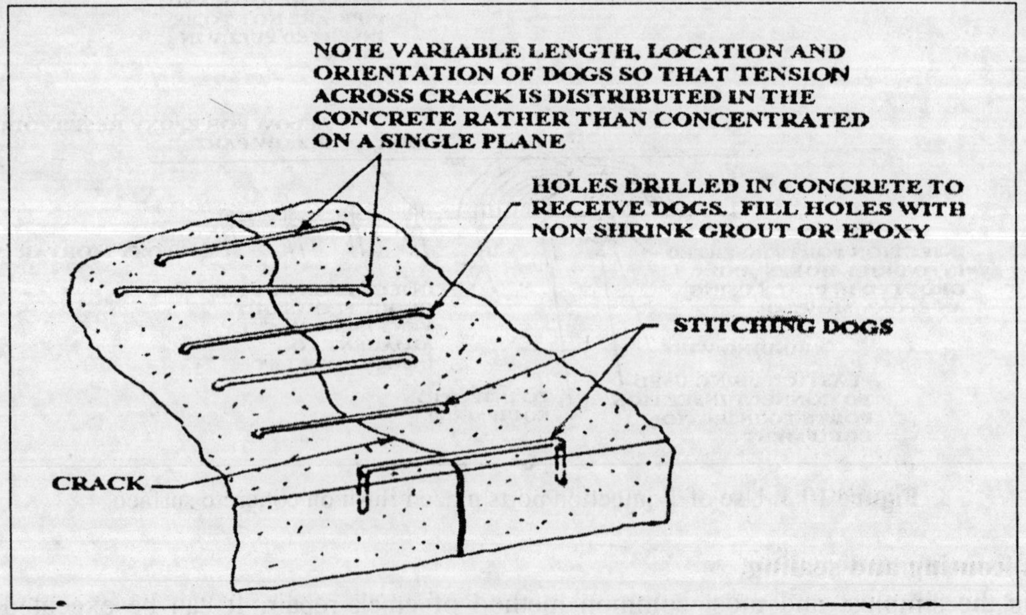

Figure 10.5. Repair of crack by stitching

A non-shrink grout or an epoxy resin based adhesive should be used to anchor the legs of the dogs. Stitching is suitable when tensile strength must be reestablished across major cracks, although stitching will not close the crack, and it is a way of stopping the movement of an active crack and thereby preventing it from spreading. Stitching dogs should be of variable length and/or orientation and so located that the tension transmitted across the crack is not applied to a single plane within the section but is spread over an area. Since stitching may accentuate restraints causing cracking, strengthening of adjacent areas of the structure to take the additional stress may be required.

10.4.4 External stressing

Development of cracking in concrete is due to tensile stress and can be arrested by removing these stresses. Cracks can be closed by external stressing by inducting compression force sufficient to overcome the tension which has caused the cracking; the compression force can be applied by use of the usual prestressing wires or rods; some form of abutment is needed for anchorage as shown in figure 10.6. The compressive force also be applied by wedging *i.e.* opening the crack and filling it with an expanding mortar, by jacking and grouting or by actually driving wedges.

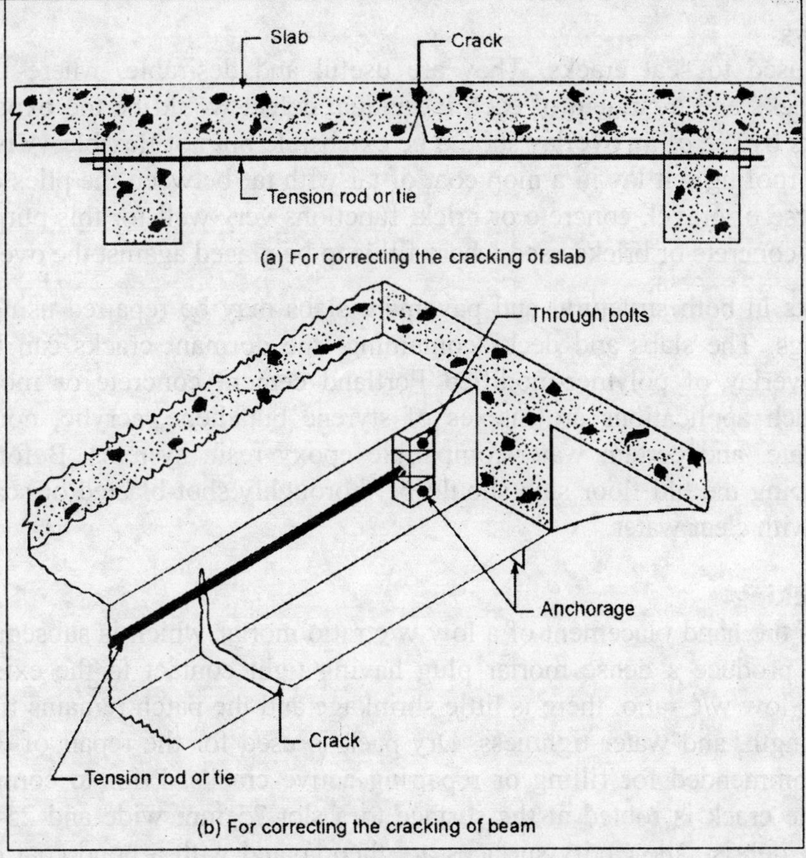

Figure 10.6. Repair of cracks by external stressing

10.4.5 Bonding

Cracks in concrete may be bonded by the injection of epoxy bonding compounds under pressure. A usual practice is to drill into cracks from face of the concrete at several locations. Water or a solvent is injected to flush out the defect. The surface is than allowed to dry. The epoxy is injected into the drilled holes until it flows out through the other holes.

Bonding with epoxies – cracks as narrow as 0.075 mm can be sealed with epoxy compounds, usually pressure injection is resorted to in sealing the cracks.

10.4.6 Blanketing

It is similar to routing and sealing but used on a large scale and applicable for sealing both active and dormant cracks and joints.

Type I – where an elastic sealant is used the sealant material is one which returns to its original shape when the externally induced stress is removed.

Type II – this is similar to the sealant chase of an elastic sealant except that the bond breaker is omitted and the sealant is bonded to the bottom as well as to the sides of the chase. The sealant is mastic rather than a compound with elastic properties. They are used where the anticipated movements are small.

10.4.7 Overlays

Overlays are used to seal cracks. They are useful and desirable, where there are large numbers of cracks and treatment of each individual defect would be too expensive. Sealing of active cracks by use of an overlay should be extensible but not flexible. A two or three ply membranes of roofing felt lay in a mop coat of tar with tar between the plies covered with a protective course of gravel, concrete or brick, functions very well for this purpose. Gravel is used for roofs, concrete or bricks used where fill is to be placed against the overlay.

Dormant cracks in both structural and pavement slabs may be repaired using bonded over lays or toppings. The slabs and decks containing fine dormant cracks can be repaired by applying an overlay of polymer-modified Portland cement concrete or mortar. Polymers suitable for such applications are latexes of styrene butadiene, acrylic, non-remulsifiable polyvinyl acetate, and certain water-compatible epoxy resin systems. Before applying an overlay or topping the old floor slab should be thoroughly shot-blasted or scabbled cleaned and saturated with clear water.

10.4.8 Dry packing

Dry packing is the hand placement of a low w/c ratio mortar which is subsequently rammed in to place to produce a dense mortar plug having tight contact to the existing concrete. Because of the low w/c ratio, there is little shrinkage and the patch remains tight, with good durability, strength, and water tightness. Dry pack is used for the repair of dormant cracks but is not recommended for filling or repairing active cracks. Prior to commencing a dry pack repair, the crack is routed at the surface to a slot 25 mm wide and 25 mm deep and cleaned of any debris. The cavity surfaces are then coated with a bond coat consisting of a thin layer of grout or mortar, just before placing the dry pack material. Dry pack mortar must be mixed just before it is to be placed. Curing of the placed material is carried out by applying a curing compound or supporting a strip of wet burlap along the length of the crack.

Relatively narrow cracks and deep holes can be rectified by dry packing; the crack is slightly undercut, the surface prepared, and mortar with just enough water for sticking is applied. Dry pack is a mortar mixture with a very low water-cement ratio, applicable for small area of repair. It is normally placed by hand. Materials commonly used dry packs are Portland cement, sand, and water.

10.4.9 Polymer impregnation

This technique consists of flooding the cleaned, dried cracked concrete surface with a monomer which is then polymerized in place, thus filling and structurally repairing the crack. A monomer system is a liquid that consists of small organic molecules capable of combining to form a solid plastic. The monomer systems used for impregnation contain a catalyst or an initiator and the basic monomer. They also contain a cross-linking agent. When heated the monomers join together or polymerize becoming a tough, strong and durable plastic that greatly enhances a number of concrete properties. For repairing fractured elements, polymer impregnation method may be used. In this case the fracture is dried first, and then it is temporarily encased in a water tight sheet metal. Now the fracture is soaked with monomer and polymerized. On polymerization of the monomer the fractured element becomes a solid mass with enhanced strength and properties.

10.4.10 Vacuum impregnation

For treating the concrete surfaces that contain a large number of cracks, vacuum impregnation may be used. The part of the structure to be repaired is enclosed with in an air tight plastic cover and then the aim from all cracks with in the cover is sacked by applying vacuum. After exhausting the air from all cracks, the monomer of resin grout is forced under one atmosphere pressure in cracks and pores of the concrete surface. On completion of impregnation, the cover is removed before the impregnate hardens. The process is extensively used to reduce the permeability of weak concrete or masonry. This process may also be used to improve the abrasive resistance of industrial concrete floor slabs.

10.4.11 Autogenous Healing

The natural process of crack repair known as autogenous healing has a practical application in closing dormant cracks in a moist environment. Healing occurs through the carbonation, calcium hydroxide in cement paste by carbon dioxide, which is present in surrounding air and water. Calcium carbonate and calcium hydroxide crystals precipitate, accumulate, and grow with in the cracks. The crystals interlace and twine, producing a mechanical bonding effect. As a result, some of the tensile strength of the concrete is restored across the cracked section, and the crack may become sealed.

10.4.12 Flexible Sealing

It is often appropriate to convert active cracks into movement joints. A recess is cut along the line of the crack and filled with a suitable flexible material. The width of the recess and strain capacity of the sealing material, determine the movement the joint will accommodate. Therefore, the sealant bond should have a suitable width and shape factor for the expected movement. A bond breaker should be used to prevent the sealant from bonding to the bottom of the recess. This allows the sealant to change shape without a concentration of stress at the bottom of the recess. The figure 10.7 shows the repair technique of an active crack by flexible sealing.

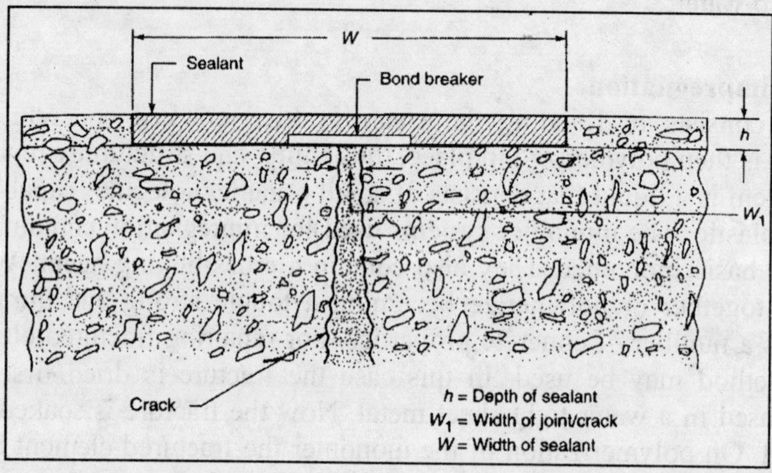

Figure 10.7. Repair of an active crack by flexible sealing

10.4.13 Drilling and Plugging

Is particularly useful for vertical cracks in retaining walls, and consists of drilling down the length of the cracks and grouting it to form a key. The method is only applicable for the cracks running in reasonably straight lines and accessible at one end. It consists in drilling a hole down the length of crack and grouting it to form a key. A hole 50 to 60 mm in diameter should be drilled, centered on and following the crack as shown in Figure 10.8.

The hole must be large enough to intersect the crack along its full length and provide sufficient repair material to structurally take the loads exerted on the key. The drilled hole is then cleaned and filled with grout. For problems concerning water tightness, the drilled hole should be filled with a resilient material of low modulus in lieu of grout. In case where both load transfer and water tightness is desired, two holes are drilled, one filled with resilient material and the second being grouted.

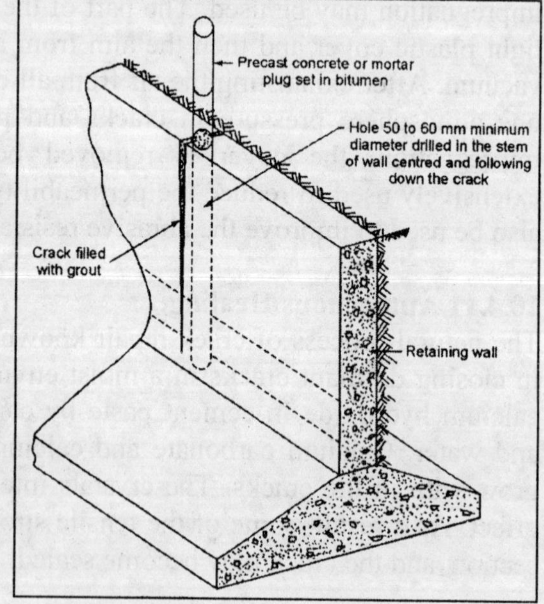

Figure 10.8. Repair of crack by drilling and plugging

10.4.14 Bandaging

A flexible strip is fixed over the crack with only the edges of the strip bonded. Where movement is not all in one plane, where there is excessive movement beyond that which can be accommodated by a recess of convenient size, or if there are factors which prohibit the cutting of a recess, a surface bandage can be used. In areas which are subject to traffic, the flexible bondage will be coated over with a wearing course.

10.4.15 Surface coatings

A wide range of surface penetrating sealers and coatings ranges from thin, purely cosmetic treatments to thick membranes can be applied to cracked concrete. If cracking has reached a stable condition, then a coating can usually be applied successfully. Low viscosity, low solids resin solutions such as epoxies, urethanes, and acrylics have been used to seal the surface of concrete in areas that are not subjected to wear. The method is effective in sealing dormant, random, multiple cracks. Bridge and parking decks and interior slabs may be effectively coated using a heavy coat of epoxy, urethane or solvent based acrylic resin. The uncured coating is usually seeded with a hard wearing aggregate. **The epoxy-based compounds are useful under three conditions:**

(a) Where an adhesive is needed to bond plastic concrete to hardened concrete or to bond rigid materials to each other
(b) For patching
(c) When a coating over concrete.

A layer of material applied to a surface, which forms a continuous membrane is termed coating. Application of suitable coatings afford the necessary protection while still allowing free passage of water vapour, is required for the protection of the structure. The coatings generally adhere to the concrete and form films after application.

Anti-carbonation coatings:

The coatings applied to concrete surface to arrest the carbonation process are known as anti carbonation coatings. There are normally based on chlorinated rubber, polyurethane resin or acrylic emulsions. The cracks which are very fine and not considered to have structural significance may be protected by applying the coating locally over cracks. These are termed conventional coatings. The coatings should have flexibility as well as ability to bridge cracks. High-build polyurethane and epoxy-polyurethane formulations have been successful.

10.4.16 Grinding

Grinding is a slow and expansive repair method, but often is a most logical solution to certain problems with concrete. Among the most common applications of grinding are the leveling of slab surface that have irregularities either from casting or wear or the removal of unsound concrete from the surface in preparation for some other treatment. Grinding is also used to remove concrete that is stained, pitted or marred with shallow cracks. Grinding is also used in repairing concrete to get an even surface.

Surface grinding:

Surface grinding can be used to repair some bulges, offsets, and other irregularities that exceed the desired surface tolerances. Excessive surface grinding, however, may result in weakening of the concrete surface, exposure of easily removed aggregate particles, or unsightly appearance. For these reasons, grinding should be performed with certain limitations. Grinding of surfaces exposed to public view should be limited in depth so that no aggregate particles more than ¼-inch in cross section are exposed at the finished surface. In no event should surface grinding result in exposure of aggregate of more than one-half the diameter of the maximum-size aggregate. Where surface grinding has caused or will cause exposure of aggregate particles greater than the fore-mentioned limits, the concrete must then be repaired by excavating and replacing the concrete.

10.4.17 Sand Blasting

Sand blasting may be a satisfactory method by itself to repair some type of damage, but it is usually used in conjunction with other techniques. Sand blasting is frequently used to remove foreign materials that could impair the bond of the patching materials to the concrete. In repairing for architectural treatment, sand blasting is quite useful.

10.4.18 Resurfacing

Protective coatings can greatly reduce the effect of such adverse conditions, and significantly improve the durability characteristics of the concrete. A variety of coatings are available, and some are tailored for greater chemical resistance, which others are formulated to resist wear and erosion. Resurfacing of a concrete surface requires thorough cleaning or removal of the existing coating. If the old coating is in good condition, then compatibility between the old and the new coating should be determined. **General resurfacing materials are:**

(a) Bituminous coatings and mastics
(b) Polyesters and vinyl esters
(c) Urethanes
(d) Epoxies
(e) Neoprene
(f) Coal tar epoxy
(g) Acrylics

Intumescent coatings:

These are specially formulated fore resistant paints applied on structural elements for the purposes of increasing their fire rating. These products are finding increased use in the repair of fire damaged buildings where they are used for fire protection of steel work.

10.4.19 Acid etching

Acid etching is a particularly efficient method of surface preparation, resulting is a sharp even finish ideal for paint application. Although sulphanic acid is recommended where stainless steel is cast into the concrete, a 10% solution of hydrochloric acid is more generally used. The acid reacts readily with the alkaline constituents of the concrete. This exposes particles of fine aggregate with in the mortar layer which form a first-class key. The acid should be sluiced off with fresh water as soon as the reaction finishes, preventing the formation of salts which might subsequently cause blistering of the paint film.

Several applications are normally necessary to achieve a suitable finish, particularly if the surface laitance is relatively intact. Where large areas are involved, there may be problems with disposal of the spent acid solution, which is a potentially harmful substance. Care should also be taken when handling the acid, because fumes are given off, and splashes will cause burns. Appropriate protection and proper ventilation is essential when using acid etch technique. Etching is especially suitable for areas demanding clean conditions because no dust is produced. Unfortunately, its most effective use is restricted to floors, owing to practical difficulties with other surfaces. Apart from this and the precautions outlined above, it is an excellent and cost effective method which sets a standard for the surface preparation of concrete.

10.4.20 Caulking

Caulking compounds are flexibilized polymeric materials which are used for two purposes:

(a) To plug irregular gaps between two rigid surfaces and
(b) To provide a dynamic bridge across the gap between two surfaces that more relative to one another.
(c) In this capacity, they perform three functions:
(d) Accommodate joint movement
(e) Prevent the infiltration of the joint by solid matter such as dust and sand, and
(f) Prevent the ingress of water into the structure.

The generic types of sealants used in industrial and domestic buildings and other engineering structures include acrylic latexes, butyls, solvent release acrylics, poly sulfides, urethanes, silicones and some flexibilized epoxies.

CHAPTER 11

Rehabilitation Techniques

11.1. INTRODUCTION

The success of repair activity depends on the identification of the root cause of the deterioration of the concrete structures. If this cause is properly identified, satisfactory repairs can be done for the improvement of strength and durability, thus extending the life of the structure, is not difficult to achieve. Repairs of the structures should be carried out as soon as the deterioration of concrete is observed. When it is found by the various tests that concrete is porous/honeycombed, it should be treated with pressure grouting, guniting/ shotcreting is another way to rehabilitate the structure, in which a dense and firmly adhesive coating is applied over the exposed surfaces. **Following steps are, however, generally used in the rehabilitation of distressed concrete structures:**

(a) Support the structural members properly as required.
(b) Remove all cracked, spalled and loose concrete.
(c) Clean the exposed concrete surfaces and steel reinforcement.
(d) Provide additional reinforcing bars, if the loss in reinforcement is more than 10%
(e) Apply shotcreting/polymer concrete for patch repair work and grouting for porous/ honeycombed concrete
(f) Apply protective coatings over the exposed/repaired surface

11.2 REPLACEMENT MORTAR

Portland cement mortar may be used for repairing defects on surfaces not prominently exposed, where the defects are too wide for dry-pack filling or where the defects are too shallow for concrete filling, and no deeper than the far side of the reinforcement that is nearest the surface. Repairs may be made either by use of shotcrete or by hand application methods, although hand application methods are generally recommended for areas subject to public view in historic preservation applications. Replacement mortar can be used to make shallow, small-size repairs to new or green concrete, provided that the repairs are performed within 24 hours of removing the concrete forms. Accomplishing successful mortar repairs to old concrete without the use of a bonding resin is unlikely or extremely difficult. Evaporative loss of water from the surface of the repair mortar, combined with capillary water loss to the old concrete, results in unhydrated or poorly hydrated cement in the mortar.

Additionally, repair mortar bond strength development proceeds at a slower rate than compressive strength development. This causes workers to mistakenly abandon curing procedures prematurely, when the mortar seems strong. Once the mortar dries, bond strength development stops, and bond failure of the mortar patch results. For these reasons, using cement mortar without a resin bond coat to repair old concrete is discouraged. A Portland cement mortar patch is usually darker than the surrounding concrete unless precautions are taken to match colours. A leaner mix will usually produce a lighter colour patch. Also, white cement can be used to produce a patch that will blend with the surrounding concrete. The quantity of white cement used must be determined by trial in an inconspicuous location on the structure.

11.2.1 Epoxy-bonded epoxy mortar

Epoxy-bonded epoxy mortar should be used where the depth of repair is less than 1-1/2 inches and the exposure conditions are such that relatively constant temperatures can be expected. Epoxy mortars have thermal coefficients of expansion that may be significantly different from conventional concrete. If such mortars are used under conditions of wide and frequent temperature fluctuations, they will cause failure just below the bond surface in the base concrete. For this reason, current practice precludes the use of epoxy mortars under conditions of frequent or large temperature fluctuations. The application of epoxy mortar to repair areas of concrete deterioration caused by corroding reinforcing steel is also not recommended. The epoxy bond coat and epoxy mortar create zones of electrical potential that are different from the electrical potential in the surrounding concrete. This difference in potential can result in the formation of a galvanic corrosion cell with accelerated corrosion at the repair perimeters.

The application of epoxy mortar to repair areas of concrete deterioration caused by corroding reinforcing steel is also not recommended. The epoxy bond coat and epoxy mortar create zones of electrical potential in the surrounding concrete. This difference in potential can result in the formation of a galvanic corrosion cell with accelerated corrosion at the repair perimeters. Epoxy mortar is properly used to make thin repairs to concrete under relatively constant temperature exposure conditions. Such applications could include tunnel linings, indoor or interior concrete, or the underside of concrete structures such as bridge decks. Applications to concrete exposed to the daily temperature fluctuations caused by exposure to direct sunlight are not appropriate for epoxy mortar repair.

11.2.2 Preparation and materials

Concrete to be repaired with replacement mortar should first have all the deteriorated or unsound areas removed. After preparation, the areas should be cleaned, roughened if necessary (preferably by wet sandblasting), and surface-dried to a saturated surface condition. The mortar should be applied immediately thereafter. Replacement mortar contains water, Portland cement, and sand. The cement should be the same type as used in the concrete being repaired. The water and sand should be suitable for use in concrete, and the same should pass through a no. 16 sieves. The cement-to-sand ratio should be 1:2 to 1:4, depending on application technique. Only enough water should be added to the cement-sand mixture to permit placing.

11.2.3 Curing

Failure to cure properly is the most common cause of failure of replacement mortar. It is essential that mortar repairs receive a thorough water cure starting immediately after initial set and continuing for 14 days. In no event should the mortar be allowed to become dry during the 14-day period following placement. Following the 14-day water cure and while the mortar is still saturated, the surface of the mortar should be coated with two coats of a wax-base or water-emulsified resin-base curing compound meeting reclamation specifications. If this curing procedure cannot be followed or if conditions at the job are such that this curing procedure will not be followed, money would be saved by using another repair material.

11.2.4 Applications

For minor restorations, satisfactory mortar replacement may be performed by hand. The success of this method depends on complete removal of all defective and affected concrete, good bonding of the mortar to the concrete, elimination of shrinkage of the patch after placement, and thorough curing. Replacement mortar repairs can be made using an epoxy-bonding agent; this technique is highly recommended.

11.3 REPLACEMENT CONCRETE

Concrete repairs made by bonding new concrete to repair areas without use of an epoxy bonding agent or mortar grout applied on the prepared surface should be made when the area exceeds 1 square foot and has a depth greater than 6 inches and when the repair will be of appreciable continuous area. Replacement concrete repairs should also be used for:

(a) Holes extending entirely through concrete sections
(b) Holes in which no reinforcement is encountered,
(c) Holes in reinforced concrete greater than one-half square foot and extending beyond reinforcement

Replacement concrete is the most common concrete repair material and will meet the needs of a majority of all concrete repairs. Replacement concrete repairs are made by bonding new concrete to the repair areas without the use of a bonding agent or portland cement grout. The replacement of concrete using pressurized form is shown in figure 11.1.The combination of a deep repair and good curing practices ensures adequate hydration water will remain at the bonding surface zone for at least 28 days, allowing the cement hydration process to develop good bond. Because the defective concrete is being replaced with high quality concrete very similar to the surrounding concrete, the repair is compatible in thermal expansion and in other physical and chemical properties with the old concrete. For this reason, in many cases, the best repair method is the use of replacement concrete. Only when an unusual increase in durability is needed, or when placing conditions or dimensions dictate otherwise, should other materials be considered.

11.3.1 Epoxy-bonded replacement concrete

Epoxy-bonded concrete is used for repairs to concrete that are between 1.5 and 6 inches thick. Shallow replacement concrete repairs, less than 6 inches thick, are subject to poor curing conditions as a result of moisture loss to evaporation and to capillary absorption by the old base concrete. Such repairs seldom develop acceptable bond strength to the old

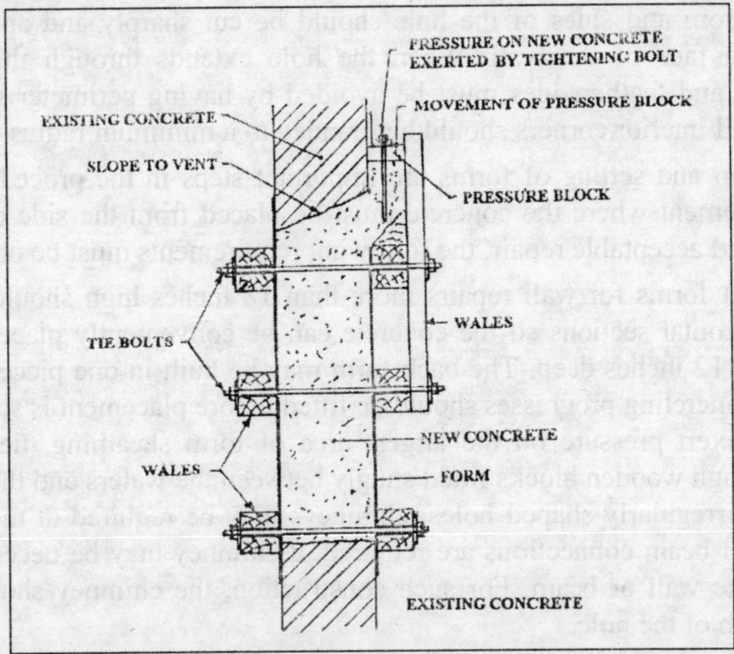

Figure 11.1. Replacing of concrete using pressurized form

concrete. The epoxy bonding resin is used to ensure a strong, durable bond between the old concrete and the replacement concrete. As with epoxy-bonded epoxy mortar, care should be exercised if epoxy-bonded concrete is to be used to repair shallow deterioration resulting from corroding reinforcement. The epoxy bond coat may create electrical potentials sufficiently different from potentials in the surrounding concrete to result in accelerated corrosion at repair perimeters.

11.3.2 Preparation and materials

To obtain satisfactory results with the replacement concrete method, preparation should be as follows:

(a) Reinforcement bars should not be left partially embedded; concrete should be removed to provide a clearance of at least an inch around each bar exposed more than one-third its circumference.

(b) The perimeter of the hole at the face should be saw cut to a minimum depth of 1 inch. If the shape of the defect makes it advisable, the remainder of the concrete removal may be chipped below the vertical saw cut and continued until a horizontal surface is obtained. The top of the hole, if on a vertical wall, should be cut on a 1:3 upward slope from the back toward the face from which the concrete will be placed. This is essential to permit vibration of the concrete without leaving air pockets at the top of the repair. In some instances, where a hole extends through a wall or beam, it may be necessary to fill the hole from both sides; the slope of the top of the cut should be modified accordingly.

(c) The bottom and sides of the hole should be cut sharply and approximately square with the face of the wall. When the hole extends through the concrete section, spalling and featheredges must be avoided by having perimeter saw cuts from both faces. All interior corners should be rounded to a minimum radius of 1 inch.

The construction and setting of forms are important steps in the procedure for satisfactory concrete replacement where the concrete must be placed from the side of the structure. To obtain a tight and acceptable repair, the following requirements must be observed:

(a) Front forms for wall repairs more than 18 inches high should be constructed in horizontal sections so the concrete can be conveniently placed in lifts not more than 12 inches deep. The back form may be built in one piece. Sections to be set as concreting progresses should be fitted before placement is started.

(b) To exert pressure on the largest area of form sheathing, tie bolts should pass through wooden blocks fitted snugly between the walers and the sheathing.

(c) For irregularly shaped holes, chimneys may be required at more than one level; when beam connections are required, a chimney may be necessary on both sides of the wall or beam. For such construction, the chimney should extend the full width of the hole.

(d) Forms should be substantially constructed so that pressure may be applied to the chimney cap at the proper time.

(e) Forms must be mortar tight at all joints between adjacent sections, between the forms and concrete, and at tie bolt holes to prevent the loss of mortar when pressure is applied during the final stages of placement. Twisted or stranded caulking cotton, folded canvas strips, or similar material should be used as the forms are assembled.

Surfaces of old concrete to which new concrete is to be bonded must be clean, rough, and in saturated surface dry condition. Extraneous material on the joint resulting from form construction must be removed prior to placement.

Concrete for repair should have the same water-cement ratio as used for similar new structures but should not exceed 0.47, by weight. Aggregate of as large a maximum size and slump as low as is consistent with proper placing and thorough vibration should be used to minimize water content and consequent shrinkage. The concrete should contain 3 to 5 percent entrained air. Where surface color is important, the cement should be carefully selected or blended with white cement to obtain the desired results. To minimize shrinkage, the concrete should be as cool as practicable when placed, preferably at about 70° F or lower. Materials should, therefore, be kept in shaded areas during warm weather. Use of ice in mixing water may sometimes be necessary. Batching of materials should be by weight; but batch boxes, if of the exact size needed, may be used. Since batches for this class of work will be small, the uniformity of the materials is important and should receive proper attention.

Best repairs are obtained when the lowest practicable slump is used. This is about 3 inches for the first lift in an ordinary large form. Subsequent lifts can be drier, and the top few inches of concrete in the hole and that in the chimney should be placed at almost zero slump. It is usually best to mix enough concrete at the start from the entire hole. Thus, the concrete will be up to 1-1/2 hours old when the successive lifts are placed. Such premixed concrete, provided it can be vibrated satisfactorily; will have less settlement, less shrinkage, and greater strength than freshly mixed concrete. Structural concrete placements should be started with an over sanded mix containing about a ¾-inch maximum-size aggregate; a maximum water-cement ratio of 0.47, by weight; 6 percent total air, by volume of concrete; and a maximum slump of 4 inches. This special mix should be placed several inches deep on the joint at the bottom of the placement. A mortar layer should not be used on the construction joints.

11.3.3 Curing

The importance of curing replacement concrete repairs cannot be overemphasized. Complete failure of repairs has been attributed to inadequate or improper curing. Because of the relatively small volume of most repairs and the tendency of old concrete to absorb moisture from new material, water curing is a highly desirable procedure, at least during the first 24 hours. When forms are used for repair, they can be removed and then reset to hold a few layers of wet burlap in contact with new concrete. One of the best methods of water curing is a soil-soaker hose laid beneath a plastic membrane covering the repair area.

When curing compound is used, the best curing combination is an initial water-curing period of 7 days (never less than 24 hours) followed, while the surface is still damp, by a uniform coat of the compound. It is always essential that repairs, even dry packed cone boltholes, receive some water curing and be thoroughly damp before the curing compound is applied. If nothing better can be devised for the initial water curing of the dry pack in cone bolt holes and similar repairs, a reliable workman should be detailed to make the rounds with water and a large brush or spraying device to keep the repaired surfaces wet for 24 hours prior to application of a curing compound. White curing compound may be used only where its color does not create objectionable contrast in appearance.

11.3.4 Application

When placing concrete in lifts, placement should not be continuous; a minimum of 30 minutes should elapse between lifts. When chimneys are required at more than one level, the lower chimney should be filled and allowed to remain for 30 minutes between lifts. When chimneys are required on both faces of a wall or beam, concrete should be placed in only one of the chimneys until it flows to the other. Attempted placement in both chimneys will result in air entrapment and/or voids in the structure.

The quality of a repair depends not only on use of low-slump concrete, but also on the thoroughness of the vibration during and after depositing the concrete. There is little danger of over-vibration. Immersion-type vibrators should be used if accessibility permits. If not, this type of vibrator can be used very effectively on the forms from the outside. Form vibrators can be used to good advantage on forms for large inaccessible repairs, especially on a one-piece back form, or attached to a large metal fitting such as hinge-based castings.

Immediately after the hole has been completely filled, pressure should be applied to the fill and form vibrated. This operation should be repeated at 30-minute intervals until the concrete hardens and no longer responds to vibration. Pressure is applied by wedging or by tightening the bolts extending through the pressure cap. In filling the top of the form, concrete to a depth of only 2 or 3 inches should be left in the chimney under the pressure cap. A greater depth tends to dissipate the pressure. After the hole has been filled and pressure cap placed, the concrete should not be vibrated without a simultaneous application of pressure. To do so may produce a film of water at the top of the repair that will prevent bonding.

Addition of aluminum powder to concrete causes the latter to expand. Under favorable conditions, this procedure has been successfully used to secure tight, well-bonded repairs in locations where the replacement material had to be introduced from the side. Time should not be allowed for settlement between lifts. When the top lift and the chimney are filled, no pressure need be applied, but the pressure cap should be secured in position so expanding concrete will be confined to and completely fill the hole undergoing repair. There should be no subsequent revibration.

Concrete replacement in open-top forms, as used for reconstruction of the tops of walls, piers, parapets, and curbs, is a comparatively simple operation. Only such materials as will make concrete of proved durability should be used. The water-cement ratio should not exceed 0.47, by weight. For the best durability, the maximum size of aggregate should be the largest practicable and the percentage of sand the minimum practicable. No special features are required in the forms, but they should be mortar tight when vibrated and should give the new concrete a finish similar to the adjacent areas. The slump should be as low as practicable, and dosage of air entraining agent should be increased as necessary to secure the maximum permissible percentage of entrained air, despite the low slump. Top surfaces should be sloped to provide rapid drainage. Manipulation in finishing should be held to a minimum, and a wood-float finish is preferable to a steel-trowel finish. Edges and corners should be tooled or chamfered. Use of water for finishing is prohibited.

Forms for concrete replacement repairs usually may be removed the day after casting unless form removal would damage the green concrete, in which even stripping should be postponed another day or two. The projections left by the chimneys normally should be removed the second day. If the trimming is done earlier, the concrete tends to break back into the repair. These projections should always be removed by working up from the bottom because working down from the top tends to break concrete out of the repair. The rough area resulting from trimming should be filled and stoned to produce a surface comparable to that of surrounding areas. Plastering of these surfaces should never be permitted.

Some replacement concrete does not require forms. Full-depth replacement of damaged or deteriorated paving or slabs involves procedures no different from those required for best results in original construction. Contact edges at the perimeter should be saw cut clean and square with the surface. If, however, needed repairs are too small for the replacement concrete method (including silica fume concrete), they should be made using the dry-pack procedure, the epoxy-bonded epoxy mortar method, or the epoxy-bonded replacement concrete method.

11.4 PREPLACED AGGREGATE CONCRETE

Preplaced aggregate concrete is an excellent repair material that has not been used much in recent years. Preplaced aggregate concrete is made by injecting portland cement grout, with or without sand, into the voids of a formed, compacted mass of clean, graded, coarse aggregate. The preplaced aggregated is washed and screened to remove fines before placing into the forms. As the grout is injected or pumped into the forms, it displaces any included air or water and fills the voids around the aggregate, thus creating a dense concrete with a high aggregate content. Because the coarse aggregate has point contact prior to grout injection, preplaced aggregate concrete undergoes very little settlement, curing, or drying shrinkage during hydration. Drying shrinkage of preplaced aggregate concrete containing 1-1/2 inch maximum size aggregate is about 200 to 400 millionths, while conventional concrete drying shrinkage containing the same size maximum aggregate is about 400 to 600 millionths.

Another advantage of preplaced aggregate concrete is the ease with which it can be placed in certain situations where placement of conventional concrete would be extremely difficult or impossible. Preplaced aggregate concrete would be extremely difficult or impossible. Preplaced aggregate concrete is especially useful in underwater repair construction. It has been used in a variety of large concrete and masonry repairs. The preplaced aggregate concrete repair the retaining wall is shown in figure 11.2. Although preplaced aggregated is

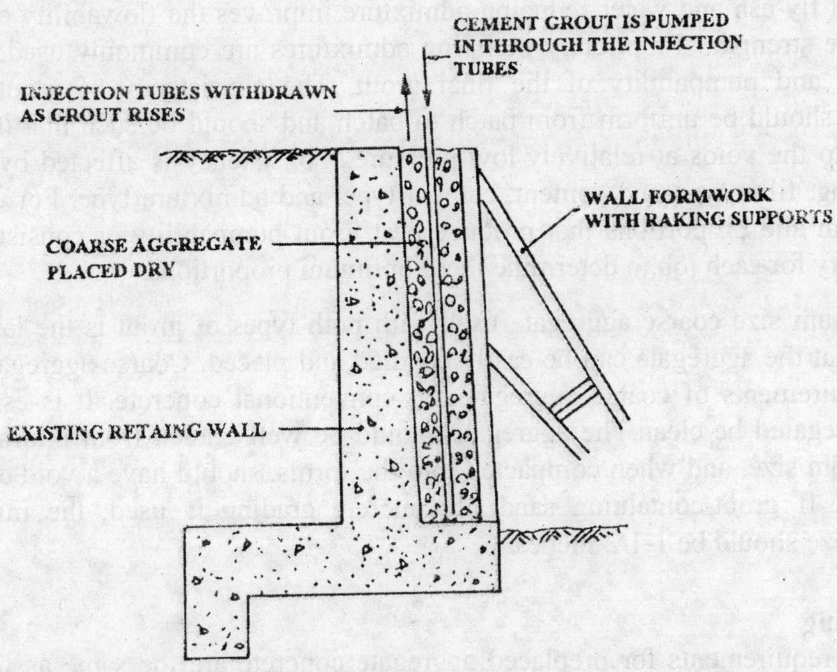

Figure 11.2. Preplaced aggregate concrete repair to retaining wall

adaptable to many special repair applications, it is essential that the work be undertaken by well-qualified personnel who are willing to follow exactly the construction procedures required for this repair material. Form work for preplaced aggregate concrete requires special attention to prevent grout loss. The construction of forms should be with workmanship better than the normally encountered with conventional concrete. Leaking forms can cause significant problems and should, by careful construction, be avoided whenever possible. The injected grout is more flowable than plastic concrete and takes slightly longer to set. Forms, therefore, must be constructed to take more lateral pressure than would be necessary with conventional concrete. Form bolts should fit tightly through the sheathing, and all possible points of grout leakage should be caulked.

11.4.1 Preparation and materials

The preparation of concrete to be repaired by preplaced aggregate concrete is identical to the preparation required for replacement concrete if the development of bond is required. Grout for preplaced aggregate concrete may be mixed with sand either of the gradation specified for conventional concrete or with fine sand, pozzolanic or fly ash fillers, water reducing admixtures, and pumping admixtures as dictate by the minimum size of the coarse aggregate. With 1-1/2 inch minimum size coarse aggregate, the sand gradation is that specified for conventional concrete. The portland cement, water, and sand are mixed using high-speed centrifugal grout mixers that produce well-mixed grouts of a creamy consistency. Best pumping characteristics will be obtained with fineness modulus between 1.2 and 2 and with the rounded shape of natural sands as opposed to crushed sands.

Addition of fly ash and water reducing admixture improves the flowability of the grout and the ultimate strength. Proprietary pumping admixtures are commonly used to increase the penetration and pumpability of the final grout. The consistency of grout for preplaced aggregated should be uniform from batch to batch and should be such that it can be readily pumped into the voids at relatively low pressure. Consistency is affected by water content, sand grading, filler type and content, cement type, and admixture type. For each mix, there are optimum and proportions that produce best grout pumpability or consistency, and tests are necessary for each job to determine these optimum proportions.

The maximum size coarse aggregate used with both types of grout is the largest available, provided that the aggregate can be easily handled and placed. Coarse aggregate should meet all the requirements of coarse aggregate for conventional concrete. It is essential that the coarse aggregated be clean.The aggregate should be well graded from minimum size up to the maximum size, and when compacted into the forms, should have a void content of 35 to 40 percent. If grout-containing sand of concrete grading is used, the minimum coarse aggregate size should be 1-1/2 inches.

11.4.2 Curing

The curing requirements for preplaced aggregate concrete are the same as for replacement concrete.

11.4.3 Application

The grout piping system used with preplaced aggregate concrete must be designed to serve at least 3 purposes: to deliver and inject grout; to provide means for determining grout level in the forms; and to serve as vents in enclosed forms for escape or air and water. Proper design and location of the grout piping system is essential for successful placement.

The grout delivery pipeline should be a recirculating system. That is, the grout delivery pipeline should extend from the grout agitator or holding tank to the grout pump, then to the injection manifold, and return to the grout agitator tank. With this type of pipeline, the grout can be kept moving and circulating in the delivery pipeline even when no grout is being injected into the aggregate. Such a system prevents stoppages and clogging of the delivery line. The delivery line should be kept as short as practicable, and the pipe size should be such that normal grout flow velocities range between 2 and 4 feet per second. All valves used in the grout piping system should be quick opening ball valves, which can be readily cleaned.

Grout injection should begin at the lowest point of the form and continue uniformly until the entire form is filled. After sufficient grout has been pumped to raise the level of grout in the form about 18 inches above the bottom outlet of the injection line, the injection line can be progressively raised, maintaining about 12 inches of embedment below the level of the grout at all times. A great deal of thought and planning is required if multiple injection lines are used. The objective is to entirely fill the form without trapping air or water. Vents must be located where needed and the injection sequence designed to promote complete filling. It is not possible to use internal vibrators to consolidate preplaced aggregate concrete. External vibrators, however, can be attached to the forms and used advantageously. External vibration will eliminate the splotchy appearance that can occur where coarse aggregate particles contact the forms. Underwater applications of preplaced aggregate concrete require additional considerations. During injection, grout pumping must continue until an undiluted flow of grout emerges from the top of the form. Formwork is usually closed at the top to prevent washout or dilution of the grout after placement if flowing water is encountered. Anti-washout admixtures might prove useful for underwater applications of preplaced aggregate concrete. Care must be taken, however, when using several different types of admixtures (*e.g.*, anti-washout, pumping aids, or high-range water reducers) that undesirable combinations are avoided. It is known, for example, that some anti-washout admixtures can significantly reduce the pumpability benefits of some high-range water reducers.

11.5 SHOTCRETE/GUNITE

Shotcreting/Guniting is mechanically applied material consisting of cement, aggregates and water. There are two guniting process, the wet mix and dry mix process. The dry mix process is more extensively used for structural repairs. The guniting process applied to damaged concrete structures is shown in figure 11.3. The cement and sand are batched are mixed in the usual way and conveyed through a hosepipe with the help of compression air. A separate pipe line brings water under pressure and the water and cement aggregate mix are passed through and intimately mixed in a special manifold and then projected at high velocity to the surface being repaired. The force of impact compacts the materials. In good quality work, a density of around 2100 kg/m^3 is achieved. For effective guniting, the nozzle should be kept 60 cm to 150 cm from the work preferable normal to the surface. Before gunite is applied, the old concrete surface is prepared properly, all the cracks treated and the new reinforcement fixed in position. Cracks wider than about 0.5 mm should be cut out and filled with hand-applied mortar or with gunite.

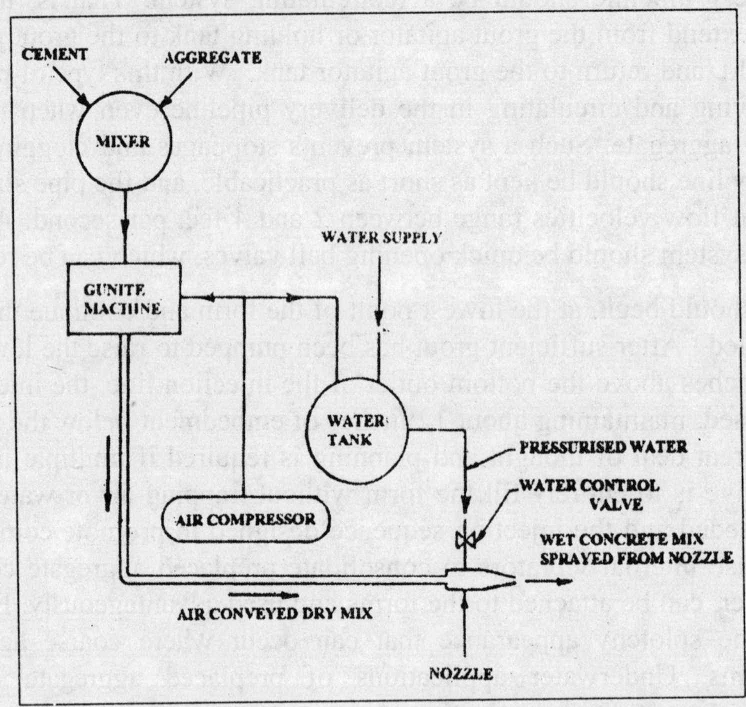

Figure 11.3. Diagrammatic representation of gunite process

Shotcrete, because it irreversibly alters the contours and surface profile of existing concrete, is generally a poor choice in historic preservation projects. The following information is included for completeness in the coverage of concrete repair methods. Shotcrete is defined as "mortar or concrete pneumatically projected at high speed onto a surface." There are two basic types of shotcrete: dry mix and wet mix. In dry-mix shotcrete, the dry cement, sand, and coarse aggregate, if used, are premixed with only sufficient water to reduce dusting. This mixture is then forced through the delivery line to the nozzle by compressed air. At the nozzle, sufficient water is added to the moving stream to meet the requirements of cement hydration. For wet-mix shotcrete, the cement, sand, and coarse aggregate are first conventionally mixed with water, and the resulting concrete is then pumped to the nozzle where compressed air propels the wet mixture onto the desired surface. The two types of shotcrete produce mixes with different water contents and different application characteristics as a result of the distinctly different mixing processes. Dry-mix shotcrete suffers high dust generation and rebound losses varying from about 15 percent up to 50 percent. Wet-mix shotcrete must contain enough water to permit pumping through the delivery line. Wet-mix shotcrete must contain enough water to permit pumping through the delivery line. Wet-mix shotcrete, as a result, may experience significantly more cracking problems due to the excess water and drying shrinkage. Advances in the development of the

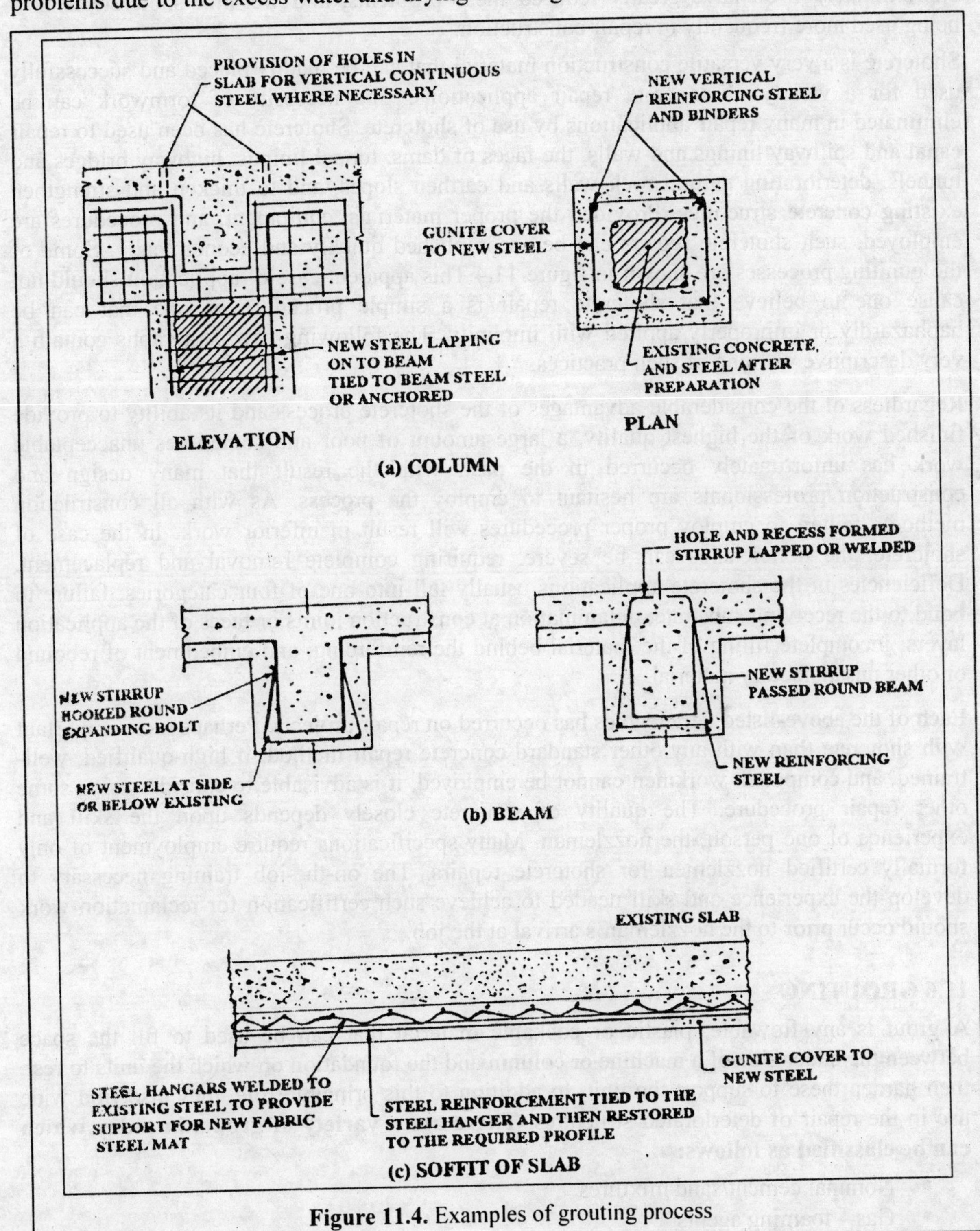

Figure 11.4. Examples of grouting process

high-range water reducing admixtures, pumping aids, and concrete pumping equipment since approximately 1960 have greatly reduced these problems, and wet-mix shotcrete is now being used more frequently in repair construction.

Shotcrete is a very versatile construction material that can be readily placed and successfully used for a variety of concrete repair applications. The necessity of formwork can be eliminated in many repair applications by use of shotcrete. Shotcrete has been used to repair canal and spillway linings and walls, the faces of dams, tunnel linings, highway bridges and tunnels, deteriorating natural rock walls and earthen slopes, and to thicken and strengthen existing concrete structures. Provided the proper materials, equipment, and procedures are employed, such shotcrete repairs can be accomplished quickly and economically. Some of the guniting processes are shown in figure 11.4.This apparent ease of application should not cause one to believe that shotcrete repair is a simple procedure or one that can be haphazardly or improperly applied with impunity. The following two paragraphs contain a very descriptive warning of such practices.

Regardless of the considerable advantages of the shotcrete process and its ability to provide finished work of the highest quality, a large amount of poor and sometimes unacceptable work has unfortunately occurred in the past, with the result that many design and construction professionals are hesitant to employ the process. As with all construction methods, failure to employ proper procedures will result in inferior work. In the case of shotcrete, the deficiencies can be severe, requiring complete removal and replacement. Deficiencies in the shotcrete applications usually fall into one of four categories: failure to bond to the receiving substrate; delamination at construction joints or faces of the application layers; incomplete filling of the material behind the reinforcing; and embedment of rebound or other unsatisfactory material.

Each of the above-listed deficiencies has occurred on repair projects. Perhaps more important with shotcrete than with any other standard concrete repair method, if high-qualified, well-trained, and competent workmen cannot be employed, it is advisable to consider using some other repair procedure. The quality of shotcrete closely depends upon the skill and experience of one person, the nozzleman. Many specifications require employment of only formally certified nozzlemen for shotcrete repairs. The on-the-job training necessary to develop the experience and skill needed to achieve such certification for reclamation work should occur prior to the nozzleman's arrival at the job.

11.6 GROUTING

A grout is any flowable, plastic or packable material that can be used to fill the space between the under side of a machine or column and the foundation on which the units to rest, then harden these to support the unit. In addition to this principle role, they also find wide use in the repair of deteriorated structures. **There are a variety of grouts in today which can be classified as follows:**

- Nominal cement/sand mixtures
- Gas – foaming agents

- Metallic aggregate grouts
- Inorganic sulfoaluminate cement or expansive additive – based grouts
- Fiber – reinforced grouts
- Organic resin grouts

11.6.1 Portland Cement Grouts:

Cracks in blocks of dams, thick concrete walls, or rock foundations for hydraulic structures may sometimes be sealed by pumping Portland cement grout into them. Grout nipples and vent pipes are installed in holes drilled into the crack and the periphery sealed by caulking or fast setting materials to prevent loss of grout. Thorough washing with water, or other cleaning agents, prior to grouting assists in obtaining a better filling of the crack.

Grout mixtures may contain cement and water or cement plus fine sand and water depending on the width of the cracks. Water reducers are used to keep the w/c ratio down so that shrinkage is reduced and strength increased. A pump is used in large scale jobs while a manual injection gun may be used for small volumes. After the crack is filled, the pressure should be maintained for several minutes to insure good grout penetration.

11.6.2 Polymer Grouts

Polymer grouts based on urethanes or acryl amides, which react with water to form solid precipitate, or a foam are used to grout fine cracks in hydraulic structures. These materials are selected for their very low viscosity and applicability in moist environments.

11.6.3 Epoxy grouting

Epoxy resin is a product of Epichlorohydrin and Bisphenol with or without additives such as plasticizers and diluents. The resin mortar may be obtained by adding fillers such as coarse sand or calcined bauxite grit. They develop excellent strength and adhesive properties, and are resistant to many chemicals. They have good chemical and physical stability; they harden rapidly and resist water penetration. They provide a toughness that couples durability with crack resistance.

11.7 RESIN INJECTION

Resin injection is used to repair concrete that is cracked or delaminated and to seal cracks in concrete to water leakage. Two basic types of resin and injection techniques are used to repair concrete: epoxy resins and polyurethane resins. Epoxy resins cure to form solids with high strength and relatively high moduli of elasticity. These materials bond readily to concrete and are capable, when properly applied, of restoring the original structural strength to cracked concrete. The high modulus of elasticity causes epoxy resin systems to be unsuitable for rebonding cracked concrete that will undergo subsequent movement. The epoxies, however, do not cure very quickly, particularly at low temperatures, and using them to stop large flows of water may not be practical. Cracks to be injected with epoxy resins should be between 0.005 inch and 0.25 inch in width. It is difficult or impossible to inject resin into cracks less than 0.005 inch in width, while it is very difficult to retain injected epoxy resin in cracks greater than 0.25 inch in width, although high viscosity epoxies have been used with some success.Epoxy resins cure to form relatively brittle materials with bond strengths exceeding the shear or tensile strength of the concrete. If these materials are used to rebond cracked concrete that is subsequently exposed to loads exceeding the tensile or shear strength of the concrete, it should be expected that the cracks will recur adjacent to the epoxy bond line. In other words, epoxy resin should not be used to rebond "working" cracks.

Epoxy resins will bond with varying degrees of success to wet concrete, and there are a number of special techniques that have been developed and used to rebond and seal water leaking cracks with epoxy resins. These special techniques and procedures are highly technical and, in most cases, are proprietary in nature. Polyurethane resins are used to seal and eliminate or reduce water leakage from concrete cracks and joints. They can also be injected into cracks that experience some small degree of movement. Such systems, with the exception of the two-part solid polyurethanes, have relatively low strengths and should not be used to structurally rebond cracked concrete. Cracks to be injected with polyurethane resin should not be less than 0.005 inch in width. No upper limit on crack width has been established for the polyurethane resins at the time this is being written.

Polyurethane resins are available with substantial variation in their physical properties. Some of the polyurethanes cure into flexible foams. Other polyurethane systems cure to semi-flexible, high-density solids that can be used to rebond concrete cracks subject to movement. Most of the foaming polyurethane resins require some form of water to initiate the curing reaction and are, thus, a natural selection for use in repairing concrete exposed to water or in wet environments. At the time this is written, there are no standard specifications for polyurethane resins equivalent to the Standard Specification for Epoxy-Resin-Base Bonding Systems for Concrete, ASTM Designation C-881.

11.8 DRY PACK AND EPOXY-BONDED DRY PACK

Dry pack is a combination of Portland cement and sand passing a No. 16 sieve mixed with just enough water to hydrate the cement. Dry pack should be used for filling holes having a depth equal to, or great than, the least surface dimension of the repair area; for cone bolt, she bolt, core holes, and grout-insert holes; for holes left by the removal of form ties; and for narrow slots cut for repair of cracks. Dry pack should not be used for relatively shallow depressions where lateral restraint cannot be obtained, for filling behind reinforcement, or for filling holes that extend completely through a concrete section. For the dry-pack method of concrete repair, holes should be sharp and square at the surface edges, but corners within the holes should be rounded, especially when water tightness is required. The interior surfaces of holes left by cone bolts and she bolts should be roughened to develop an effective bond; this can be done with a rough stub of 7/8-inch steel-wire rope, a notched tapered reamer, or a star drill. Other holes should be undercut slightly in several places around the perimeter. Holes for dry pack should have a minimum depth of 1 inch.

11.8.1 Preparation and application

Application of dry-pack mortar should be preceded by a careful inspection of the hole, which should be thoroughly cleaned and free from mechanically held loose pieces of aggregate. One of three methods should be used to ensure good bond of the dry-pack repair. The first method is the application of a stiff mortar or grout bond coat immediately before applying the dry-pack mortar. The mix for the bonding grout is 1:1 cement and fine sand mixed with water to a fluid paste consistency. All surfaces of the hole are thoroughly brushed with the grout, and dry packing is done quickly before the bonding grout can dry. Under no circumstances should the bonding coat be so wet or applied so heavily that the dry-pack material becomes more than slightly rubbery. When a grout bond coat is used, the hole to be repaired can be dry. Presoaking the hole overnight with wet rags or burlap prior to dry packing may sometimes give better results by reducing the loss of hydration water, but there must be no free surface water in the hole when the bonding grout is applied.

The second method of ensuring good bond starts with presoaking the hole overnight with wet rags or burlap. The hole is left slightly wet with a small amount of free water on the inside surfaces. The surfaces are then dusted lightly and slowly with cement using a small dry brush until all surfaces have been covered and the free water absorbed. Any dry cement in the hole should be removed using a jet of air before packing begins. The hole should not be painted with neat cement grout because it could make the dry-pack material too wet and because high shrinkage would prevent development of the bond that is essential to a good repair. A third method of ensuring good bond is use of an epoxy bonding resin. Epoxies bond best to dry concrete. It may be necessary to dry the hole immediately prior to dry packing using hot air, a propane torch, or other appropriate method. The concrete temperature, however, should not be high enough to cause instant setting of the epoxy or to burn the epoxy when it is applied. After being mixed, the epoxy is thoroughly brushed to cover all surfaces, but any excess epoxy is removed. Dry-pack mortar is then applied immediately, before the epoxy starts to harden. The epoxy must be either fluid or tacky when dry packing takes place. If it appears that the epoxy may become hard before dry packing is complete, fresh fluid epoxy can be brushed over epoxy that has become tacky. If the epoxy becomes hard, it must be removed before a new coat is applied. The epoxy ensures a good bond between the dry-pack repair and the old concrete. It also reduces the loss of hydration water from the repair to the surrounding concrete, thus assisting in good curing; however, the epoxy-bonded dry pack still requires curing as discussed below. Where appearance is not important, epoxy has sometimes been used on the surface in place of a curing compound. This procedure is not recommended.

11.9 SPRAYED CONCRETE

Sprayed concrete is a mixture of cement, aggregate and water projected at high velocity from a nozzle in to place against an existing structure or formwork where it is compacted by its own velocity to produce a dense homogenous mass. Sprayed concrete in placed by either dry or wet process. Concrete should be cut away from corroded steel at the ends of defective patches until atleast 75 mm of uncorroded steel in exposed. The exposed steel should be cleaned by abrasive blasting or other appropriate means to remove all rust and scale. The repair should be encased in the specified minimum thickness of sprayed concrete. When the defects are in isolated area and patchwork repair are justified the area of sprayed concrete. Rectangular patches are preferable to irregular shapes.

Generally, a welded steel fabric is used, or steel fibres are incorporated in case of sprayed concrete without polymer admixtures to minimize the risk of cracks of sufficient size developing (which allows penetration of air and water) by encouraging the development of a large number of very fine, insignificant cracks. A typical fabric for this purpose would be of 3 to 4 mm diameter high tensile steel bars in the form of 75 to 100 mm mesh. With such a fabric the sprayed concrete thickness should be 50 mm minimum which will provide adequate cover to the fabric itself. The fabric reinforcement should be securely fixed by the nails driven into plugs set in parent concrete and bent over to grip it, with spacers to hold it at least 12 mm from the surface. Fixing points should be spaced at sufficiently close intervals such that the mesh does not belly out during the concrete spraying. The sprayed concrete thickness may be reduced in case of non-rusting steel fabric. The minimum thickness may be 30 to 50 mm. Chases at least 20 mm wide x 20 mm deep should be cut with pneumatic tools at the perimeter of the area to be sprayed, into which the edge will tuck to provide a sound finish at that point. Light timber profiles fixed securely in correct position should always be used as a guide to thickness and to provide eyeable lines on all main arises.

The aggregate to be used in sprayed concrete should be clean, well graded from 10 mm to fines, but without an excess of fines, clay, silt or dust. Typical mix proportions (aggregate-cement ratios) are 3.5: 1 or 4: 1, and typical strength of 30 MPa at 28 days. Construction joints are formed at a slope wherever possible, not at right angles as in cast concrete. The face of sprayed concrete is normally carried forward at a slope so that a construction joint of this kind is formed naturally. The interface of the joint must be cleaned of rebound, overshoot and laitance so that a good bond is achieved at the joint. Where defects are in isolated areas and patchwork repairs are justified the repairs should extend at least 300 mm on to the sound concrete at the perimeter of the defective area and terminate in chases. Rectangular patches are preferable to irregular shapes. Prior to commencing spraying, the interface must be prepared and wetted so that it does not absorb water from the sprayed concrete, but at the same time is not so wet that there is excess free water on it.

The adequacy of plant and of the properties of the material to be sprayed is confirmed by a pre-construction test panel. It consists in placing a test box 600 or 750 mm square and 100 mm thick on the surface simulating the construction conditions of proposed work and then spraying the concrete. Concrete cores of suitable age are taken from the test panels for testing for the specified standards of crushing strength, porosity, water penetration resistance, etc. However, previous experience with similar plant and materials is often accepted in lieu of such pre-construction tests except on very big jobs. Test panels also help to judge the competence of nozzlemen. Fabric mesh or bar reinforcement should be checked for secure fixing, correct positioning and overlaps, spacers and cleanliness. Profile guides, whether timber formers or stretched wires must be firm, tight and correctly positioned.

The spraying should be carried out with the nozzle held approximately at right angles to the interface, and at such a distance that the concrete compacts effectively. The nozzle should be fanned from side to side, and up and down so that layer is evenly built up over the area in front of the nozzleman. If there is any unevenness in the spray of material, the nozzleman must turn the nozzle away from the work until the spray becomes even again. Where the full thickness has to be built up in more than one layer, it is not advisable to apply more than 50 mm or so at a time on vertical surfaces, or 25 mm or so overhead. The surface of each layer may be lightly trimmed with the edge of a steel float, and must be wetted again once set before applying the next layer. Curing of the sprayed concrete is even more important than that of conventional cast concrete because thinner section may make water loss easier and more serious. The curing method may include a fine water spray, wetted hessian and curing compounds.

11.10 SLAB JACKING TECHNIQUE

The technique of pressure injection of grouts for the purpose of raising or stabilizing faulty concrete pavement floors, and other slab on grade surfaces, is termed slab jacking. It is referred to as slab jacking when lifting or leveling is involved, or simply pressure grouting where void filling is the sole objective. In addition to highway maintenance, slab jacking is presently used for a wide variety of remedial work such as repair or expansion joints and settlement damage in concrete slabs.

11.10.1 Advantages

The main advantages are:

(a) Cost effectiveness – grout leveling is frequently the most economical method.
(b) Down time – it is generally faster than other methods or repair.
(c) Interruption of other operations – a grouting schedule can be planned to cause minimum interference with other activities.
(d) Accessibility – the grout pump and other large equipment can be placed several hundred feet from the actual work area.
(e) Surface maintenance - for concrete pavements, the repair maintains the surface texture and appearance, provides a smoothes riding surface, and extends the useful life of the concrete surfacing.
(f) Load cavity – lifting of under – slab voids increases the load carrying ability of the slab.

11.10.2 Disadvantages and limitations

Disadvantages and limitations are:

(a) Most slabs that have undergone significant settlement contain extensive cracks. Although properly performed slab jacking will not cause new cracks, in many cases existing cracks, such as those in the bottom of significant sag that are in compression, will tend to open as the say is raised.
(b) Where small areas are involved or where use of the improvement is of little importance, lost may exceed that of replacement or other type of repair.

11.11 TREMIE CONCRETE

The most common method of placing concrete underwater is by tremie. A tremie consists of a vertical pipeline, topped by a hopper, which is long enough to reach the lowest point to be concreted from a working platform above the water. The tremie method is widely used for casting mass concrete in caissons, cofferdams seals and bridge piers.

11.12 CHLORIDE EXTRACTION AND REALKALISATION OF CONCRETE

This process involves the application of an external potential which causes chloride ions to be moved away from the reinforcement. In this case, the overlay is of sprayed cellulose fibre saturated with an alkaline solution. Chloride ions migrate towards the anode and are removed with the overlay. To install a system, electrical connections are first established with the reinforcement. Typically one contact point is made for each $50m^2$ of concrete surface. Wooden battens are attached to the concrete surface using plastic fixings and a layer of cellulose fibre saturated with electrolyte is applied by spray. The temporary anode is installed on the wooden battens using plastic plugs and bedded into the fibre layer. Anodes are usually in the form of mesh which can be easily bent and cut to shape. Mesh may be of titanium or steel. Titanium mesh is inserting and can be reused; steel mesh corrodes during the application of current and may cause stains on the surface which are difficult to remove. Finally, a further layer of cellulose fibre is sprayed over the mesh and connections are made to a DC power source. Connection is usually made to the anode every $10m^2$. The voltage is applied such that the external mesh becomes its positively charged anode and hence attracts the negatively charged chlorides ions. They move through the concrete and in to the fibre layer. At the same time, hydroxyle ions are produced at the negatively charged reinforcement thus increasing the PH at this location.

$$2H_2O + O_2 + 4e^- = 4OH^-$$

The alkaline solution is the fibre layer may be drawn into the concrete by a process of electro-osmosis, again increasing the pH. Replenishment of the alkalis is a quicker process than extraction of chlorides. Extraction of chlorides is more difficult because some of the chloride may be bound up in reaction products with the cement hydrates. The bound chlorides are in dynamic equilibrium with the chlorides dissolved in the pore water. As the chlorides in solution are removed by the electrolytic process, they are replaced by chlorides from the reaction products. The rate of which chlorides can be extracted is controlled by the rate at which the decomposition of the hydrated occurs. To overcome this effect, current has been applied intermittently in some cases. The rest period allows time for more chloride to be liberated from the reaction products and these are extracted when the current is reapplied giving an increase in the overall efficiency of operation.

11.13 CATHODIC PROTECTION

Impressed current cathodic protection (CP) is an electrical process of controlling the corrosion of embedded reinforcing steel. The process involves pumping an electrical current to the reinforcing steel, thus, preventing the bar from corroding. An electrical direct current (DC) volt power source is connected between the reinforcing steel requiring protection and an installed anode system. Electrical power is pumped into the anode system, which distributes the electrical current to the surrounding concrete. The electrical current moves toward the electrically connected reinforcing steel. The concrete between the anode and the reinforcing steel acts as an electrolyte (material capable of transmitting electrical current). In order for the concrete to carry electrical current, adequate moisture levels are required. With top side systems involving protective overlays, it is important that the materials used provide for electrical current flow. All impressed current cathodic protection (CP) systems require constant monitoring and adjustment.

11.14 MEMBER REPLACEMENT

Replacement is time consuming and expensive. It can cause extensive disruption in the use of the building on various floors because of the need for temporary propping. However, the procedure is effective provided proper care is taken to achieve continuity of reinforcement.

CHAPTER 12
Strengthening Techniques

12.1. INTRODUCTION

A significant number of facilities were constructed during the first half of the 20th century using reinforced or prestressed concrete materials. Now, at the beginning of the next century, many of these buildings have reached the end of their planned service life, and deterioration in the form of steel corrosion, concrete cracking, and spalling is observed frequently. In addition, many of these structures were built to carry loads that are significantly smaller than the current needs. Because of these factors, many structural engineers are faced with the challenge of evaluating and implementing effective and economical repair and strengthening programs.

Unfortunately, there is no single solution that offers a simple, straightforward method for all repair and strengthening projects. Further, the processes of repair and retrofit of existing structures are complicated because most of these structures are occupied, and much of the mainstream construction community's expertise is centered on new construction. However, success can be achieved if the repair and strengthening systems are tailored to serve a structure's intended use without interfering with its occupants or function. The key to success is a combination of the different design skills and application techniques — structural strengthening and structural repair — necessary for such projects. As such, the engineer must relay his or her expertise in using mechanical and structural behavior principles to develop a comprehensive retrofit solution.

Strengthening measures are required in structures when they are required to accommodate increased loads. Also, when there are changes in the use of structures, individual supports and walls may need to be removed. This leads to a redistribution of forces and the need for local reinforcement. In addition, structural strengthening may become necessary owing to wear and deterioration arising from normal usage or environmental factors.

12.1.1 Need for strengthening

Concrete structures need to be strengthened for any of the following reasons:

1. Load increases due to higher live loads, increased wheel loads, installations of heavy machinery, or vibrations.
2. Damage to structural parts due to aging of construction materials or fire damage, corrosion of the steel reinforcement, and/or impact of vehicles.
3. Improvements in suitability for use due to limitation of deflections, reduction of stress in steel reinforcement and/or reduction of crack widths.
4. Modification of structural system due to the elimination of walls/columns and/or openings cut through slabs.

5. Errors in planning or construction due to insufficient design dimensions and/or insufficient reinforcing steel.

12.1.2 Terms of repair

Concrete experts commonly use the terms structural repair and strengthening to describe building renovation activities. Although the two terms sound similar, they refer to slightly different concepts. Structural repair describes the process of reconstruction and renewal of a facility or its structural elements. This involves determining the origin of distress, removing damaged materials and causes of distress, as well as selecting and applying appropriate repair materials that extend a structure's life.

Structural strengthening, on the other hand, describes the process of upgrading the structural system of an existing building to improve performance under existing loads or to increase the strength of structural components to carry additional loads. For upgrade projects, design engineers must deal with structures in which every element carries a share of the existing load. The effects of strengthening or removing part or all of a structural element — such as penetrations or deteriorated materials — must be analyzed carefully to determine their influence on the global behavior of the structure. Failure to do so may overstress the structural elements surrounding the affected area, which can lead to a bigger problem and even localized failure.

12.1.3 Structural concrete repair

Although durable, buildings constructed using reinforced and prestressed concrete have a finite service life. When exposed to harsh environments, de-icing salts, and other chemicals, these structures may experience significant deterioration, which typically occurs in the form of steel corrosion, concrete spalls, delamination, and cracks.

Interestingly, one of the most severe and widespread problems in concrete is the internal damage caused by the corrosive action of external chlorides on reinforcing or prestressing steel embedded in concrete. Corrosion problems are caused by a corrosion-process by-product (rust) that expands up to eight times its original volume. This expansion creates internal pressure, which causes the concrete to crack and spall, resulting in a reduction of the effective area of steel reinforcement and reduced structural capacity of the affected member. If not addressed at early stages, corrosion will continue to grow rapidly, ultimately creating a safety issue of falling concrete and loss of strength.

Figure 12.1. Unsuccessful Concrete Repair

Figure 12.1 shows an unsuccessful concrete repair in structural member. The repair materials started to delaminate just a few short months after the first repair was completed. Adequate surface preparation could have prevented this repair failure. The assessment, design, and implementation of a durable repair to an existing structure are indeed more complex than for new construction. In addition to the unknown state of existing structural materials, the degree to which repair materials and the existing material will act as a composite and share loads must be addressed. Prior to establishing a repair strategy, the concrete-repair expert must diagnose the problem's root cause, which enables prescribing repairs that are long lasting and durable. Failing to follow this process may result in a frustrating, but common, cyclical outcome known as "repairing the repair."

12.1.4 Structural repair techniques for reinforced concrete

Structural repair describes the process of reconstruction and renewal of a facility or its structural elements. This involves determining the origin of the distress, removing damaged materials and cause of distress, as well as selecting and applying appropriate repair materials that extend the structure's useful life. In addition to proper repair techniques, regularly scheduled and budgeted maintenance can have significant impacts on the required repairs outcome.

Essential preparation:

A crucial step in achieving a durable repair is surface preparation. Multiple close-up observations of spalled concrete that had been repaired previously indicate that in many cases repairs have failed not because of material or technology, but because of the poor quality of surface preparation. The care with which deteriorated concrete is removed and the concrete and steel reinforcement surfaces are prepared often will determine whether a repair project will turn out to be successful. Equipment selection is important, because the concrete-removal method should not weaken or crack the surrounding sound concrete. Also care should be exercised to avoid further damage to the reinforcing and prestressing steel. This is achieved by using only industry-specified, lightweight demolition equipment. The effect of concrete removal on the structural integrity should be investigated carefully. A temporary shoring system may be required to relieve the loads on repaired elements in cases where removal of concrete or corroded reinforcing steel is significant enough to affect the structure's load-carrying capacity.

Once concrete has been removed, the reinforcing steel should be carefully inspected to determine whether the steel should be simply cleaned, repaired, or replaced. The final step in concrete and steel surface preparation is cleaning using abrasive or water-blasting techniques to remove loose materials and achieve an open-pore structure for the exposed concrete substrate and to remove the rust from the steel. This results in an appropriate surface to bond the new repair material and stop the corrosion process.

Cost of concrete repair versus time:

The cost of repairs becomes exponentially more expensive when preventative maintenance is not scheduled regularly or budgeted for a concrete facility. There are three observed phases describing the natural evolution of the concrete deterioration process and the influence of maintenance on this process: preventative maintenance, repair, and replacement. During the preventative maintenance phase, the owner may spend a fixed, annual maintenance cost to install systems, such as protective coatings, to slow down the deterioration process. Money spent in this phase will delay the ingress of aggressive materials, thus delaying the start of active deterioration (repair phase). In the repair phase, the concrete deterioration has begun, and the repair cost curve increases exponentially over time. The reason for the rapid increase in cost is that once aggressive materials that cause deterioration have permeated sufficiently into the concrete (a process that may take 20 to 30 years), the deterioration rate is rapid and irreversible. Throughout the replacement phase, a "wholesale" deterioration occurs throughout the structure at such a rapid rate that repair costs may exceed the costs of replacing the entire structure. However, total replacement of the structure may not be an option because of interruption to the function of the structure. Incurring additional costs in the early years of a structure's life to ensure well-protected concrete, and addressing deterioration problems as soon as they are observed would delay excessive deterioration and may increase significantly the service life of a structure.

12.2 STRUCTURAL CONCRETE STRENGTHENING

Many buildings that originally were constructed for a specific use now are being renovated or upgraded for a different application that may require higher load-carrying capacity. As a result of these higher load demands, existing structures need to be reassessed and may require strengthening to meet heavier load requirements. In general, structural strengthening may become necessary because of code changes, seismic upgrade, deficiencies that develop because of environmental effects (such as corrosion), changes in use that increase service loads, or deficiencies within the structure caused by errors in design or construction. The structural upgrade of concrete structures can be achieved using one of many different upgrading methods such as span shortening, external composites, externally bonded steel, external or internal post-tensioning systems, section enlargement, or a combination of these techniques. Similar to concrete repair, strengthening systems must perform in a composite manner with an existing structure to be effective and to share the applied loads. The strengthening is generally done by adopting (i) Jacketing Technique (ii) Externally Bonding Technique.

Strengthening or stiffening of members such as columns and girders is usually achieved either by replacing poor quality or defective material by better quality material or by adding additional material to the member. In either case, the new material will usually be reinforcing steel, high quality concrete, thin steel plates and straps or various combinations of these materials. The main difficulty in this type of operation is to achieve continuity of structural action between the original material and the new material. The various techniques of bolting, gluing, dowelling and keying have been developed to provide positive force transfer and composite action.

Column strengthening:

One of the simplest and most effective methods for strengthening a column in an existing building is to partially loaded the column by jacketing between floors and then insert two or more props to carry portion of the axial load. The props are usually rolled steel sections which may subsequently be encased in concrete to improve fire protection and appearance. The disadvantages of this method are that considerable floor space is lost, and that the props may not be effective in transferring moment unless positive connection details are introduced at both ends, for example in the form of end plates bolted through holes drilled in the floors.

Strengthening and stiffening of beams and girders:
Commonly used methods for strengthening flexural members include:

(a) Provision of additional concrete on the compressive face
(b) Addition of tensile reinforcement with a cast-in place or gunited cover
(c) Bolting or providing steel plates or straps to the surface of the member

The main problem of strengthening, stiffening and repair operation is to ensure good composite action between old and new material. Bolting of new concrete or steel to the existing member is effective but time consuming and expensive. For this reason, gluing procedures have been developed which provide strong and durable connections between concrete and surface straps and plates.

Strengthening and stiffening of slabs:
The simplest and cheapest procedure of strengthening and stiffening slabs is by providing additional props either in span or in the vicinity of existing columns and walls. However, this solution will often be unacceptable for architectural reason, even through props placed quite close to existing columns may be sufficient to bring the structure to an acceptable standard of serviceability.

12.2.1 Jacketing technique:
Jacketing is the process of fastening a durable material over concrete and filling the gap with a grout that provides needed performance characteristics. The materials used for jacket are metals, rubber, plastics, ferrocement and concrete. A typical jacketing process is shown in

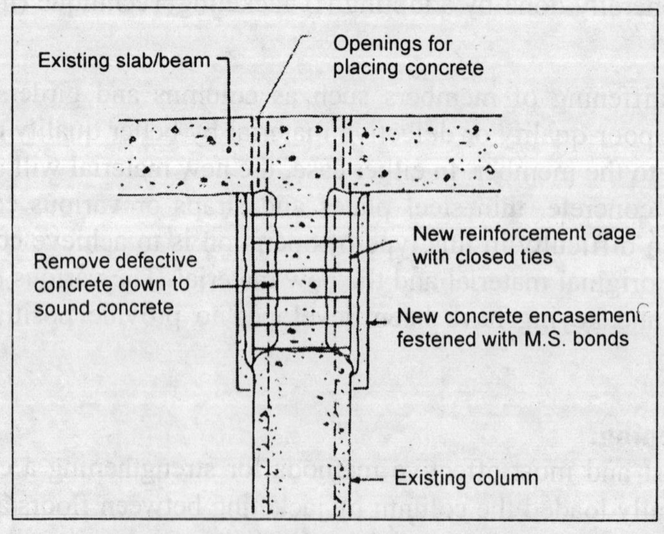

Figure 12.2. Process of jacketing technique

figure 12.2. This restores structural values, protects the reinforcements from exposure to the harmful elements and improves the appearance of the original concrete. Jacketing materials may also be secured to concrete by means of bolts, screws, nails or adhesives by bond with the existing concrete, or by gravity. The method of securing employed, will depend up on the pressure, the material used and the positioning of the jacketing material. Fibre glass reinforced plastics, ferrocement and other materials such as poly propylene can be used for jacketing.

The column jacket can also be used for increasing the punching shear strength of column-slab connections by using it as a column capital. When the jacket is provided around the periphery of the column, it is termed a collar. In most of the applications, the main function of the collar is to transfer vertical load to the column. Circular reinforcement can be used for load transfer. The practice of transferring load through dowel bars embedded into columns or shear keys has a disadvantage in that they require drilling of holes for dowels or cutting shear keys which are costly and time consuming, and can damage the existing column. Reinforcement encircling the column can be used to transfer the load through shear friction. The expansion of collar as it slides along the roughened surface causes the tensioning of circular reinforcement resulting in radial compression, which provides normal force needed for load transfer. The shear transfer strength is provided by both frictional resistance to sliding and dowel action of reinforcement crossing the crack.

The collar can also be used as mid-column bearing surface, acting as circumferential beam to distribute the concentrated load around the column. The collar is subjected to shear and bending along the collar circumference as well as direct bearing stress under concentrated load. Thus in addition to shear transfer reinforcement, the collar should be provided with reinforcement for shear and moment within the collar. The repair can be used as an alternate load path from the column to the collar and then to the connecting structural component. Column collars can be provided below the slab to act as column capital to improve punching shear strength of the slab column connection.

12.2.2 Externally bonding technique

In this, suitable laminates are pasted externally in the flexural zone to improve upon the strength and deformation characteristics of distressed structural members. The laminates may be fabricated from any one of the following material: Steel, Stainless steel, FRP, Ferrocement and SIFCON and SIMCON. The laminates are bonded to the distressed structural elements using epoxy resin. **Design criteria suggested by researchers worked in plate bending techniques:**

(a) The plate/laminates width to thickness ratio should not be greater than 50.
(b) The central axis depth should not be greater than 0.4 times the effective depth of member.

Strengthening with external bonding of plates has the following limitations:

(a) Difficulty in manipulating the plates in field conditions
(b) Deterioration of the bend steel concrete interface

(c) Lack of proper formation of joints

A typical process of plate bonding technique is shown in figure 12.3 using steel plates.

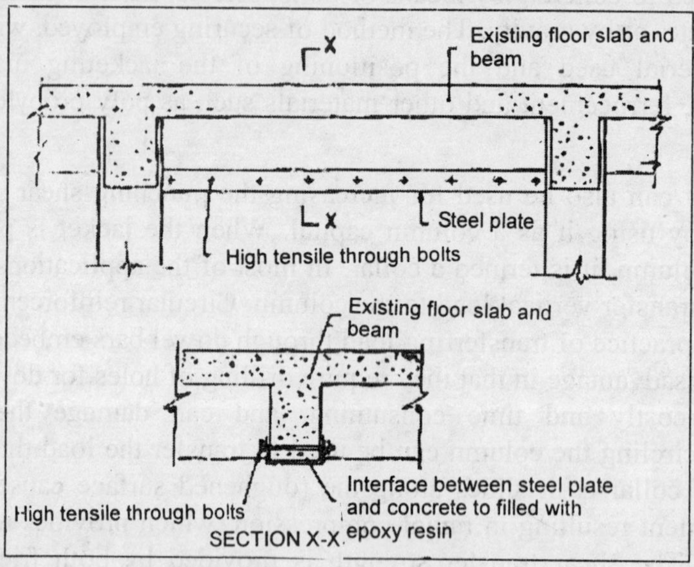

Figure 12.3. Process of plate bonding technique by steel plates

12.2.3 Externally bonded mild steel plates

The technique of gluing mild steel plates to the soffit of reinforced concrete beams and slabs can be applied to improve the structural performance of the existing structures. It increases the strength and rigidity and also reduces the flexural crack width in the concrete. The technique can be used to enhance the load carrying capacity either due to increase in load requirements or design or construction deficiencies. This technique has many advantages when applied to bridges. The plates can be easily handled and placed, this allowing the procedure to be carried out relatively quickly with minimal disruption to traffic. However, this technique is not recommended for application to bridges with unformed concrete or those affected by corrosion unless the source of the problem is rectified. The solution of an appropriate adhesive and surface preparation of the bonding surface is the most critical parameters for the successful implementation of this technique. The practical considerations to facilitate the placement of plates include the temperature requirements for mixing, applying and curing adhesive should be consistent with the existing conditions in the field and the adhesive should develop its strength rapidly.

Bonded steel elements - Strengthening concrete members by using bonded steel plates was developed in the 1960s. With this method, steel elements are glued to the concrete surface by a two-component epoxy adhesive to create a composite system and improve shear or flexural strength. The steel elements can be steel plates, channels, angles, or built-up members. In addition to epoxy adhesive, mechanical anchors typically are used to ensure that the steel element will share external loads in case of adhesive failure. The exposed steel elements must be protected with a suitable system immediately following installation. And regardless

Figure 12.4. Externally bonded steel plates with FRP

of the specified corrosion protection system, its long-term durability properties and maintenance requirements must be considered fully. A roof was strengthened with FRP and steel plates externally as shown in figure 12.4.

The roof consisted of prestressed concrete, hollow planks. Installation of the skylights required cutting openings in the planks that would reduce their load-carrying capacity. This issue was resolved by designing a hybrid strengthening system composed of FRP fabric and steel elements. The externally bonded FRP strengthened the planks adjacent to the one to be cut, while the steel elements tied the plank to the adjacent ones, thus creating a new unit consisting of three planks with adequate capacity. In addition to the fast application of this system, this was a cost effective solution that also was aesthetically pleasing.

12.3 STRENGTHENING WITH EXTERNAL REINFORCEMENT

Fibre composites are becoming increasingly important in the construction Industry with a number of applications in many areas. The fibres such as glass, carbon and aramid can be introduced in a certain position, volume and direction in the matrix to obtain the maximum efficiency. Other advantages offered by the fibre reinforced and greater efficiency, in construction when compared with the more conventional materials. **Some important applications of FRP are as follows:**

(a) Use of FRP castings to confine new concrete beams and columns or to strengthen beams in seismic area.
(b) Development of prestressing tenders made of parallel alignment of continuous fibres.
(c) Strengthening of concrete beams with uni-directional fibre reinforced concrete sheets bonded on their tension faces by the use of epoxy adhesives.

Composites:

Fiber reinforced polymer (FRP) systems are high-strength, lightweight reinforcement in the form of paper thin fabric sheets, thin laminates, or bars that are bonded to concrete members with epoxy adhesive to increase their load carrying capacity. These systems have been used extensively in the aerospace, automotive, and sport-equipment industries, and now are becoming a mainstream technology for the structural upgrade of concrete structures. Important characteristics of FRPs for structural repair and strengthening applications include their non-corrosive properties, speed and ease of installation, lower cost, and aesthetic appeal.

As with any other externally bonded system, the bond between the FRP system and the existing concrete is critical, and surface preparation is very important. Typically, installation is achieved by applying an epoxy adhesive to the prepared surface, installing the FRP reinforcement into the epoxy and, when required, applying a second layer of the epoxy adhesive. After curing, the FRP composite will add capacity to the element because it has a tensile strength up to 10 times that of steel.

Figure 12.5 shows a schematic for the structural strengthening of a utility tunnel. The utility tunnel roof originally functioned as a pedestrian walkway. A new dormitory structure required the walkway to be the primary access for emergency vehicles. Analysis of the tunnel's top slab revealed that it did not have adequate strength to carry loads from fire trucks and other emergency vehicles. A structurally efficient, easy-to-install, and cost-effective strengthening option was achieved by using externally bonded FRP sheets. The strengthening solution consisted of carbon FRP sheets bonded to the bottom of the slab, serving as additional bottom tension reinforcement. In addition, the overhanging portions of the slab were strengthened using carbon FRP bars epoxy-bonded in grooves made on the slab's top side. The latter technique is more appropriate than FRP sheets since the bars were bonded below the surface, thereby avoiding traffic damage to the externally bonded reinforcement.

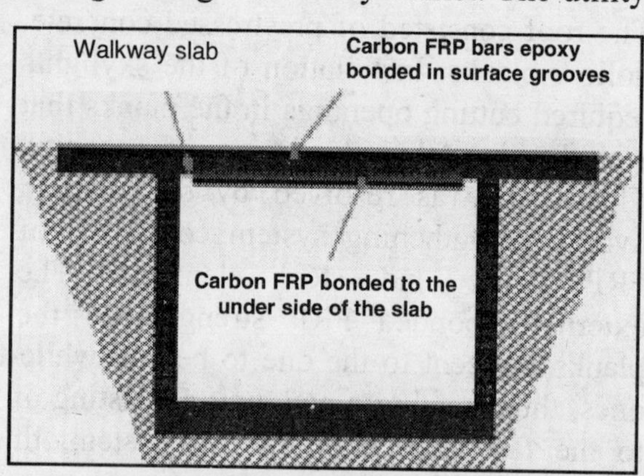

Figure 12.5. Strengthening with carbon FRP sheets

In addition to FRP, steel reinforced polymer composites (SRP) may be used as externally bonded reinforcement. This steel-based, innovative strengthening system (known as Hardwire) was developed and first introduced to the market in 2002. Hardwire is a low-cost, reinforcement system consisting of ultra-high-strength steel wires that are twisted together to form reinforcing steel cords approximately 0.035 inches in diameter. The steel wires have a tensile strength equal to about 10 times more than the strength of conventional structural steel, and the same in elastic modulus. This strengthening system can be applied using epoxy or cementitious materials and can be used to increase the shear and flexural capacity of structural elements.

Figure 12.6. Short spanning technique

12.4 SHORT SPANNING

Span shortening is accomplished by installing additional supports underneath existing members. Appropriate materials for span shor-tening include structural steel members and cast-in-place reinforced concrete members, which are simple to install. Connections can be designed easily using bolts and adhesive anchors. An example of this upgrading method is shown in figure 12.6.

The structural steel system shown was installed on a parking deck to shorten the span and carry part of the load, transferring it to the existing supporting system. On the down side, such applications may result in loss of space and reduced headroom.

12.5 EXTERNAL POST-TENSIONING

The external post-tensioning technique has been used effectively to increase the flexural and shear capacity of both reinforced and prestressed concrete members since the 1950s. With this type of upgrading, active external forces are applied to the structural member using post-tensioned (stressed) cables to resist new loads. Because of the minimal, additional weight of the repair system, this technique is effective and economical, and has been employed with great success to correct excessive deflections and cracking in beams and slabs, parking structures, and cantilevered members.

The post-tensioning forces are delivered by means of standard prestressing tendons or high-strength steel rods, usually located outside the original section. The tendons are connected to the structure at anchor points, typically located at the ends of the member. End-anchors can be made of steel fixtures bolted to the structural member, or reinforced concrete blocks that are cast into place. The desired uplift force is provided by deviation blocks, fastened at the high or low points of the structural element. Prior to external prestressing, all existing cracks are epoxy-injected and spalls are patched to ensure that prestressing forces are distributed uniformly across the section of the member.

Figure 12.7 illustrates an external post-tensioning system used to strengthen prestressed double tees damaged by vehicular impact. Four double tee stems on an overpass located were damaged when the driver of an over-height truck failed to observe the posted height restriction. The four stems suffered excessive concrete cracking and spalling, and damage occurred to some of the internal prestressing steel.

Figure 12.7. External post-tensioning technique

Proposed solutions included replacing the damaged double tees with new ones and installing a steel frame underneath for support. Both options would render the overpass out of service for a longer-than-desired period. The option of an external post-tensioning system was more economical, required less time to complete, and allowed for a strengthening system that provided active forces. Therefore, it was more compatible with the existing construction. After all cracks were injected, the sides of the stems were formed and new concrete was cast to restore the integrity of the stems. The strengthening system then was installed, and — after the concrete cured — the external strands were stressed according to the engineer-specified forces.

12.6 SECTION ENLARGEMENT

This method of strengthening involves placing additional bonded reinforced concrete to an existing structural member in the form of an overlay or a jacket. With section enlargement, columns, beams, slabs, and walls can be enlarged to increase their load-carrying capacity or stiffness. A typical enlargement is approximately 2 to 3 inches for slabs and 3 to 5 inches for beams and columns.

Figure 12.8 depicts details of a section enlargement used to increase the capacity of a main girder in a university parking garage. The girder was re-evaluated because of a change in the required loading and found to be deficient in flexure and shear. To correct the deficiency, additional flexural and shear steel were added. The entire beam then was formed and a 4-inch jacket of concrete was cast to enlarge the section.

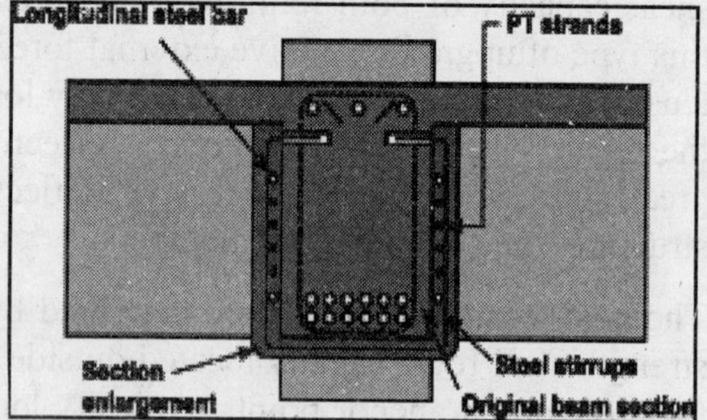

Figure 12.8. Section enlargement technique

12.7 STRENGTHENING BY SIMCON

The experimental results demonstrate that SIMCON exhibits improved properties in tension, compression, flexure and shear even when comparatively low fiber volume fraction fiber-mats are used. Furthermore, since fiber-mats are pre-packed in the plant, distribution and orientation of fibers can be more accurately controlled, than is the case with short discontinuous fiber HPFRCs. These characteristics allow for the manufacturing of a unique cement-based fiber composite that can have different yet easily controllable properties in the longitudinal and transversal directions. These material characteristics are desirable in repair/ retrofit of structural elements such as columns, which require a high increase in strength and toughness in the transverse direction while increasing only ductility but not strength in the longitudinal direction (*i.e.*, "moment-carrying" direction).

The investigations also demonstrate that SIMCON has considerable potential for both seismic repair/retrofit, as well as the development of novel, high-performance composite structural systems. In a retrofit situation continuous SIMCON fiber-mats, delivered in large rolls, can be easily installed by wrapping around members to be rehabilitated. The procedure of rehabilitation for beam – column joint stepwise as shown in figures 12.9.

In new construction of high-performance composite frames SIMCON is well suited for manufacturing high strength, high ductility, and thin stay-in-place formwork elements that eliminate the need for secondary and most of the primary reinforcement. The presence of a SIMCON layer led to:

(a) Both improved structural performance and durability of the member, and

(A)

(B)

(C)

Figure 12.9. SIMCON wrapping process

(b) Optimization of member dimensions, amount of reinforcement and member weight.

First Step: Re-bars wrapped in SIMCON are put along the column to provide moment continuity through the joint region, as well as replacement of concrete with SIMCON in the anchorage region of the discontinuous bottom beam reinforcement.

Second Step: Additional layers of SIMCON mat are added to increase moment capacity at the column and beam zones facing the joint.

Third Step: The entire column (and portion of adjacent beams) is jacketed with SIMCON. Formwork is put next and mat is injected with a high-strength slurry (say 14,000 psi).

Manufacturing of SIMCON is based on the use of widely available construction equipment and building expertise, and can thus be relatively easily introduced into the field without major re-training and changes in existing construction practices. Hence, this novel type of HPFRCC provides some unique new ways of developing durable and cost-effective high-performance infrastructural systems, essential for the economic well-being of the nation in the next century. Slurry Infiltrated Mat Concrete (SIMCON) can be used in: (1) seismic retrofit, (2) the development of a novel, partially-cast-in-place High Performance Composite Frame, and (3) the development of a "self-stressing" SIMCON stay-in-place formwork that can provide active confinement after the core of the member has been cast-in-place.

12.8 DAM SAFETY: CONCRETE REPAIR TECHNIQUES

Concrete is an inexpensive, durable, strong and basic building material often used in dams for core walls, spillways, stilling basins, control towers, and slope protection. However, poor workmanship, construction procedures, and construction materials may cause imperfections that later require repair. Any long-term deterioration or damage to concrete structures caused by flowing water, ice, or other natural forces must be corrected. Neglecting to perform periodic maintenance and repairs to concrete structures as they occur could result in failure of the structure from either a structural or hydraulic standpoint. This in turn may threaten the continued safe operation and use of the dam.

12.8.1 Considerations

Floor or wall movement, extensive cracking, improper alignments, settlement, joint displacement, and extensive undermining are signs of major structural problems. In situations where concrete replacement solutions are required to repair deteriorated concrete, it is recommended that a professional engineer be retained to perform an inspection to assess the concrete's overall condition, and determine the extent of any structural damage and necessary remedial measures.

Typically, it is found that drainage systems are needed to relieve excessive water pressures under floors and behind walls. In addition, reinforcing steel must also be properly designed to handle tension zones and shear and bending forces in structural concrete produced by any external loading (including the weight of the structure). Therefore, the finished product in any concrete repair procedure should consist of a structure that is durable and able to withstand the effects of service conditions such as weathering, chemical action, and wear. Because of their complex nature, major structural repairs that require professional advice are not addressed here.

12.8.2 Repair methods

Before any type of concrete repair is attempted, it is essential that all factors governing the deterioration or failure of the concrete structure are identified. This is required so that the appropriate remedial measures can be undertaken in the repair design to help correct the problem and prevent it from occurring in the future. The following techniques require expert and experienced assistance for the best results. The particular method of repair will depend on the size of the job and the type of repair required.

A. **The Dry-pack method:** The dry-pack method can be used on small holes in new concrete which have a depth equal to or greater that the surface diameter. Preparation of a dry-pack mix typically consists of about 1 part Portland cement and 2 1/2 parts sand to be mixed with water. You then add enough water to produce a mortar that will stick together. Once the desired consistency is reached, the mortar is ready to be packed into the hole using thin layers.

B. Concrete replacement: Concrete replacement is required when one-half to one square foot areas or larger extend entirely through the concrete sections or where the depth of damaged concrete exceeds 6 inches. When this occurs, normal concrete placement methods should be used. Repair will be more effective if tied in with existing reinforcing steel (rebar). This type of repair will require the assistance of a professional engineer experienced in concrete construction.

C. Replacement of unformed concrete: The replacement of damaged or deteriorated areas in horizontal slabs involves no special procedures other than those used in good construction practices for placement of new slabs. Repair work can be bonded to old concrete by use of a bond coat made of equal amounts of sand and cement. It should have the consistency of whipped cream and should be applied immediately ahead of concrete placement so that it will not set or dry out. Latex emulsions with Portland cement and epoxy resins are also used as bonding coats.

D. Shotcrete (Gunite): A popular concrete replacement technique for repairing large areas of severely deteriorated concrete and spalled vertical and overhead faces is the use of pneumatically placed concrete or shotcrete. Shotcrete consists of a mixture of moistened cement and fine aggregate (sand) that is sprayed onto the repair area under pressure. Professional assistance and sophisticated equipment are needed to apply shotcrete.

E. Prepacked aggregate with intruded mortar: This special commercial technique has been used for massive repairs, particularly for underwater repairs of piers and abutments. The process consists of the following procedures:

(a) Removing the deteriorated concrete,
(b) Forming the sections to be repaired,
(c) Prepacking the repair area with coarse aggregate, and
(d) Pressure grouting the voids between the aggregate particles with a cement or sand-cement mortar.

F. Synthetic patches: One of the most recent developments in concrete repair has been the use of synthetic materials for bonding and patching. Epoxy-resin compounds are used extensively because of their high bonding properties and great strength. In applying epoxy-resin patching mortars, a bonding coat of the epoxy resin is thoroughly brushed onto the base of the old concrete. The mortar is then immediately applied and troweled to the elevation of the surrounding material.

Conclusion: Before attempting to repair a deteriorated concrete surface, all unsound concrete should be removed by sawing or chipping and the patch area thoroughly cleaned. A sawed edge is superior to a chipped edge, and sawing is generally less costly than mechanical chipping. **Before concrete is ordered for placing, adequate inspection should be performed to ensure that:**

(a) Foundations are properly prepared and ready to receive the concrete,

(b) Construction joints are clean and free from defective concrete,
(c) Forms are grout-tight, amply strong, and set to their true alignment and grade,
(d) All reinforcement steel and embedded parts are clean, in their correct position, and securely held in place
(e) Adequate concrete delivery equipment and facilities are on the job, ready to go, and
(f) Capable of completing the placement without addition unplanned construction.

12.9 GUIDELINES FOR SEISMIC REHABILITATION OF EXISTING BUILDINGS

Advances in earthquake-related technology during the past few decades have led to a realization that seismic risk to life and property can be reduced significantly by improving seismic performance of existing seismically deficient buildings. Detailed post-earthquake investigations of building failures have provided engineers with considerable information concerning the details of building design and construction that enhance earthquake resistance.

12.9.1 Seismic vulnerability

The vulnerability of a building subjected to an earthquake is dependent on seismic deficiency of that building relative to a required performance objective. The seismic deficiency is defined as a condition that will prevent a building from meeting the required performance objective. Thus, a building evaluated to provide full occupancy immediately after an event may have significantly more deficiencies than the same building evaluated to prevent collapse. Depending on the vulnerability assessment, a building can be condemned and demolished, rehabilitated to increase its capacity, or modified so that the seismic demand on the building can be reduced. Thus, structural rehabilitation of a building can be accomplished in a variety of ways, each with specific merits and limitations related to improving seismic deficiencies.

12.9.2 Common seismic deficiencies

Regardless of the evaluation method used, failure to meet the stipulated performance objective implies certain seismic deficiencies. These deficiencies are described below:

Global strength:

Global strength typically refers to the lateral strength of the vertically oriented lateral force-resisting system. For degrading structural systems characterized by a negative post yield slope on the pushover curve, a minimum strength requirement may apply. In certain cases, the strength will also affect the total expected inelastic displacement and added strength may reduce nonlinear demands into acceptable ranges. A deficiency in global strength is common in older buildings either due to a complete lack of seismic design or a design to an early building code with inadequate strength requirements. If prescriptive equivalent lateral force methods or linear static procedures have been used for evaluation, inadequate strength will directly relate to unacceptable demand-to-capacity ratios within elements of the lateral force-resisting system.

Global stiffness:
Global stiffness refers to the stiffness of the entire lateral force-resisting system although the lack of stiffness may not be critical at all levels. For example, in buildings with narrow walls, critical drift levels occur in the upper floors. Conversely, critical drifts most often occur in the lowest levels in frame buildings. Stiffness must be added in such a way that drifts are efficiently reduced in the critical levels. Although strength and stiffness are often controlled by the same existing elements or the same retrofit techniques, the two deficiencies are typically considered separately. Failure to meet evaluation standards is often the result of a building placing excess drift demands on existing poorly detailed components.

Configuration:
This deficiency category covers configuration irregularities that adversely affect performance. In codes for new buildings, these configuration features are often divided into plan irregularities and vertical irregularities. Plan irregularities are features that may place extraordinary demands on elements due to torsional response or the shape of the diaphragm. Vertical irregularities are created by uneven vertical distribution of mass or stiffness between floors that may result in concentration of force or displacement at certain levels. In older existing buildings, such irregularities are seldom taken into consideration in the original design and, therefore, normally require rehabilitation measures to mitigate.

Load path:
A discontinuity in the load path, or inadequate strength in the load path, may be considered overarching because this deficiency will prevent the positive attributes of the seismic system from being effective. The load path is typically considered to extend from each mass in the building to the supporting soil. For example, for a panel of cladding, this path would include its connection to the supporting floor or floors, the diaphragm and collectors that deliver the load to components of the primary lateral force resisting system (walls, braces, frames, etc.), continuity of these components to the foundation, and finally the transfer of loads between foundation and soil. Many load path deficiencies are difficult to categorize because the strength deficiency may be considered to be part of another element. For example, an inadequate construction joint in a shear wall could be considered a load path deficiency or a shear wall deficiency in the category of global strength.

Inadequate component detailing:
Detailing, in this context, refers to design decisions that affect a component's or system's behavior beyond the strength determined by nominal demand, often in the nonlinear range. An example of a detailing deficiency is poor confinement in concrete gravity columns. Often in older concrete buildings, the expected drifts from the design event will exceed the deformation capacity of such columns, potentially leading to degradation and collapse. Although the primary gravity load design is adequate, the post elastic behavior is not, most often due to inadequate configuration and spacing of ties. Identification of detailing deficiencies is significant in selection of mitigation strategies because acceptable performance often may be achieved by local adjustment of detailing rather than by adding new lateral force-resisting elements.

In the case of gravity concrete columns, acceptable performance can be achieved by enhancing deformation capacity by adding confinement rather than by reducing global deformation demand by adding lateral force-resisting elements.

Diaphragm deficiencies:
The primary purpose of diaphragms in the overall seismic system is to act as a horizontal beam spanning between lateral force-resisting elements. Diaphragm deficiencies include such factors as inadequate shear or bending strength, stiffness, or reinforcing around openings or reentrant corners. Inadequate local shear transfer to lateral force-resisting elements or missing or inadequate collectors are categorized as load path deficiencies.

Foundation deficiencies:
Foundation deficiencies can occur within the foundation element itself, or due to inadequate transfer mechanisms between foundation and soil. Element deficiencies include inadequate bending or shear strength of spread foundations and grade beams, inadequate axial capacity or detailing of piles and piers, and weak and degrading connections between piles, piers and caps. Transfer deficiencies include excessive settlement or bearing failure, excessive rotation, inadequate tension capacity of deep foundations, or loss of bearing capacity due to liquefaction.

Other deficiencies:
Deficiencies that do not fit into one of the categories described above may include:

(a) Geologic hazards
(b) Adjacent buildings
(c) Deteriorated structural materials

12.9.3 Strategies for rehabilitation schemes

Technical Considerations:
Selected techniques must eliminate deficiencies, preferably more than one deficiency. First one should consider enhancing existing elements such as shear walls, moment frames, and bracing frames. The deformation compatibility between new elements and existing elements must be considered. In some cases, the application of base isolators or damping devices is the most efficient way to eliminate deficiencies.

Non-technical considerations:
The solution chosen for rehabilitation is almost always dictated by building-user oriented issues rather than by merely satisfying technical demands. There are five basic issues that are of concern to building owners or users:

(a) Construction cost,
(b) Seismic performance,
(c) Short term disruption of occupants,
(d) Effects on long-term functionality of building, and
(e) Aesthetics, including consideration of historic preservation.

All of these characteristics are always considered, but an importance will eventually be put on each of them, either consciously or subconsciously, and a combination of weighting factors will determine the scheme chosen.

Cost:

Construction cost is always important and is balanced against one or more other considerations deemed significant. However, sometimes other economic considerations, such as the cost of disruption to building users, or the value of contents to be seismically protected, can be orders-of-magnitude larger than construction costs. Thus, cost may be the only criterion applied when choosing among equivalent rehabilitation options.

Seismic performance:

Prior to the emphasis on performance based design, perceived qualitative differences between the probable performance of difference schemes would be used to assist in choosing a scheme. Specific performance objectives are often set prior to the development of schemes. Objectives that require a limited amount of damage or continued occupancy will severely limit the retrofit methods that can be used and may control the other issues.

Short-term disruption of occupants:

When seismic rehabilitation is done at the time of major building remodeling, disruption issue is minimized. However, in cases where the building is partially or completely occupied, this parameter commonly becomes dominant and controls the design.

Effects on long-term functionality of building:

This characteristic is often judged less important than others. The planning flexibility is only subtlety changed. However, it can be significant in building occupancies that need open spaces such as retail spaces and parking garages.

Aesthetics:

In historic buildings, considerations of preservation of historic fabric usually control the design. Performance objectives are controlled by limitations imposed by preservation. In non-historic buildings, aesthetics is commonly stated as a criterion, but is often sacrificed, particularly in favor of minimizing cost and disruption to tenants.

12.9.4 Rehabilitation techniques

Different building types require different mitigation techniques for a specific seismic deficiency. Depend on building types and associated seismic deficiencies, alternative recommendations are made to satisfy the performance objective of rehabilitation. Rehabilitation techniques are being developed for common building types. They are:

Wood light-frames: One- and two-detached dwellings of one or more stories in height.

Multi-story, multi-unit residential wood frames: Large residential buildings with commercial space at the ground floor.

Steel moment frames: Buildings consist of steel beams and columns, and lateral forces are resisted by moment frames.

Steel braced frames: Buildings consist of frame assemblies of steel beams and columns. Lateral forces are resisted by diagonal steel members placed in selected bays.

Steel frames with infill masonry shear walls: Buildings are normally older buildings that consist of gravity frames with unreinforced masonry, tightly infilling the space between columns.

Concrete moment frames: A complete system of concrete beams and columns. Lateral loads are resisted by cast-in-place moment frames.

Concrete shear wall buildings (Bearing wall systems): Usually all concrete with flat slab or precast plank floors and concrete bearing walls. Little, if any, of the gravity loads are resisted by beams and columns. Lateral loads are resisted by shear walls.

Concrete shear wall buildings (Gravity frame systems): Buildings have columns and beams or columns and slabs that essentially carry all gravity loads. Lateral loads are resisted by concrete shear walls surrounding shafts, at the building perimeter, or isolated walls placed specifically for lateral resistance.

Concrete frames with infill masonry shear walls: Buildings are normally older buildings that consist of essentially complete gravity frame assemblies of concrete columns and floor systems. The floors can be of a variety of concrete systems including flat plates, two way slabs, and beam and slab. Exterior walls are constructed of unreinforced masonry, tightly infilling the space between columns horizontally and between floor structural elements vertically, such that the infill interacts with the frame to form a lateral force resisting system.

Tilt-up concrete shear wall buildings: Buildings are constructed with perimeter concrete walls cast on the site and tilted up to form the exterior of the building. The majority of these buildings are one story with wood roof framing; however, a good number of multi-story buildings also exist with composite deck floors and a wood or steel framed roof.

Precast concrete frames with shear walls: Buildings consist of concrete columns, girders, beams and/or slabs that are precast off the site and erected to form a complete gravity load system. This building type has a lateral force-resisting system of concrete shear walls, cast-in-place or precast. In California, precast floor T-beams or hollow core planks are covered by a cast-in-place topping slab, reinforced to provide diaphragm action. In other areas, methods of joining floor sections vary, and include use of welded insert plates.

Reinforced masonry bearing wall buildings (Similar to tilt-up concrete shear wall buildings): Building are constructed with reinforced masonry (brick cavity wall or concrete masonry unit) perimeter walls with a wood or metal deck flexible diaphragm.

Reinforced masonry bearing wall buildings (Similar to unreinforced masonry bearing wall buildings): Buildings are multistory, and typically has interior concrete masonry unit walls and shorter diaphragm spans.

Reinforced masonry bearing wall buildings (Similar to concrete shear wall buildings with bearing walls): Building consist of reinforced masonry walls and concrete slab floors that may be either cast-in-place or precast. This building type is often used for hotel and motels and is similar to the concrete bearing wall type.

Unreinforced masonry bearing wall buildings: Buildings consist of unreinforced masonry bearing walls, usually at the perimeter and usually brick masonry. The floors are typically of wood joists and wood sheathing supported on the walls and on interior post and beam construction.

Rehabilitations methods are recommended for the following five categories of techniques as appropriate:

(a) Add new elements,
(b) Enhance existing elements,
(c) Improve connections between elements,
(d) Reduce demand, and
(e) Removal deficient elements.

Seismic rehabilitation techniques not necessarily related to a specific building type such as those related to diaphragms, foundations, and nonstructural components and significant global techniques that could be applied to any building, such as seismic isolation or addition of damping are included in this category.

Regardless of the experience and experimental knowledge gained in more than 100 years of reinforced concrete construction, structures require repair and/or strengthening because of natural causes, human error, and change in loading conditions. Further, it is important to recognize that concrete repair and strengthening is a scientific art form that involves the use of conventional, cement based materials, as well as new techniques and materials. It is crucial that structural engineers recognize that strengthening assessment and design is infinitely more complex than new construction. Typically, challenges arise because of unknown factors associated with the structural state — such as continuity, load path, and material properties — as well as the size and locations of existing reinforcement or prestressing. The degree to which the upgrade system and the existing structural elements share the loads also must be evaluated and addressed properly in the upgrade design, detailing, and implementation procedure. The importance of detailing and its direct effects on the effectiveness and durability of structural upgrades cannot be overemphasized. In fact, inadequate detailing is one factor that can lead to the total failure of a structural repair system.

OBJECTIVE TYPE QUESTIONS

1. The selection of repair technique is based on the following objectives
   (a) To improve the functional requirements
   (b) To improve the durability
   (c) To restore the load carrying capacity
   (d) All the above

2. The preparation of surface for repair consists of
   (a) Complete removal of unsound material
   (b) Removing cracks from the surface
   (c) Providing rough but uniform surface
   (d) All the above

3. The best way of removing laitance from the surface of concrete is
   (a) Sand blasting
   (b) Acid etching
   (c) Grinding
   (d) Cleaning with caustic soda

4. The repair of large and deep patches of deteriorated concrete in a structure, the material filling in done by
   (a) Dry packing
   (b) Pre packing
   (c) Concrete replacing
   (d) Grouting

5. Deep and wide cracks in concrete members can be repaired by
   (a) Guniting
   (b) Grouting
   (c) Epoxy injecting
   (d) Mortar replacing

6. Epoxy injection technique is used for
   (a) Sealing narrow cracks in structural members
   (b) Scaling actively leaking cracks
   (c) Repairing hydraulic structures
   (d) All the above

7. For sealing cracks in concrete structures by epoxy, the minimum width of routing should be
   (a) 3 mm
   (b) 9 mm
   (c) 6 mm
   (d) 15 mm

8. Identify the incorrect statement of the following
   (a) The stitching of cracks is adopted to re-establish the tensile strength across major cracks.
   (b) The stitching of cracks is adopted to prevent the crack from further propagation.
   (c) The stitching of cracks is adopted to stiffen the structure.
   (d) The stitching of cracks is adopted to strengthen the cracks, which have a tendency to close as well as to open.

9. Sulphur impregnated concrete can be used
   (a) For repairing fractured members
   (b) As a practical and expansive substitute of polymer impregnated concrete
   (c) For repairing slab and decks containing fine dormant cracks
   (d) None of the above

10. Identify the correct statement of the following
    (a) For polymer concrete usually methyl methacrylate is used
    (b) For polymer concrete latexes of SBA is used
    (c) For polymer concrete 99.9% pure sulphur is used
    (d) All the above

11. Jacketing technique
    (a) The process of fastening a durable material over concrete surface and filling the gap with grout
    (b) It increases the section of an existing member by encasement a new concrete.
    (c) It is used for compression members as columns & piers
    (d) All the above

12. Autogenous healing
    (a) It can not take place to continuous saturation
    (b) It is natural process of crack repair developing in the presence of moisture
    (c) It is particularly affective in cycles of drying and remoisturising
    (d) All the above

13. Identify the incorrect statement of the following
    (a) The choice of under water repair is influenced by the available method of access
    (b) The placement method and formwork adopted must be able to minimize the mixing of water to the repair material
    (c) The under water repair requires adequate preparation and special technique
    (d) None of the above

14. The technique of epoxy injection is used foı
    (a) Sealing actively leaking cracks
    (b) Repairing hydraulic structures
    (c) Sealing narrow cracks
    (d) All the above

15. For scaling cracks in concrete structures by using epoxy, the minimum routing should be
    (a) 20 mm (b) 15 mm
    (c) 9 mm (d) 6 mm

16. The bonded over lays of polymer modified Portland cement concrete or mortar
    (a) Should have minimum thickness of 75 mm
    (b) Should be used for 72 hrs
    (c) Can be used for repairing slabs and decks containing fine dormant cracks
    (d) Should be mixed, placed and finished with in 30 min.

17. Polymer impregnation of concrete is not effective in the following cases except
    (a) When cracks contain moisture
    (b) When cracks are dry
    (c) When cracks are fine
    (d) When monomer evaporates

18. The best fire resistance aggregate is
    (a) Sand stone
    (b) Limestone
    (c) Brick aggregate
    (d) Blast furnace aggregate

19. Addition of bonded steel plates t o beam increases their,
    (a) Stiffness
    (b) Cracking
    (c) Deformations
    (d) All the above

20. The plate bonding technique can be used
    (a) To increase deflection
    (b) To enhance the load carrying capacity
    (c) To improve crack width
    (d) All the above

21. A reduction in plate thickness on plate bonding technique will lead to
    (a) Premature brittle failure
    (b) Premature shear failure
    (c) Premature bond failure
    (d) All the above

22. The technique of grooving is used for
    (a) Sealing of leaking cracks
    (b) Sealing shallow pits or cracks
    (c) Scaling narrow cracks
    (d) None of the above

23. Improving the abrasion resistance of concrete floors by
    (a) Polymer impregnation
    (b) Resin injection
    (c) Plate bending
    (d) Grinding

24. The reliable and commercially valid techniques of leak sealing consulted with
    (a) Leaking expansion joints
    (b) Leaking construction joints
    (c) Seepage through random cracks
    (d) All the above

25. Cracks may be considered problemative, if they are
    (a) Aesthetically unacceptable
    (b) Make the structure non water tight
    (c) Affect the durability of the structure
    (d) All the above

26. The objective of the rehabilitation techniques is
    (a) To improve durability
    (b) To improve the appearance
    (c) To improve the structural integrity
    (d) All the above

27. Dry pack is used for the repair of
    (a) Dormant cracks
    (b) Filling
    (c) Growing cracks
    (d) All the above

28. A bond breaker should be used to prevent the sealant
    (a) From bonding to the top of the recess
    (b) From bonding to the bottom of the recess
    (c) From binding to the middle of the recess
    (d) None of the above

29. The inherent ability of concrete to heal cracks with in itself is termed,
    (a) Stiffening
    (b) Softening
    (c) Autogenous healing
    (d) Accelerated healing

30. Shotcrete contains well–graded aggregate of
    (a) 10 mm or less in size
    (b) 20 mm or less in size
    (c) 10 mm or more in size
    (d) None of the above

31. The quality and strength of tremie concrete is dependent on
    (a) Proper placement
    (b) Proper mix design
    (c) Proper temperature
    (d) All the above

32. Attempts to vibrate the tremie concrete causes
    (a) Bleeding
    (b) Honeycombing
    (c) Excessive laitance
    (d) 36. None of the above

33. Repair of concrete structures is decided upon the factor
    (a) The cause of damage
    (b) Type of damage
    (c) The extent of damage
    (d) All the above

34. Repair of concrete structures is carried out the first stage
    (a) Removal of damaged concrete
    (b) Pre treatment of surfaces and reinforcement
    (c) Pre treatment of reinforcement
    (d) None of the above

35. For effective Guniting, the nozzle should be kept ------------ from the work preferable normal to the surface.
    (a) 20 cm to 50 cm
    (b) 60 cm to 150 cm
    (c) 160 cm to 180 cm
    (d) None of the above

36. The natural process of crack repair known as
    (a) Autogenous healing
    (b) Accelerated healing
    (c) Stitching
    (d) None of the above

37. Dormant cracks in both structural and pavement slabs may be repaired using
    (a) Coatings
    (b) Toppings
    (c) Screening
    (d) All the above

38. The coatings applied to concrete surface to arrest the carbonation process are known as
    (a) Carbonation coatings

(b) Membrane coatings
(c) Anti carbonation coatings
(d) None of the above

39. Mortar or concrete conveyed through a pressure hose
(a) Shot Crete
(b) Grout
(c) Dry pack
(d) All the above

40. ------------------------- is a mortar mixture with a very low water –cement ratio
(a) Repacked
(b) Replaced
(c) Dry pack
(d) Gunite

41. Under water repair technique is called
(a) Preplaced aggregate concrete
(b) Pretreated concrete
(c) Shot Crete
(d) Sprayed concrete

42. ------------------ is used in repairing concrete to get an even surface
(a) Repairing
(b) Grinding
(c) Replacing
(d) None of the above

43. Use of vacuum impregnation is
(a) Reducing the permeability of the concrete
(b) Increasing permeability of the concrete
(c) Reducing the impermeability of the concrete
(d) None of the above

44. Concrete projected at high velocity from a nozzle in to place against an existing structure is called
(a) Ready mix concrete
(b) Grouted concrete
(c) Sprayed concrete
(d) All the above

45. The format of surface seals
(a) Fill out surface imperfections with resin based gront
(b) Apply recommended primer
(c) Apply two coats of high – build paint
(d) All the above

46. Defective concrete pipeline joints due to
    (a) Excessive settlements in pipe line
    (b) Defective manufacture
    (c) Settlement of the pipe bedding
    (d) None of the above

47. Point leakage through concrete pipe walls due to
    (a) Excessive settlements in pipe line
    (b) Defective manufacture
    (c) Settlement of the pipe bedding
    (d) None of the above

48. Cracks in the concrete pipes due to
    (a) Excessive settlements in pipe line
    (b) Defective manufacture
    (c) Settlement of the pipe bedding
    (d) None of the above

49. The method of concrete placement is suited to underwater or tidal work
    (a) Replacement concrete
    (b) Preplaced aggregate concrete
    (c) Sprayed concrete
    (d) All the above

50. Floors that are cast directly on top of a sub-base is called
    (a) Ground floors
    (b) Suspended floors
    (c) Concrete floors
    (d) All the above

51. ---------------- treatment will seal the upper part of the crack against the ingress of moisture, chlorides and carbon dioxide.
    (a) Flexible sealing
    (b) Toppings
    (c) Fed crack sealing
    (d) None of the above

52. It is often appropriate to current active cracks into movement joints
    (a) Flexible sealing
    (b) Toppings
    (c) Fed crack sealing
    (d) None of the above

53. -------------- may be used to seal cracks and very useful where there are large numbers of cracks and expansion joint is not possible
    (a) Toppings

(b) Screeds
(c) Overlays
(d) None of the above

54. The technique used for a wide variety of remedial work such as repair of expansion joints and settlement damage in concrete slabs
(a) Slab jacking technique
(b) Jacketing technique
(c) Collar technique
(d) None of the above

55. The most common method of placing concrete underwater is by
(a) Roller
(b) Tremie
(c) Screedy
(d) None of the above

56. --------------- are intermediate between coatings and pore blocken
(a) Membranes
(b) Sealers
(c) Collars
(d) All the above

57. The process involves the application of an external potential which causes chloride ions to be moved away from the reinforcement
(a) Chloride extraction
(b) Realkalisation
(c) Cathodic protection
(d) None of the above

58. The repair of cracks can be achieved with the following techniques
(a) Replaced concrete
(b) Sprayed concrete
(c) Routing and sealing
(d) None of the above

ANSWERS

1. d	2. d	3. b	4. c	5. b	6. a	7. c	8. c	9. b	10. a
11. d	12. b	13. d	14. c	15. d	16. c	17. b	18. d	19. a	20. b
21. d	22. b	23. a	24. d	25. d	26. d	27. d	28. a	29. c	30. a
31. d	32. c	33. d	34. a	35. b	36. a	37.b	38. c	39. a	40. c
41. a	42. b	43. a	44. c	45. d	46. a	47. b	48. c	49. b	50. a
51. c	52. a	53. c	54. a	55. b	56. b	57. a	58. c		

SHORT ANSWER QUESTIONS

1. How do you get economical rehabilitation?
2. How do you select repair methods?
3. What are the objectives followed in repair procedure?
4. What are factors depends on repair procedure?
5. What are the stages carried out in repair procedures?
6. What are the steps involved in the surface preparation for repair?
7. Write about Autogenous Healing?
8. What are the design criteria suggested by researchers worked in plate bending techniques?
9. Specify the limitations of plate bending technique.
10. Explain shortly about Acid etching.
11. Write shortly about caulking.
12. What is the use of vacuum impregnation?
13. Define sprayed concrete.
14. How do you repair reinforcement corrosion damaged concrete by sprayed concrete?
15. Outline the format of surface seals.
16. Write about leak sealing in concrete pipelines.
17. What is meant by preplaced aggregate concrete?
18. What is ground floor slabs?
19. What is a suspended floor?
20. Define toppings & screeds.
21. Write about gravity – fed crack sealing.
22. Define flexible sealing.
23. What is the function of overlays?
24. How do you repair dormant cracks?
25. How do you repair leaking cracks?
26. What is the use of preplaced aggregate concrete?
27. List the limitations of preplaced aggregate concreting method.
28. Define slab-jacking technique.
29. What are advantages of slab jacking technique?
30. Write down the disadvantages & limitations of slab jacking technique.
31. Define the term shotcrete.
32. What is the purpose of using shortcrete?
33. What are the advantages of shotcrete?
34. Write down the disadvantages of shotcrete.
35. What is Tremie concrete?
36. Define Sealers.
37. Write about chloride extraction and realkalization of concrete.
38. How do you repairs cracks in concrete?
39. Classify the types of repair.

40. Explain the corrective measures for cracks.
41. Explain the externally bonded mild steel plates as strengthening technique.
42. Explain the procedure of strengthening with external reinforcement
43. What is the function of overlaps?
44. What is the purpose of surface coatings?
45. What is meant by anti-carbonation coatings?
46. How are coatings to protect cracked concrete?
47. How to leak sealing applied?
48. Explain leak sealing conventional method.
49. Explain leak sealing injection techniques.
50. What are the factors decided the rehabilitation techniques?
51. Explain shotcrete/Gunite.
52. What is meant by dry pack?
53. Write about preplaced aggregate concrete.
54. What are the stages carried out in repair process?
55. What are the aspects comprises in pretreatment?
56. How do you remove loose material on the surface of concrete?
57. Explain the Grinding process.
58. What is the purpose of sand blasting?
59. What is the use of concrete replacement system?
60. What is the use of mortar replacement system?
61. What is meant by grouting?
62. What is the use of prepacked concrete?
63. How to use polymer impregnation in repair of fractured elements?
64. Explain vacuum impregnation process in concrete repair?
65. Explain resin injection technique.
66. What is the function of externally bonded mild steel plates?
67. What is meant by durable repair system?
68. Define the procedure repair to repair.
69. List out the safety measures in dames.
70. Name the rehabilitation techniques for earthquake damage.

THEORY QUESTIONS

1. Explain with neat sketches about the various techniques used in the repair of cracks.
2. Explain in detail about the plate bonding strengthening technique.
3. Explain with sketches about the fibre sheet bonding strengthening technique.
4. What is ferrocement? Explain about ferrocement jacketing.
5. Explain the procedure of epoxy coating of rebars.
6. With the help of a simple sketch, explain guniting/shotcreting.
7. How do you strengthen the various structural elements? Explain in detail with sketches.
8. Explain in detail about acid etching and resurfacing.
9. Explain with sketches the retrofitting of damaged RC members by polymer modified concrete and ferrocement.
10. Explain with neat sketches about the jacketing techniques for the repair of seismically affected beam – column joint in a building.
11. What are the different types of plates/laminates used for strengthening of structural elements. Discuss the different strengthening techniques and their relative merits.
12. Give a complete procedure of epoxy injection to structural crack repair in RC structural elements.
13. Explain the classification of cracks and how the cracks can be measured and reported.
14. Explain coating systems adopted in concrete repairs.
15. Explain the replacement concrete and preplaced aggregate concrete techniques.
16. Explain grouting and dry packing concrete techniques.
17. Explain the considerations in structural repairs in reinforced concrete structures.
18. Explain strengthening techniques with external reinforcement.
19. Explain section enlargement and short span strengthening techniques.
20. Explain external post-tensioning strengthen techniques.
21. Explain the durable repair system of concrete structures.

PART V
MAINTENANCE AND DEMOLITION

CHAPTER 13

Necessity and Classification of Maintenance

13.1 INTRODUCTION

Maintenance Engineering is defined as the work done to keep the civil Engineering structures and work in a condition so as to enable them to carry out the functions for which they are constructed. **The maintenance of structure is done to meet the following objectives:**

(a) Prevention of damages due to natural agencies and to keep them in good appearance and working condition.

(b) Repair of the defects occurred in the structure and strengthen them, if necessary.

The maintenance work is broadly classified as:

(a) Preventive maintenance
(b) Remedial maintenance
(c) Routine maintenance
(d) Special maintenance

13.1.1 Preventive maintenance

The maintenance work done before the defects occurred or damage developed in the structure is called preventive maintenance. It includes thorough inspection, planning the programme of maintenance and executing the work. It depends upon the specifications, condition and use of structure.

13.1.2 Remedial maintenance

It is the maintenance done after the defects or damage occurs in the structure. It involves the following basic steps.

(a) Finding the deterioration
(b) Determining the causes
(c) Evaluating the strength of the existing structures
(d) Evaluating the need of the structure
(e) Selecting and implementing the repair procedure

13.1.3 Routine maintenance:

It is the service maintenance attended to the structure periodically. The nature of work done and interval of time at which it is done depends upon specifications and materials of structure, purpose, intensity and condition of use. It includes white washing, patch repair to plaster, replacement of fittings and fixtures, binding of road surface.

13.1.4 Special maintenance:

It is the work done under special condition and requires sanction and performed to rectify heavy damage. It may be done for strengthening and updating of the structure to meet the new condition of usage or to increase its serviceability. It may include particular or complete renewal occurring at long interval, such as removal of floors, roofs etc.

13.2 NECESSITATION OF THE MAINTENANCE

The causes which necessitate the maintenance effects the service and durability of the structure are as follows:

(a) Atmospheric Agencies
(b) Normal wear and tear
(c) Failure of structure

13.2.1 Atmospheric agencies

Rain:

It is the important source of water, which affects the structure in the following ways: **Physical–**

- Dissolving and carrying away minerals as it is universal solvent
- Expansion and contraction - The material is subjected to repetitive expansion and contraction while they become wet and dry and develops the stresses
- Expansion of water - The variation of temperature causes the expansion and contraction of absorbed water and affects the micro- structures of the materials
- Erosion - Transportation and attrition and abrasion of the material is quite evident effect of the water

Chemical - The water available in nature contains acids and alkali and other compound in dissolved form acts over the material to give rise, which is known as chemical weathering.

- **Wind** - It is the agent, which transports the abrasive material and assists the physical weathering. Its action is aggravated during rains and when it is moving with high speed it may contain some acidic gases like CO2 fumes which may act over the material and penetrates quite deeply in the material and structure.
- **Temperature** - The diurnal, seasonal and annual variation of the temperature, difference in temperature in two parts of the material and the surface of material causes expansion and contraction. By this movement of the material bond and adhesion between them is lost when it is repeated. This is responsible for the development of cracks and the rock may break away into small units.

Exploitation or peeling off the shell takes place if exterior layer are heated externally with respect to internal layers. The temperature variation may also cause change in the structure and chemical composition of the material.

13.2.2 Normal wear and tear

During the use of structure it is subjected to abrasion and thereby it looses appearance and serviceability.

13.2.3 Failure of structure

Failure is defined as behaviour of structure not in agreement with expected condition of stability or lacking freedom from necessary repair or non-compliance with desired use of and occupancy of the completed structure. In field it may result in visual collapse of the structure or even suspension of the services *e.g.* the collapse of towers, sliding or over-turning of dam, settlement of foundation, crushing of columns etc. **The causes of failure may be broadly grouped as:**

- **Improper design** - due to incorrect, insufficient data regarding use, loading and environmental conditions, selection of material and poor detailing
- **Defective construction** - poor materials, poor workmanship, lack of quality control and supervision
- **Improper use of structure** - overloading, selecting the structure for the use for which they are not designed such as deteriorating environment due to impurities from industrial fuel burning, sea water minerals, chemicals, storage of chemicals for which they are not designed.
- **Lack of maintenance** - Lack of upkeep, proper protection, precaution and preservation, deteriorates the structure, which may result in the failure.

13.3 INSPECTION PERIODS

The maintenance programme is normally drawn taking climatic conditions under consideration the year is divided into the following periods:

(a) Pre – monsoon period
(b) Monsoon period
(c) Post – monsoon period

For drawing up the maintenance programme, it is very essential to inspect the structure to get first-hand information regarding defects and their causes.

Pre – monsoon inspection is made to decide the maintenance programme to be done before monsoon such as cleaning of drains, checking of roof leakage, collection of material and equipment require during monsoon repairs, bridges and other structure which are not easily approachable during the monsoon are inspected fully and repaired, if necessary.

Monsoon inspection is needless to mention that the emergency work carried out in monsoon period, e.g. canal or dam breach, washing away of roads, railway tracks, collapse of roof, and failure of walls.

Post – monsoon inspection is made to repair the damage caused by water and draw up the programme of repair according to the priorities. All replacement renewal and rehabilitation work are carried out.

13.4 BACKGROUND OF MAINTENANCE

Maintenance is the continuous protective care of the fabric, contents and setting of a place. A maintenance plan is the most cost-effective way to ensure the ongoing value of any asset, particularly a registered place. The plan should include regular inspections, prompt attention to repairs and a cleaning program. Advantages of a maintenance plan include:

(a) The property is organised and maintained in a systematic rather than haphazard way;
(b) The standard and presentation of the property can be maintained; and
(c) Emergency repairs are minimised.

There is no such thing as a maintenance-free building and serious problems can develop when maintenance is neglected. Spending small amounts of time and money on inspections and repairs at regular intervals is more effective than undertaking extensive repair work and replacing items that have deteriorated due to neglect.'

Why is it important to take care with maintenance in a registered building?

(a) To protect valuable historic information and to maintain the value of an historic asset;
(b) To ensure that the fabric is properly conserved; and
(c) To retain the significance of the place through appropriate management.

There are three ways to obtain an approval to undertake building maintenance works in a registered place.

13.4.1 General exemption

Under General Exemption issued for all places certain building maintenance can be carried out at any time if the conditions of the exemption and the guidelines are followed. Building maintenance approved under General exemption includes:

(a) Work that is the continuous protective care of existing material;
(b) Work to retain the condition of existing fabric or the operation of an element;
(c) Removal of surface deposits, organic growths or graffiti by the use of low pressure water, neutral detergents and mild brushing and scrubbing as appropriate for the surface;
(d) Clearing of paths and drains; and
(e) Cleaning in accordance with the cleaning guideline

Building maintenance work does not include removal of or damage to existing materials or the introduction of new materials.

13.4.2 Apply for an exemption certificate

When a building maintenance project is not covered by General Exemption and does not cause detriment to the cultural heritage significance, submit an exemption certificate application seeking approval.

Examples of building maintenance work that may require an exemption certificate application include:

(a) Cleaning sandstone or limestone;
(b) Removal of paint from a masonry surface that was not originally painted;
(c) Removal of difficult stains on masonry; and
(d) Installation of additional wall vents to sub-floor spaces to facilitate cross ventilation.

13.4.3 Apply for a Development Permit

If the proposed maintenance works are likely to impact on the significance of the place, lodge a development application through the Integrated Development Assessment System. Examples of building maintenance projects that require a development application include:

(a) High pressure abrasive cleaning of stone monuments, walls, features or pathways; and

(b) Applying a coating to an uncoated stone monument or stone/brick fence pier.

13.5 MAINTENANCE PROCESSES

Having listed the maintenance objectives and their measures the next logical step is to learn the means of achieving them. The maintenance processes that help achieving the listed objectives are discussed briefly.

(a) Design for Maintainability (DfMt)
(b) Preventive maintenance (PM)
(c) Predictive maintenance (PdM)
(d) Reliability centered maintenance (RCM)
(e) Reactive maintenance
(f) Spares management
(g) Maintenance logistics support
(h) Total productive maintenance (TPM)
(i) Organizing for maintenance
(j) Computerized maintenance management system (CMMS)
(k) Decision support systems & tools
(l) Statutory requirements

13.5.1 Design for maintainability

Maintenance is a non-value-adding activity; for it reduces the availability of equipment for production at the same time costs resources to maintain it. Ideal equipment is the one which is available for production all the time and doesn't cost anything to maintain. Such ideal equipment has to be conceived even at the drawing board. Various design principles have been developed to aid designing equipments that will have minimum or even zero maintenance.

13.5.2 Preventive maintenance

Equipment and its parts are designed to perform their intended functions under certain working conditions. When these conditions change for worst the equipment will fail. A periodic check of these conditions and the parts is what is known as preventive maintenance. A well-planned PM can bring down the breakdowns considerably. On the other hand a poor PM will not only increase the maintenance cost but also breakdowns.

13.5.3 Predictive maintenance

PM schedules are normally fixed based on the past history of failures or manufacturer's recommendations. Neither of them truly represents the expected life of a part. There is always a possibility of replacing an under-utilized part or not replacing an over-used part. An effective maintenance should replace / repair a part when it is required, neither earlier nor later. This is possible only if the maintenance is based on the condition of the part. PdM is one such process in which the condition of the part is monitored using special equipments and the maintenance action is decided based on the monitoring results. This process may require initial investments but will pay for itself in long-run.

13.5.4 Reliability centered maintenance

Reliability is the probability that equipment will perform its intended function under given operating conditions for a defined period. RCM is centered on improving this reliability of the equipment. Activities start from design stage and makes use of the past failure history and condition monitoring results. RCM is a subject by itself and is rooted in statistics.

13.5.5 Reactive maintenance

How much ever PM or PdM is adapted, eliminating breakdown is not possible due to continuously varying working conditions. Maintenance carried out to repair such breakdowns is known as reactive maintenance. The skills required to attend to breakdowns are special and the maintenance department should equip themselves for that. Expertise is required in mechanical elements, hydraulics, pneumatics, electrical and electronics. Pooling such skill is very difficult hence organizations rely on the machine manufacturers for breakdown maintenance. This forms a part of the annual maintenance contract.

13.5.6 Spares management

For a cost effective maintenance, spares management is to be done judiciously. Excessive stocking will lead to money block and deterioration on the other hand under stocking will lead to prohibitive downtimes. Extensively researched models are available to arrive at the optimum stocking; however, we are restricting our discussion to the practical models.

13.5.7 Maintenance logistics support

Spares management is from the user point of view, whereas logistics is from the equipment supplier's point of view. Maintenance logistics provides for the initial planning, funding, and controls which help to assure that the end user will receive a system that will not only meet performance requirements, but one that can be expeditiously and economically supported throughout its programmed life cycle.

13.5.8 Total productive maintenance

TPM is a Japanese concept aimed at involving users in maintaining their machines. Maintenance personnel support such activities in addition to performing specialized functions, which includes DfMt initiatives. TPM has been successfully adapted in many companies and they have reported substantial gains in almost all business and operational parameters. Implementation of TPM is an involved exercise and requires top management's commitment.

13.5.9 Organization for maintenance

The various processes discussed above indicate the quantum and quality of work involved for an effective maintenance. Management must not only provide human resources to accomplish such activities but also organize them appropriately with respect to the functional needs. Organizational structure differs depending upon the type of industry, size of the organization, process choice, and activity levels etc. The types of structures and applicability are discussed in the subsequent section

13.5.10 Computerized maintenance management system

Now a days computers form an essential part of maintenance function. Generally they are used to store the data about the machines, failure history, maintenance schedules and spare parts management. In specialized industries, computers are used in on-line condition monitoring and even to trouble shoot. Specialized software systems are available for limited applications; however, large business houses rely on ERP packages such as SAP as they provide an integrated system that aligns with the rest of the business operations 3 TPM Club of India offers live details about TPM activities in India.

13.5.11 Decision support systems and tools

Decision support systems are required in maintenance at various levels and stages, which are listed below:

(a) New equipment procurement
(b) Planning maintenance strategy such as PM, PdM etc.
(c) Maintenance budgeting
(d) Spares management
(e) Logistics support for the sold equipment, and
(f) Residual life and equipment disposal

While mathematical models are available to aid decision, the popular ones are decision trees.

13.5.12 Statutory requirements

To inculcate a sense of safety The Industrial Ministry has formulated certain basic requirements for the factories. The compliance to these statutes is normally managed by the maintenance department with the help of administration department. The popular and common statutes are:

(a) Factories act that approves the factory layout, fire safety precautions and safety of electrical installations
(b) Pressure vessel certification to audit compressors, boilers etc.
(c) Building safety statute that inspects the safety of the building construction
(d) Pollution control statutes that audits noise level, air quality, water quality and effluent and waste disposal

The 12 maintenance processes explained above help achieving the maintenance objectives.

CHAPTER 14
Maintenance Procedure

14.1 BUILDING MAINTENANCE

The cleaning method that should be considered as part of an overall maintenance plan for a registered place. It is not an exhaustive guide but aims to assist owners to care appropriately for the fabric of a registered place. The cleaning process done under the following headings:

(a) Background
(b) Maintenance plan
(c) Caution
(d) Testing
(e) General principles for cleaning
(f) Regular cleaning guide
(g) Recommended frequency for regular cleaning
(h) Graffiti

14.1.1 Background

Cleaning is a vital part of maintenance. Maintenance cleaning does not involve the removal of any fabric or the introduction of new materials. Many of the cleaning activities mentioned are common sense and are already followed by most owners and occupants of a registered place.

14.1.2 Maintenance plan

A maintenance plan should be part of the management of a registered building. A regular cleaning program is an essential part of a maintenance plan. When buildings are neglected, defects can occur which may result in extensive and avoidable damage to the fabric or equipment. Cleaning is work that is undertaken at regular intervals to prevent failure and unnecessary deterioration of a property. It contributes to the general upkeep of the place.

14.1.3 Caution

When cleaning any registered place there will be a large number of variable and unpredictable factors involved. When selecting a cleaning method the test is – will the method damage the fabric? If a test sample indicates that the cleaning will cause damage seek advice from your local office and/or a conservation consultant before proceeding.

14.1.4 Testing

All cleaning methods can be detrimental to a building or element of a building if carried out incorrectly or inappropriately. Test small areas before carrying out the full job. Be aware that the full range of surface variations cannot be discovered in a small area so allow for flexible treatments on the larger surface.

14.1.5 General principles for cleaning

(a) Ensure that the cleaning is necessary. Over cleaning of older buildings can cause problems.
(b) Maintenance plan should identify areas requiring special care, *e.g.* special finishes not normally found in modern buildings; fragile areas like rusted metal; and areas containing fabric of potential danger like asbestos.
(c) Select the cleaning method that is appropriate for the job and the condition of the substrate.
(d) Understand the dangers of some cleaning methods and guard against them. Test the proposed method.
(e) Generally avoid strong alkalis or strong acids or any method that is abrasive.
(f) Remove graffiti as quickly as possible.
(g) Use cleaning as an opportunity to check the condition of finishes.
(h) Seek advice if in doubt.
(i) Always adhere to safety advice and environmental guidelines.

14.1.6 Regular cleaning guide and intervals

The following Table 14.1 shows the regular cleaning guide for structural elements and the Table 14.2 given the time interval for regular cleaning.

Table 14.1. Regular cleaning guide

Item	*Cleaning*
Windows & Doors:	
Anodised aluminium	Clean with warm soapy water applied with a cloth or chamois. Dry and polish with a soft dry cloth.
Bronze	Apply thin coating of paraffin oil; allow drying thoroughly and polishing with a soft cloth.
Ordinary glass	Wash with dilute alkaline solution (*e.g.* a proprietary window cleaner) or clean water and wipe with damp chamois.
Leadlight or Historic glass	Wash with softly sprayed distilled or deionized water and a soft cloth or soft brush.
Lacquered metal	Clean with warm soapy water applied with a cloth or chamois. Dry and polish with a soft dry cloth.
Painted timber	Wash with warm water and weak detergent using a soft brush. Rinse with clean water.
Stainless steel & Painted	Clean with warm soapy water, dry with a chamois and

Item	*Cleaning*
steel	polish with a soft cloth.
Hardware:	*(e.g. nails, screws, bolts, fasteners, locks, latches, hinges, door/ window fittings)*
Anodised aluminium	Clean with warm soapy water applied with a cloth or chamois. Dry and polish with a soft dry cloth.
Brass	Apply thin coating of metal polish. Allow to dry and polish with a soft cloth. OR leave to dull naturally.
Bronze	Apply thin coating of paraffin oil. Allow to dry and polish with a soft cloth.
Ceramic	Clean with warm soapy water applied with a cloth. Dry with a soft dry cloth.
Chromium plate	Clean with warm soapy water applied with a cloth or chamois. Dry and polish with a soft dry cloth.
Lacquered metal	Clean with warm soapy water applied with a cloth or chamois. Dry and polish with a soft dry cloth.
Stainless steel	Clean with warm soapy water, dry with a chamois and polish with a soft cloth.
Walls:	
Cement glaze	Wash down with warm water and weak detergent. Rinse with clean water and wipe with a chamois.
Face brick	Brush down with a stiff dry brush.
Glazed brickwork	Wash with warm water and weak detergent. Rinse with clean water.
Glazed wall tiles	Wash with warm water and weak detergent. Rinse with clean water and dry with a soft cloth.
Granite	Wash with warm water and weak detergent. Rinse with clean water and dry with a soft cloth.
Marble	Wash with warm soapy water. Rinse with clean water and dry with a soft cloth. Apply a thin coating of wax furniture cream and polish with a soft cloth.
Painted timber (ext.)	Wash with warm water and weak detergent using a soft brush. Rinse with clean water.
Painted timber (int.)	Wipe with damp cloth.
Plaster	Wash with neutral pH soap in water using dampened soft cotton cloth. Rinse with clean water. Do not saturate.
Plastics laminates	Remove dirty marks with polish. Wipe over with a damp cloth and dry with a soft cloth.
Sandstone	Cleaning agents suitable for sandstones vary with the type

Item	Cleaning
	of sandstone. Seek advice from a stone conservator before cleaning sandstone.
Timber *	Wipe with damp cloth. (*varnished or stained timber)
Vitreous enamel	Wash with warm water and weak detergent. Rinse with clean water and polish with a soft cloth.
Wallpaper	Wallpapers have a range of finishes. Seek advice before cleaning.
Floors:	
Carpets, Rugs	Vacuum in the direction of the pile with low-level suction. Periodically vacuum the back of the rug. If liquids are spilt on the fabric, clean the area immediately by soaking up as much liquid as possible with clean absorbent towel or cloth. Blot gently with soda water or a mix of half white vinegar and cold water to neutralize the spot. Repeat if necessary. Do not rub. Place clean absorbent material over the area, weight down until dry. Do not use commercial carpet cleaners. For stubborn stains consult a carpet cleaning specialist experienced in older carpets.
Hardwood (waxed)	Apply paste or liquid wax polish, rub into surface with a dry cloth and when dry buff with a polishing machine. Remove accumulations of polish by dry rubbing with a medium steel wool pad or wet rub with a medium steel wool pad, dipped in liquid polish and when dry, buff with a polishing machine.
Linoleum	Remove dirty marks with a damp cloth or mop. Scuff marks which cannot be removed by this method may be removed by light rubbing with a fine steel wool pad dipped in liquid polish. Apply paste or liquid wax polish, rub into surface with a dry cloth and when dry buff with a polishing machine. A water/wax emulsion polish gives greater non-slip properties. Apply with a damp cloth or mop and when dry buff with a polishing machine. Remove accumulations of polish with detergent in hot water, rinse with clean water and when dry re-apply polish or dry rub with a fine steel wool pad and re-apply polish. Do not use alkaline cleaners.
Oilcloths	Clean by hand with a warm, soapy sponge. Dry with a soft cloth.
Polished softwood	Wash with a solution of water and vinegar or methylated spirits.
Rubber	Apply water/wax emulsion polish with a damp cloth or mop

Item	*Cleaning*
	and when dry lightly buff with a polishing machine. Remove accumulations of polish with paste cleaner using warm water, rinse with clean water and when dry re-apply polish.
Tessellated tiles	Damp mop or wipe with damp cloth.
Terrazzo	Lightly scrub with warm water using a gritty cleansing powder, rinse with clean water and squeegee to remove surplus water.
Tiles	Wash with warm water and detergent and rinse with clean water. Squeegee to remove surplus water.
Verandah floors	Sweep. Damp mop for grime. Take care with washing shot edge boards to avoid water dripping through to spaces below.
Vinyl	Sweep or vacuum to remove dust. Wipe over with a damp mop using warm water.
Paintwork internal: Gloss paint, Semi-gloss paint, Flat paint, Emulsion paint	Brush and wipe down surfaces with damp rags or sponges using warm water and household detergent or sugar soap. DOES NOT use soap powders. Change water and rinse rags and sponges regularly. Rinse thoroughly with clean water and wipe down with chamois.
Lime wash/washable distemper	Brush lightly with a dry fine-hair broom.
Furnishings:	
Fabric upholstery	Vacuum clean and brush with a stiff upholstery brush.
Painted timber	Wash with warm water and weak detergent. Rinse with clean water and polish with a soft cloth.
Polished timber	Apply wax furniture cream and polish with a soft dry cloth.
PVC upholstery	Clean with warm soapy water, applied with a soft brush. Rinse with clean water and dry with a soft cloth.
Bathrooms:	
Chromium plate	Clean with warm soapy water applied with a cloth or chamois. Dry and polish with a soft dry cloth.
Copper	Apply thin coating of metal polish. Allow to dry and polish with a soft cloth.
Mirrors	Wash with clean water and wipe with damp chamois.
Sanitary fittings	Clean with porcelain cleaning powder and warm water. Rinse with clean water and polish with a soft cloth.

Item	*Cleaning*
Stainless steel	Clean with warm soapy water, dry with a chamois and polish with a soft cloth.

Table 14.2. Recommended frequency for regular cleaning

Item	*Task*	*Frequency*
Furniture:		
Fabric upholstery	Vacuum clean	Monthly
Mirrors and glass	Dust	Daily
	Wash	Monthly
Painted wood and metal	Dust/Wash	Daily/Annually
PVC covered upholstery	Dust/Wash	Daily/Monthly
Plastic	Dust	Daily
Polished wood and metal	Dust/Polish	Daily/Fortnightly
Floors:		
Carpets and rugs	Vacuum clean	Weekly
Concrete	Sweep	Daily
Hardwood	Sweep/Polish	Daily/Monthly
Linoleum	Sweep and lightly buff	Daily
Rubber	Sweep/Wash	Daily/Weekly
Terrazzo	Sweep/Wash	Daily/Weekly
Tiles	Sweep/Wash	Daily/Weekly
Verandah floors	Sweep/mop	Weekly/as needed
Walls:		
Face brickwork	Brush down	Annually
Glazed tile	Dust/Wash	Weekly/Monthly
Glazed brickwork	Dust/Wash	Weekly/Six-monthly
Paint(gloss, emulsion, semi-gloss, flat oil)	Dust	Weekly
Polished hardwood	Dust/Polish	Weekly/Annually
Ceilings:		
Washable distemper	Dust	Monthly
Paint (gloss, semi-gloss, flat, emulsion)	Dust/Wash	Monthly/Annually
Door furniture:		
Metal finishes	Dust/Clean	Daily/Weekly
Sanitary fittings:		
Ceramics	Clean	Daily
Window glass	Clean	Monthly

14.1.7 Graffiti

Begin cleaning as soon as possible after the graffiti appears. Work on a test section first. For masonry, it is usually best to remove graffiti by treating the whole of the affected area to the margins of the individual blocks. On a painted surface, if the remover works by softening the paint film, it is better to restrict the treatment to the actual paint surface. If a graffiti inhibitor has been applied to the surface follow the instructions of the manufacturer. For acrylic, epoxy or alkyd paint graffiti, use a paint remover with a bristle brush or medium-pressure water jet. Remove spirit pen and aerosol paint graffiti by scrubbing with a solution of detergent and warm water.

Poultice may be the best method of removing some graffiti. A poultice consisting of absorbent material or clay powder (shredded paper, paper towel, kaolin or fuller's earth) mixed with a liquid (solvent or other remover) to form a paste is applied to the graffiti. The poultice should be kept moist and left on as long as necessary to drain the graffiti stain out. As the paste dries it absorbs the graffiti stain. Pre-mixed poultices are available in a paste or powder form. The masonry must be pre-wet before applying an alkaline cleaning agent, but not when applying a solvent. Rinse thoroughly after the graffiti stain has been removed.

14.2 STEEL WORK MAINTENANCE

Steel is very much affected by atmospheric agencies, especially when it comes in contact with water, the slow but continuous oxidation of steel results in its corrosion or rusting. **Some of the methods adopted for protecting corrosion are given below:**

(a) **Painting**: this is the most common method for protecting steel and other metals, but great care is needed in the selection of type of paint. Red lead and alluminium paints are mostly used for iron and steel work. The life of painting is limited depending on the impurity in atmosphere.

(b) **Encased into concrete**: if the steel is to come in contact with water *e.g.* storage reservoirs, to prevent from rusting, it is better to encase it in concrete.

14.2.1 Painting of steel work

The surface of iron and steel should be free from scales, rust and grease before painting. Scales and rust can be cleaned with hot water so that remaining acid is washed off. Sometimes it is also dipped in hydrated lime solution after washing. Grease of cold rolled steel should be removed by using any petroleum solvent. Grease can be removed by hot alkaline solution also. Phosphoric acid is applied over the surface to protect it from rusting. First coat is applied after preparing the surface. Mainly red lead or lead oxide in linsed oil is used in first coat. Second coat is given only after first coat has dried. Paint or finishing coat must be durable and elastic. If it is desired to have black colour then graphite can be used. For white colour zinc oxide or white lead can be used. Lead or zinc paint should not be applied directly on the surface or iron as encourages galvanic action which destroys paint.

14.2.2 Repainting of steel work

Scales and rust be cleaned before repainting the surface of iron and steel. If some grease is present on the surface then it should be leaned by lime and water. After the complete clearing, one or two coats of paint are applied in the same way. If it is desired to remove the old paint completely then following method may be adopted.

(a) **Burning -** Paint may be burnt with a blow lamp suitable for this purpose. After burning the surface can be scrapped and cleaned. This method can not be used for wood.

(b) **By paint solvent -** Paint can be dissolved by any solvent and then removed.

(c) **By stipso -** It is a patented product for removal of paint. Paint can be removed very quickly by this method from any type of surface. Stipso spread over the painted surface and allowed to remain for sometime, after that the surface is washed by cold water.

14.2.3 Defects of painting

If painting is not done properly then many types of defects occur as follows.

(a) Cracking and scaling
(b) Chalking or rubbing off the clothes
(c) Exudation
(d) Discolouration
(e) Alligatoring
(f) Wrinkling
(g) Washing
(h) Blistering and peeling
(i) Checking
(j) Mildew
(k) Running and sagging

Cracking and scaling:

This defect is due to the brittleness of paint film. If thickness of coat is more its effect is still worse. Cracking occurs before scaling in many cases due to the expansion and contraction of wood under paint. Due to cracks moisture creeps in and consequently scaling occurs. Paint which gives elastic film does not suffer from this defect.

Chalking or rubbing off the clothes:

If paint is deficient in oil or if first coat has not been given then base is separated slowly. Due to this if one comes in contact with painted surface then paint sticks to hand or clothes.

Exudation:

Knots exude resin and hence the painted surface is spoiled. Knotting should be done to prevent it.

Discolouration:

Discolouration of paints occurs slowly. For example colour of white lead changes if it comes in contact with sulphur. Proper selection of paint should be done to avoid this defect.

Alligatoring:
The essential feature of alligatoring is the sliding of one layer of paint over another. This occurs due to marked dissimilarity of paints. To avoid this defect, it is essential that the properties of all coats should be similar. Second coat should be applied only after first coat has dried.

Wrinkling:
If the thickness of paint is very large or if there is more of oil in paint then this defect occurs. Under portion of paint does not dry due to large thickness and later on it shrink after drying. It is better to give 1 to 3 thin coats than one thick coat.

Washing:
This defect occurs when low-grade paint is used or the paint contains water. If there is presence of SO_2 gas in atmosphere then this defect occurs even in good paints.

Blistering and peeling:
This is caused by excess moisture vaporizing in back of the paint film. Due to this pressure is created and blistering and peeling occurs. Painting should not be done just after heavy rain. Blistering can occurs due to strong sun also. If pores of wood contain moisture while painting; then peeling will occur.

Checking:
If finishing coat is harder as compared to first coat then this defect occurs. Due to this the appearance of painted surface is spoiled.

Mildew:
Mildew thrives in a place where warmth is and moisture; especially if there is darkness. Zinc oxide has proved to be very good for checking mildew growth. Phenyl mercury oleate can also be used for purpose.

Running and sagging:
Painting over a smooth and glossy surface results in this defect because paint does not stick to it. Hence the surface should be rubbed by sand paper application of paint. Coat should be thin.

14.3 WOOD WORK MAINTENANCE

Our country is rich in forests and enormous timber species are grown. Timber has been in very common use for engineering purpose, since ancient times. Main criteria in the design and construction of timber structures are the choice of species with regards to strength and durability. Unless wood is given an approved preservative treatment it contains a high percentage of naturally durable heartwood it will decay. **The agencies responsible for deterioration of timber are:**

(a) Moisture

(b) Temperature
(c) Insects – such as termite, beetles and marine borers

Moisture and temperature on timber causes:

(a) Dimensional changes
(b) Shape changes
(c) Mould growth

Moisture effects due to:

(a) Alternate dry and wet conditions
(b) Dark and damp conditions
(c) Lack of ventilation
(d) Fungi developing stains and rots

14.3.1 Decay prevention

The following points are the antitermite measures to protect woodwork:

(a) No soil wood contact – protection can be obtained by isolating wood from the soil by use of shields of metal or plastic between the wood and the footing.
(b) No vulnerable wood near cracks in concrete
(c) No undue moisture accumulation about susceptible wood
(d) No possible entry points near sources of warmth
(e) Ensure that the wood remains dry in buildings
(f) Use only wood that is naturally durable

Mode of eradication of termite:

The successful eradication of termite depends on the careful assessment of the infestation, and the thoroughness with which treatments are carried out. Treatment of the entire building is often better than treatment only at points of observed infestation. Chemicals are now available in eradicating termites. A termite eradication scheme can be dividing into three main operations, which can be carried out simultaneously.

(a) Structural repairs
(b) Nest treatment
(c) Creation of poisoned soil barriers

Inspection and maintenance:

The effect of termite eradicating chemicals usually last several years but even then monthly routine inspection should be carried out in the even of reoccurrence of attack, preventive treatments should be immediately undertaken. It would help in controlling termite infestation. If the neighbourhood of the building is kept clear of waste wood stumps and all plant debris.

Precautions:

All the chemicals mentioned are toxic to human beings and animals. They should therefore be handled and used with care. Hand gloves, gumboots and respirators if available, should be used. Contamination of food and drinking water with the chemicals should be strictly avoided. If accidentals spilled on the body or cloth, the effected parts should be immediately washed with warm soap water. Empty containers of the chemicals and surplus solutions should be disposed of in holes dug in the ground.

14.3.2 Repainting old woodwork

Basic principles of repainting are:

(a) Preparation of surface
(b) Application of prime coat
(c) Application of final coat

Before repainting the old painted surface must be thoroughly cleaned off., if greasy, with lime and water, rubbed with pumice and all cracks, dents and holes filled with putty. If the ole paint has blistered, the surface must be rubbed down and scraped.

14.4 INSPECTION OF BUILDING

Inspection of a building and any other civil engineering structure is the routine duty of person in-charge maintenance. They are supposed to inspect the same job at a specific period of interval. Inspection means keen analytic and dynamic observation regarding the change in condition of the structure, deterioration, reason and the causes, failure and concluding the remedies for the same.

The inspections should be made on the following points:

(a) Condition of wall paint
(b) Condition of paint on woodwork and grill
(c) Condition of flooring
(d) Roof leakage, leakage, if any
(e) Condition of service fittings
(f) Drainage from terrace or pitched roof
(g) Growth of vegetation
(h) Structural defects like Crack, Settlement, and Deflection

14.4.1 Condition of wall paint

The paint is acted upon weathering action and protects main structure against weathering the condition of wall paint will indicate the necessity of maintenance and method, way and case of the structure.

14.4.2 Conditions of paint on woodwork and grill

For protection of wood and iron, the painting is very essential. The main defect in painted surface is occurred in form of cracks or blisters. They are due to atmospheric action and alternate wet and dry condition.

14.4.3 Condition of flooring

It consists of Cracks, Dampness, Dusting, Pot holes, and Crowning.

(a) **Cracks -** The most probable cause of cracking in floor may lack of adhesion of the granolythic to the concrete sub-floor, settlement of sub-floor. Previously, these cracks appear thin and small, after some time, these cracks became heavy.

(b) **Dampness -** The dampness in floor is caused due to the following reasons:

Inadequate drainage

Rise in watertable

Use of bad material in filling

The remedial measures may be:

- ❖ Change of filling materials
- ❖ Proper sub-soil water drainage
- ❖ Use of water proofing material in concrete flooring

(c) **Dusting -** Dusting is the appearance of dust particles. Actually the sand and cement particles, which cannot stand against the abrasion or as these are not in proper bond into the main floor. This may be due to segregation, wrong placing and cementing or its composition of concrete and insufficient curing of the floor.

(d) **Pot Holes -** Potholes are the small depressions in the floor at the time of laying finest cement mortar. Sometimes air bubble left in between sub-base and floor and the walking of persons, which particular pot goes down from a hole, this hole is known a Pot Hole and this hole increases successively. Potholes may also be caused by the presence of hard small particles in floor, if it is not maintained properly it further develops in the bigger pothole.

(e) **Crowning -** It is bulging of the tile floors. It may be due to sagging and also because of relative movements of tiles and sub-floor, because of temperature and moisture.

14.4.4 Roof leakage, leakage if any

This may be due to defective drainage of rainwater, from roof or due to construction joint. By the leakage of roof, dampness may occur, which may spoil the structure. To inspect the crack and leakage before the monsoon and defects arrested in roof covering.

14.4.5 Condition of service fittings

The service fittings like hot and cold water supply pipe, sewage pipe, drain water pipe and other service fittings should be inspected. Whether there is some leakage, chocking, breakage and needs some repair and replacement.

14.4.6 Drainage from terrace or pitched roof

The water may remain stagnant on the terrace due to improper slope, insufficient drainage, local depression and wrong position of drainpipe. Leakage also takes place through defect in roof covering. In case of pitched roof, it may give rise due to cracks, breaking of the roof tile and improper drainage and their jointing.

14.4.7 Growth of vegetation

The humid environment and ill – ventilated room and continuous presence of dampness the growth of vegetation occurs. It is harmful for the structure, because it acts biochemically over the plaster.

14.4.8 Structural defects

The structural defects like settlement of foundations, cracks in masonry walls, beams, slabs and other elements of the building needs repair. Deflection and sagging of members like beams, slabs, purlins, rafters, and lintels must also be observed and needs any procedure for improving. Even bulging, which is failure of member under compression, should be attended immediately and the member may be strengthened.

14.5 ROUTINE BUILDING MAINTENANCE

The following are few of the routine maintenance of the building, which are attended either at a fixed interval or attended as and when needed and provisions made in the annual repair fund.

(a) **Whitewashing or colour washing** - These washings come under routine maintenance, because they are usually done at a fixed interval of time to maintain the building properly.

(b) **Distempering** - It is done once in a year or two years.

(c) **Plaster and roof repairs** - According to the necessity, it is undertaken in floor for holes, cracks, etc. and for plaster crack filling and patch repair.

(d) **Roof repair for leakages** - Roof repairs are done before monsoon. If there is concrete roof, having a source of leakage, we can use cement slurry as well as cement concrete.

(e) **Replacement of fittings and fixtures** - The door fittings, sanitary fixtures and defective service pipes are repaired and replaced in order to protect the building and render efficient services.

14.6 DEPARTMENTAL PROCEDURE FOR REPAIRS OF BUILDINGS

The departmental procedure consists of estimate for ordinary repairs, estimate for special repairs, and control of expenditure on ordinary repairs, repair orders and maintenance allowance for building.

14.6.1 Estimate for ordinary repairs

The ordinary repair estimate should be prepared. The cost of repairs is first fixed as a lump sum when the building is constructed or purchased, and is usually taken as the following percentage of the capital cost of the building;

First class building	1.5%
Second class building	2%
Third class building	3%

The above percentage is revised, when the building is altered in such a way as to affect the cost of ordinary repairs, *e.g.* when substantial additions or alterations are carried out. The sanctioned estimates for ordinary repairs to buildings will be current for one working year ending on 31st March. In the case of irrigation buildings, when the cost of repairs is dubitable to the allotment for maintenance of irrigation works, the working year for estimates for repairs ends on the 30th June. Repairs to public buildings are classed under two categories, *viz.* Ordinary and Special. Ordinary repairs are repairs that are carried out as a matter of routine with (*i*) year by year (annual) or (*ii*) at longer intervals (Periodical)

14.6.2 Estimates for special repairs

Estimate for special repairs will be prepared in detail for each building separately and remain current till the completion of the repairs.

Some of the special repairing works in a building are given below:

(a) A broken W.C. seat to be replaced
(b) Opening is to be made in existing wall
(c) Repairs of sagged mangalore tiled roof
(d) Replacement of broken S.W. sewer pipe
(e) Cleaning of choked residential sewer line
(f) Replacement of broken W.C. gully trap

14.6.3 Control of expenditure on ordinary repairs

In April each year, the Superdent Engineer will place at disposal of the Executive Engineer the total grant for ordinary repairs to all the buildings in the division.

(a) 30% for the payment of work establishment chowkidars, ordinary monsoon repairs, white washing and other repairs, which remain constant from year to year.
(b) 70% for payment on repairs to be specified in repair order.

14.6.4 Repair orders

Each repair order will be prepared in duplicate and will show the work to be done and the estimated cost. A separate repair order will be prepared for each building, or for each group of buildings under the same minor head of account, *viz.*, Education, Medical, Forest etc.

(a) After sanctioning and allotting a serial number to the repair order.
(b) During the execution of the repairs covered by the repair order, the subordinate will record on the back of the form the details of the work done and the actual cost of the labour employed and the materials used.
(c) Immediately the work is completed, the subordinate will record on the form the actual cost of each item, sign the certificate and send the completed form to the superior.
(d) The original and duplicate repair orders will be filed by the superior in an indexed file. Separate files may be opened, if necessary for large buildings or a group of buildings.
(e) The superior will check the completed repair orders on the spot and verify if the changes for labour and materials are reasonable, and will then sign the certificates given on the form.
(f) The repair orders are not account forms and will not be subject to audit, they are for use by executive officers, and should be posted in such a way that superior officer can readily check, on the spot, the work actually done.

(g) In the work abstracts and register of works the expenditure should be booked under the one head "Ordinary repairs" and not separately under "Annual and Periodical".

14.6.5 Maintenance allowance for building

For all building either residential or non-residential an amount of Rs. 3.50 per sq./m. of plinth area is allowed. An additional amount is Rs. 1.50 is allowed for hospital building. This amount is only for annual repairs. Amount for repair is obtained as and when necessary depending on the condition of the building.

14.7 INTEGRAL MAINTENANCE OF BUILDING

Maintenance of a building is the process of keeping it in proper condition structurally and functionally. Integral maintenance embraces buildings with all internal and external services and furniture as well as all appurtenant equipment inside, *viz.*, lifts, air-conditioners, heating equipment, water coolers and refrigerators. It thus covers all the fittings, fixtures and equipment; external services like electricity, water-supply, sewage disposal, drains and internal roads on which the comfort of the occupants depends. No norms can be laid down for maintenance of private buildings as these entirely depend on the owner's resources and interest in his property.

14.7.1 Role of maintenance engineer

Integral maintenance demands a blend of engineering knowledge, experience managerial skill and public relations. The Maintenance Engineer should possess ingenuity insight and be capable of innovation and improvisation. He has to manage the resources of materials and labour and tackle problems pertaining to different types and levels of users. **The role of the maintenance engineer:**

(a) To serve the users for whom the assets have been created

(b) To report with users by meeting in various cross-sections frequently and by understanding and anticipating their genuine problems

(c) To adapt a positive approach to the needs and problems of the users and give a fair deal to all

(d) To inculcate a feeling of ownership in the occupants of residential buildings so that they take care of the building, its fittings and fixtures, thus reducing the need for repairs

(e) With prompt attention to rectification of electric, water supply and sewage defects will promote good will

14.7.2 Scope of maintenance

Integral maintenance covers:

(a) Petty repairs, replacements and structural repairs of buildings, white and colour washing, distempering and painting at prescribed intervals

(b) Repair and renewal of furniture

(c) Operation, periodical maintenance, repairs renewals of machinery and equipment for electric and water-supply, air conditioning, refrigeration, vehicles and sewage installations

(d) Repair of roads, culverts and resurfacing the roads

(e) Repair and renewal of fences, boundary walls, drains and miscellaneous items

14.7.3 Scale of maintenance funds

The annual grants are generally made on the basis of a fixed percentage of the capital cost. In case of buildings the amounts are laid down by the financial authorities separately for permanent and temporary buildings at fixed rates on the plinth area. The rates vary depending on the importance of the building *e.g.* offices, hospitals and residential accommodation for different ranks of personnel. It is impossible to carry out effective maintenance with the meager annual grants at a fixed percentage of the capital cost, when prices of materials and labour are continually rising. The building cost index went up by 100% between 1966 and 1975 and nearly 300% between 1975 and 1985. It is logical to link the scales to the building cost index by a suitable formula and also to connect it to the residual life of the building, as the quantum of repairs necessarily increases with the age of building.

14.7.4 Steps to reduce repairs and replacement

(a) Use of quality materials and fittings and adoption of good specifications during construction

(b) Delicate or intricate architectural frills merely for sake of looks should be avoided Geographical and climatic conditions need consideration in the choice of materials

(c) The engineer has to keep abreast with modern techniques and developments for economical and faster construction.

(d) The tendency to follow the easiest course and traditional pattern should be discouraged and engineers must face the challenge for innovation.

(e) It is essential to avoid wasteful expenditure on wholesale renewal of floors, drains etc.,

(f) To lay down priorities for items of maintenance to be done in a year

(g) To lay down what items should be executed only after obtaining clearance from the maintenance engineer

(h) Greater attention should be paid to the reliability, maintainability and suitability for the end use to avoid frequent breakdowns or costly replacements

14.7.5 Normal break-up

All maintenance tasks involve:

For buildings:

(a) Periodical services,

(b) Repairs through contracts,

(c) Petty repairs and

(d) Replacements by departmental labour

For installations:

(a) Operations involve labour, fuel and lubricants
(b) Periodical maintenance involves labour and materials
(c) Repair, renewal and overhaul
(d) Payment to other agencies, if bulk supply of power and water is received from them

For furniture:

(a) Requiring polishing,
(b) Rescanning of chairs,
(c) Restringing of charpoys,
(d) Repairs and replacements

14.7.6 Management tools for effective maintenance

The maintenance engineer should follow the principle of management by objectives. The tools at the disposal are:

(a) Data
(b) Management of resources (labour and materials)
(c) Information and control and
(d) Public relations

For high standard of maintenance, all data are under maintenance charge of the engineer. It is helpful to plan management of resources and to anticipate trouble spots.

Management of materials is the vital aspect of maintenance. All plans, organization of labour and standing orders will go waste if materials are not readily available. Without a proper inventory control, it is not possible to ensure smooth maintenance. If proper and modern tools in adequate number are not available, the management of labour and materials cannot give desired results. Similarly equipment like telescopic ladders, repair kits for electrical repairs, sewage line cleaning equipment, safety equipment for workmen as well as accessibility of rooftops and windows in tall building are necessary. Handcarts for transport of petty stores and mechanical transport for labour and stores and supervisory staff should be available since mobility is essential for economy and efficiency.

Management of labour is the most delicate and difficult task. In Government department there are large labour forces in maintenance organizations. As they are not paid according to wages prevalent in the labour market, it is difficult to recruit highly skilled workmen. Since incentives are lacking and punishment of defaulters is difficult, if not impossible, the morale is not high. The engineer should evolve a procedure for obtaining regular information of the state of maintenance of the different categories of amenities. For example:

(a) A daily report on the state of service fittings like air conditioners including standby equipment
(b) A weekly report on
    - The number of complaints received during the previous week in respect of each residential colony
    - The number and description of complaints which could not be attended

- Reasons, whether due to non-availability of materials, labour or any other problem

(c) A weekly report on electric supply, water supply, central air-conditioning and sewage installations and vehicles that periodical maintenance tasks have been carried out

(d) A weekly report in the condition of street lights

(e) Fortnightly report

- That all external water, electric and sewage lines have been inspected physically and that there are no leakages or defects unattended; that no overflow in overhead reservoirs and service tanks is noticed;
- Repair, state of furniture-number of important items category-wise received in the workshop for repair and recanning, how many completed and the reasons for not completing the remaining items.
- State of maintenance stores-whether any item is below the minimum level or out of stock; what items have been received during the fortnight.

Preventive maintenance is essential to avoid breakdowns and costly repairs. All buildings, fittings, fixtures, furnitures and all machinery and equipment, including electric, water and sewage lines should be inspected thoroughly at regular intervals and timely action taken to rectify defects. Fans should be oiled once a month in summer and overhauled in winter according to a prearranged programme. All heating appliances like geysers should be thoroughly checked in summer to insure that they are in working condition before the cold season starts.

CHAPTER 15

Safety in Maintenance and Demolition

15.1 SAFETY IN MAINTENANCE

Safety depends to a great extent on the individual safety performance. Safety is one of the most important aspects of modern management today. The workers must be taught into a multitude of situations without much help from prescribed rules. An accident is defined as an unforeseen event, which causes personal injury or property damage. Accident defines it even more broadly as "any occurrence that interrupts or interferes with the orderly progress of the activity in question."

15.1.1 Causes of accidents

Some of the causes of accidents are as follows:

(a) Lack of intelligence
(b) Lack of experience and training
(c) Effects of drugs and drinks
(d) Fault of machines and equipment
(e) Environmental factors
(f) Physical handicaps

15.1.2 Remedies to avoid accidents

(a) Physical fitness of labour should be judged before allotting maintenance work
(b) Proper instruction to handle lifting devices and machines should be given
(c) For hard jobs the time period of work should be reduced
(d) Centering and scaffolding should be strong enough
(e) Clear instructions should be given regarding procedure to handle heavy loads
(f) Avoid unnecessary crowding, at work place
(g) Proper ventilation in the working place
(h) Educate workers on special work to be entrusted
(i) Provide warning notices at proper places

15.1.3 Accident prevention

An accident prevention programme should be initiated on every project in order to reduce the cost of construction. The following are the objects of accident prevention programme:

(a) To reduce the human life scarified
(b) To lessen the temporary and permanent injuries to workers
(c) To avoid loss of materials resulting from accident
(d) To reduce the amount of time lost as a result of accident

(e) To safeguard against loss of damage to equipment
(f) To hold the expense of workmen's compensation to a minimum

15.2 CONSTRUCTION ACCIDENTS

They may be grouped as follows:

(a) Uncontrolled contact between men and equipment or between men and materials, such as cranes, trucks and material storage
(b) Failure of temporary structures such as forms, scaffolds ramps, ladders, cofferdams, etc.
(c) Inherent engineering hazards, such as the use of explosives, presence of injurious gases, toxic dusts, etc.
(d) Unsafe practices of individual workers or personal hazards resulting from the carelessness of works.

15.2.1 Safety programme for construction

Most of fundamentals of safety apply to all kinds of construction. Safety programme should be made an integrated part of the operations of each construction company. The following measures when vigorously promoted by management have proved effective in reducing the accident rate on construction:

(a) Interest in safety on the part of top management
(b) Adequate and capable full time safety personnel
(c) Publicity of safety programme
(d) Development of a safety programme for each job
(e) Indoctrinate new employees
(f) Promote good house-keeping
(g) Provision of adequate first-aid facilities
(h) Assistance from the insurance agencies

15.3 SAFETY IN BUILDING MAINTENANCE

For dismantling and maintenance of various engineering structures it is essential to complete the operations safely and effectively. Monumental structures that are well worth considering required scheduled controls and periodic maintenance procedures to retain the traditional cultural values of the community. Numerous projects of modern building system are being constructed with special architectural and new modern facilities due to the increasing demands for improvements to the quality of life, and sense of well being at the work place. The buildings in urban areas have to changing harmony with the need of the age. Moreover, the changing work patterns with modern, productivity-enhancing equipment lead to new conditions to which the building has to be adopted. Under these circumstances, it is natural to consider demolition work to be as important as this construction work for the sound development work of urban areas. **The need of dismantling/demolition work:**

Demolition work is inevitable for many developmental works as follows, when:

(a) The structure or a part of it is massive and unsafe to perform any kind of function

(b) The structure or a part of it needs extensive structural modifications to drive superior advantages
(c) The structure may be safe, but not suitable for modern requirements
(d) The substructure or the foundation bed removal is essential for development works
(e) The structures severely suffered during natural hazards such as earthquake, landslide, fire, cyclone, floods, etc.
(f) The structures severely damaged during the war time.

15.3.1 Precautions prior to dismantling

(a) If some part of the structures is to be dismantled for repairs have got hidden damages, caused by fire, earth quakes, flood, necessary measures to prevent accidental collapse using shorting, bracing etc., shall be provided.
(b) Prominent danger signs should be placed all around the property and all entrances to the building should be barricaded except for the workers. At night time lights shall be placed on or above all barricades.
(c) All service connections such as electricity, system, gas and water should be disconnected outside the property line or at intermediate points, for safety of workers.
(d) All necessary safety appliances like steel helmets and goggles, gloves etc., required for maintenance of different structures should be provided to the workers prior to starting of dismantling and maintenance work.
(e) For safety of public, safety distances should be clearly marked and prominently sign posted. All main roads should be kept open, where necessary diversions for pedestrians should be constructed for safety.
(f) As far as possible no demolition work should be carried out during nights. If such night work has to be done, additional precautions by way of red warning signals, working lights and watchmen should be provided to avoid any injury to workmen and public.

15.3.2 Precautions during dismantling

(a) Materials of fragile nature like glasses should be removed before commencement of work
(b) To prevent harm to workmen dust should be controlled by suitable means
(c) All openings should be boarded up
(d) Proper ventilation and lighting shall be provided at the working places
(e) During emergency easy exit should be provided for quick evacuation of the workers
(f) Stacking of debris and materials should be within safe limits of the structural members

15.3.3 Dismantling sequence

The dismantling work should always proceed systematically storey-by-storey from top to bottom. In upper floor all work should be completed and approved by the maintenance Engineer-in-charge prior to disturbance to any supporting member on the lower floor.

15.3.4 Dismantling wall

(a) Wall should be removed part by part in reasonably level courses. They should not be allowed to fall in large masses, which endangers the adjoining properly or exceeds the safe loads of the floor below. Debris should be removed at frequent intervals.
(b) If the walls are less than 30 cm thick and dangerous to work by standing over them, suitable staging should be provided for the workers.
(c) Workers standing on scaffolding clear of the arch should demolish arches.
(d) For unsound walls adequate lateral bracing should be provided.
(e) In case of exterior walls of multistory structures, scaffolding platform of sufficient strength to prevent injuries to workers below and public should be provided. Overcrowding and storage of material on catch platforms should be avoided.
(f) Safety belts or ropes should be used by workmen when working at higher levels.

15.3.5 Dismantling floors

(a) Prior to removal of masonry or concrete floor adequate support, centering should be provided.
(b) Planks of sufficient strength should be provided to give workmen firm support to guard against any unexpected floor collapse.
(c) No person should be allowed to work in an area directly underneath and access to such area should be barricaded.
(d) Debris may be dropped through holes in the floor. In such a case precautions should be taken to avoid overloading of the floor.
(e) Debris should be removed at the earliest to ensure safe and adequate working space.

15.4 DEMOLITION MANAGEMENT

A construction removal and demolition company is comprised of team of engineers, project managers and demolition technicians providing effective and efficient work. The demolition work is carried out through cutting, drilling, and rigging in a safer way, for the environment and its employees. With well-planned and precise drawings of the background of the site and projects are able to complete any designated job at a better and fairer price by demolition contractors. With routine check ups at the site, that are able to provide a safer work environment and with better equipments. Hence, contractors are able to remove construction materials and demolish structures within a set time frame and with lesser accidents.

Demolition Services:

(a) Diamond wire sawing demolition service for heavily reinforced concrete, solid stainless steel and rock.
(b) Blade sawing demolition service for steel-reinforced concrete, pre-cast concrete, asphalt, stonewalls and cobblestone street beds.
(c) Core drilling demolition service for steel-reinforced concrete, pre-cast concrete, asphalt, masonry, metal and rock.
(d) Parapet/curb cutting demolition service for steel-reinforced concrete, pre-cast concrete and asphalt.

15.4.1 Selective Demolitions

With combination of special machinery, skills and experience offer selective demolition of any kind or anywhere. Understand the physical and environmental risk of any demolishing tasks especially selective demolition. Understand the time constraint and the value of budget as well. Therefore, conduct a detail research before initiating a task of any magnitude, to study the risk and threat factors and come up with the solution, beneficial to all.

Hydro Demolition:

A hydro Demolition service is one of a kind in this industry. Experience with wire sawing helps to make cuts of limitless size. When all other options failed or became too expensive, but diamond wire cutting system and innovative hydro demolition techniques enabled the project to be completed with in budget.

Marine Demolition:

Marine Demolition is one of the most risky jobs in this demolition industry. It is possible to make on many marine explorations to cut or remove underwater metal structures and to complete marine demolition needs.

Underwater Demolition:

Underwater Demolition could include jobs such as blasting a hill, making way for river or for a dam or blasting riverbed for tunnel. These types of underwater demolition jobs require special machinery, well-sketched plan, and experienced divers. Successful demolition requires good communication among the people doing the job. This is especially true for underwater demolition. Along with communication, coordination of work is necessary as well to create a safer working environment. To understand that a minor mistake could cause a huge physical and financial loss for any projects. Therefore, we take extra care before and during the work.

Bridge Demolition:

Demolition of a bridge requires great deal of careful observation before the demolition work. The bridge demolition work should not disturb the traffic flow. If the bridge is currently being used, you will need to construct an artificial bridge to detour the oncoming traffic.

Building Demolition:

There are basically two ways to demolish a building. One is the old fashion manual way and the other is by exploding with dynamite. The second is used mostly these days to lower the cost and save time. However, this solution is easier if the building is secluded. If the building that needs to be demolished is within the city limit then the preferable way would be the manual way. This process may take a longer time but it is environmentally safer. But due to time and budget constraints, use of explosives may be the preferred choice.

Nuclear Demolition:

Nuclear plants have been used as power generators. These sensitivity plants need to be maintained on timely basis. Maintenance may include demolition of certain section. Demolishing heavily reinforced structures and sensitive areas make the choice for demolition of nuclear related demolition includes; such nuclear tasks as decommissioning, commodity removal, head and reactor vessel replacement.

Runway Demolition:
As we all know air traffic has grown to be busier day by day. With growing airplane traffic, old runways demolition and construction material removal need to be done quickly and without hindering the airline operations at the airport. With expertise on cutting and coring techniques to tackled projects at a lightning speed.

15.5 CONCRETE DEMOLITION

The following figure 15.1 shows a concrete demolition procedure.

Figure 15.1. Concrete demolition

15.5.1 Methods of demolition

(a) Pneumatic and Hydraulic Breakers
(b) Dismantling
(c) Pressure Bursting
(d) Explosives
(e) Ball and Crane for Demolishing Masonry and Concrete Structures

15.5.2 Demolition tools and materials

(a) Hydraulic Splitter
(b) Expansive Demolition Agents
(c) Demolition Hammer
(d) Rotary Hammer
(e) Chipping Hammer
(f) Pavement Breaker
(g) Mounted Breakers
(h) Hydraulic Hammer
(i) Pneumatic Hammer

(j) Hydraulic Concrete Crusher
(k) Diamond Wire Saw

15.5.3 General information regarding demolition permits

The following is for GENERAL INFORMATIONAL USE ONLY. This is a very general overview of the permitting process for demolition projects. The actual process may vary widely between regions of the country, counties and municipalities.

(a) When is a demolition permit required?
(b) Why is demolition permits required?
(c) Who can apply for a demolition permit?
(d) What does the demolition permit cost?
(e) What is involved in the typical demolition permit process?
(f) What is involved in the demolition inspection process?

Safety:

(a) Prior to the Start of Demolition
(b) Hand Tool Hazards
(c) Power Tool Hazards
(d) Machine-Mounted Tool Hazards
(e) Clothing and Dress Tips
(f) Recycling Concrete

15.6 REVIEW OF ADVANCED DEMOLITION TECHNIQUES

Mechanisation of demolition works on reinforced concrete structures started in late fifties with the introduction of pneumatic hand hammer breakers and steel balls. Subsequent development of diamond cutters, flame jetting methods, C-shaped concrete crushers, and expansive chemicals have lead to more sophisticated process of civil engineering works. Initial methods of demolition have caused many serious problems of environmental pollution hazards in public life.

Expansive demolition agents have been widely used to demolish rocks and concrete structures. To design the demolition works using expansive demolition agents rationally, it is important to estimate the expansive pressure exactly and to determine its action on the inner surface of a borehole. The new pressure transducers developed can directly measure the expansive pressure in a borehole. Several variables affecting the nature of expansive pressure in a borehole were investigated with the transducers. They are

(a) Expansive pressure is transmitted in a manner similar to that of fluid.
(b) The influence of the physical restraint of surrounding materials is very small.
(c) The expansive pressure depends on thermal conditions including ambient temperature, and the degree of hydration of CaO.

The traditional approach to demolish concrete structures or to reduce the sizes of large rocks or boulders has typically included the use of **explosives**. The resulting explosions are associated with the obvious risks posed by shock waves and fly rock. These, along with other detrimental side effects of using explosives, have increased an interest in the use of alternative methods to demolish rock and concrete structures. Soundless **chemical demolition agents** (SCDAs) have proven to be viable substitutes for the use of explosives. SCDAs are powdery materials that will expand considerably when mixed with water. This expansion, when occurring under confinement, generates significant expansive pressures. These pressures are sufficient to break up rock and concrete when the SCDA is confined in a borehole or a series of boreholes. Experimental results show that the amount of mixing water and the ambient temperature are the most important variables in influencing the generation of SCDA expansive pressures.

Hydro-demolition is a relatively new method of removing select portions of a hardened concrete structure. By using the erosive power of high-velocity water streams, hydro-demolition equipment breaks up concrete by disintegrating the cement matrix between aggregates. The demolishing effect can be tightly controlled to a desired level of removal, ranging from light scarification of the surface to deep penetration of the structural element. The use of the hydro-demolition process has several advantages over conventional concrete removal methods, such as jackhammering. These advantages include a reduction in new damage caused by the removal process automation, which produces a very consistent level of removal energy over large areas; the ability to seek out and remove weak or deteriorated locations at various depths; and a rough, high-quality bonding surface for repair materials. These characteristics are favourable for construction projects that involve rehabilitation of corrosion-damaged reinforced concrete structures, particularly bridge decks.

15.6.1 Review of guidelines

Design for explosive demolition has four points for critical examination, namely,

(a) Structural stability under preliminary weakening of critical links
(b) Location, size, shape and triggering time of charge for plastification of structure
(c) Effect of ground impact on people and adjacent property, and
(d) Mechanization to clear debris at the site

The construction and design details of the building have to be well understood, before the design for explosion starts. This will provide the information for partial delinking of the major structural components without the loss of stability. Depending on the type of structural element such as, mass concrete, reinforced concrete beam or column, steel girder, steel plate, etc. charge shape, size, pattern of holes and delay time can be arrived.

Dismantling of massive reinforced concrete or steel structures by blasting technique is an effective and an elegant method. The developmental programme can be quickly initiated by clearing the unwanted structures by blasting technique. The developments have facilitated adoption of this popular technique a safer and economical method.

The data collected from previous blast demolition works, helps to improve the guidelines. It is essential to conduct an engineering survey, hold prejob safety meetings, and make sure that the public is aware of this operation and its consequences. The buildings likely to have high risk of damage due to blasting operation, and the buildings with medium/less risk of damage are to be identified prior to the operation. The equipment used should be checked for its operation. The local bodies and police personnel should be informed about necessary safety steps to be taken. Safety of people and protection of surrounding property has to be given top priority.

15.6.2 Modern technology for demolition

Build Aids offers sound solution to any technically demanding project through diamond cutting technology. In numerous construction and renovation projects, diamond-cutting technology offers the competitive edge in concrete demolition and removal. Architects, engineers, and general contractors agree that diamond-cutting technology is cost effective and time-efficient. Often used in conjunction with more conventional impact demolition techniques, diamond cutting and drilling methods provide various benefits, which can make up the difference between a successful project and an inefficient one.

With agents all over India working in tandem with qualified personnel, each acting as a regional center of execution, build aids is ideally placed to service the local market with its access to the latest technology.

Equipped with the most advanced development in drilling and sawing technology, Build Aids serves customers all over India with a host of services that include:

(a) Floor sawing
(b) Hand sawing
(c) Diamond core drilling
(d) Wire sawing
(e) Controlled demolition by blasting

Diamond tools can cut any type of building material and hence are used widely in the construction industry for new construction or for demolition operations.

Typical applications include:

(a) Partial demolition of RCC structures
(b) Sawing floors for vibration isolation and jointing
(c) Drilling holes in both RCC and brick work
(d) Cutting expansion and control joints
(e) Large, bulk concrete removal
(f) Bump grinding and grooving of uneven concrete surfaces
(g) Rectification/removal of damaged foundation bolts of machinery

Build aids offers the following services to the customers/clients-core drilling, stitch drilling, and diamond wire sawing and controlled demolition system.

OBJECTIVE TYPE QUESTIONS

1. Preventive Maintenance depends upon
   (a) Specification of structures
   (b) Condition of structures
   (c) Use of structures
   (d) All the above

2. A maintenance done after the defects or damage is called
   (a) Remedial Maintenance
   (b) Routine maintenance
   (c) Special maintenance
   (d) All the above

3. The routine maintenance done by the fund provided annually
   (a) 2 % of the cost of construction
   (b) 1.5 % of the cost of construction
   (c) 3 % of the cost of construction
   (d) None of the above

4. The main aim of maintaining any structure is to
   (a) Improve its appearance
   (b) Utilise the funds provided
   (c) Stabilize the structure
   (d) Utilise services of incharge maintenance

5. Which one of the following is preventive maintenance
   (a) Filling up cracks in masonry
   (b) Painting of doors and windows
   (c) Repair of pot holes on road
   (d) Removal of flooring

6. Tile turning of tiled roof in the month of may is the example of
   (a) Special repair
   (b) Pre-monsoon repair
   (c) Remedial repair
   (d) Post-monsoon repair

7. A roof has fallen in rainy season. Its maintenance will be classified under which of the following maintenance:
   (a) Special repairs
   (b) Routine
   (c) Pre-monsoon
   (d) None of the above

8. The life of painting to steel work is limited depending on the
   (a) Impurity in atmosphere
   (b) Temperature in atmosphere
   (c) Humidity in atmosphere
   (d) None of the above

9. No cracks are visible on the exposed surface of a verandah slab, but it leaks sufficiently during rains. The possible reason is
   (a) Less thickness of slab
   (b) Porosity of slab material
   (c) Under-reinforcement of slab
   (d) Excessive thickness of slab

10. Steel structure is likely to come in contact with water. To prevent rusting
   (a) Apply bitumen coat
   (b) Apply anti-corrosive coat
   (c) Varnish paint
   (d) Slacked lime coat

11. One of the ill effects of exposure of reinforcement is
   (a) Reduction in tensile strength of steel
   (b) Reduction in steel section
   (c) Reduction in bond section
   (d) Gradual decrease in RC section

12. The most effective method of removing old paint from structure is
   (a) Washing with water
   (b) Scrubbing with gunny bags
   (c) Use of oxy- acetylene flame
   (d) Polishing with grinding stone

13. Which one of the following methods will be the best method to make RC slab leak proof
   (a) Plaster the roof with cement mortar
   (b) Lime koba layer over RC roof with proper slopes
   (c) Soap solution to be sprinkled over the RC roof
   (d) Bitumen paint on whole roof surface

14. The most appropriate cement for repairing cracks in RCC work is
   (a) Portland cement
   (b) Expansive cement
   (c) Quick settling cement
   (d) Low heat cement

15. Which one of the following is used to protect the steel from rusting
   (a) Zinc oxide

(b) White lead
(c) Molten lead
(d) Red lead

16. Steel reinforcement in an RCC work is exposed. This will result in
(a) Reduction in steel sectional area
(b) Bad appearance
(c) Development of shrinkage cracks
(d) Settlement of structure

17. Most suitable method of protecting steel columns of a multistoried building against rusting is to
(a) Paint with aluminium paint
(b) Paint with red oxide
(c) Paint with bitumen paint
(d) Encase in concrete

18. Deflection was noticed in a newly constructed properly designed RCC beam. The defect may be due to
(a) Very long span
(b) Diagonal cracks near edge of support
(c) Insufficient concrete cover
(d) Early removal of formwork

19. The defect due to the brittleness of paint film is
(a). Discolouration
(b) Exudation
(c) Cracking and scaling
(d) Chalking or rubbing

20. The defect due to marked dissimilarity of paints is
(a) Exudation
(b) Alligatoring
(c) Wrinkling
(d) None of the above

21. The defect due to more of oil in paint is
(a) Wrinkling
(b) Exudation
(c) Scaling
(d) Rubbing

22. The defect due to painting over a smooth and glossy surface is
(a) Cracking and scaling
(b) Chalking or rubbing
(c) Running and sagging

(d) None of the above

23. The defect due to excess moisture vaporizing in back of the paint film is
(a) Wrinkling
(b) Washing
(c) Running
(d) Blistering

24. Termite control is essential to maintain the serviceability of
(a) Timber structure
(b) Steel structure
(c) Masonry structure
(d) RCC structure

25. Termite make the timber
(a) Crack
(b) Porous
(c) Moist
(d) Wet and dry rot

26. A portion of wall near kitchen plinth is seen always damp. This type of defect is not noticed in the rest of the building. The probable reason is
(a) Use of bad materials of construction
(b) Bad workmanship
(c) Effect of atmospheric agencies
(d) Improper drainage

27. Mosaic tiles flooring gets damp during monsoon season. Which one of the following remedial measures will be most effective?
(a) Change the mosaic tiles and do the flooring again
(b) Apply a coat of water proofing compound over the tiles
(c) Replace the plinth filling with murum and boulders and reconstructed mosaic tiles flooring again
(d) Put a layer of 4cm thick cement concrete over the mosaic tiles flooring

28. The door shutter of newly constructed college building becomes reshaped. One of the reasons for the same may be
a. Poor workmanship
b. Use of cheap quality of wood
c. Use of cheap quality of iron fittings
d. Use of unseasoned wood

29. There is seepage of water on both sides of load bearing wall of a building with flat roof. What remedy will you suggest?
(a) Remove plaster on the wall and replaster with rich mix
(b) Inspect the roof for cracks over the wall and seal the crack

(c) Paint both the surfaces of the wall with oil paint
(d) Plaster both the surfaces of wall with cement mortar mixed with water proofing compound

30. Sinking of floor of a domestic building is due to
(a) Loose filling of plinth by excavated earth
(b) Insufficient curing of floor
(c) Dampness in floor
(d) Thick wearing surface material

31. The maintenance of structure is done to meet the
(a) Prevention of damages
(b) Repair of the defects
(c) Strengthen the structure
(d) All the above

32. One of the maintenance works is
(a) Preventive maintenance
(b) Long term maintenance
(c) Short term maintenance
(d) None of the above

33. The causes which necessitate the maintenance
(a) Atmospheric Agencies
(b) Normal wear and tear
(c) Failure of structure
(d) All the above

34. The defect, improper use of structure means
(a) Improper loading
(b) Improper workmanship
(c) Improper material
(d) Improper climate

35. The maintenance programme is normally drawn taking climatic conditions are
(a) Pre – monsoon period
(b) Monsoon period
(c) Post – monsoon period
(d) All the above

36. One of the method adopted for protecting corrosion is
(a) Painting
(b) Distempering
(c) Varnishing
(d) Cleaning

37. Grease of cold rolled steel should be removed by using
    (a) Alkaline solution
    (b) Petroleum solvent
    (c) Phosphoric acid
    (d) All the above

38. If it is desired to remove the old paint completely following method may be adopted
    (a) By burning
    (b) By paint solvent
    (c) By stipso
    (d) All the above

39. Main criteria in the design and construction of timber structures
    (a) Strength and durability
    (b) Elastic and plastic
    (c) Hardness and toughness
    (d) All the above

40. The agencies responsible for deterioration of timber are
    (a) Moisture
    (b) Temperature
    (c) Insects
    (d) All the above

41. Moisture and temperature on timber causes
    (a) Cracks
    (b) Peeling
    (c) Dimensional changes
    (d) None of the above

42. Moisture effects on timber due to
    (a) Coloured conditions
    (b) Lack of ventilation
    (c) Climate condition
    (d) None of the above

43. The anti termite measures to protect woodwork
    (a) Soil wood contact
    (b) Vulnerable wood near cracks
    (c) Not use durable wood
    (d) None of the above

44. One of the termite eradication scheme
    (a) Structural assessment
    (b) Nest treatment
    (c) Non creation of poisoned soil barriers

(d) All the above

45. Condition of flooring consists of
(a) Dampness
(b) Dusting
(c) Crowning
(d) All the above

46. The dampness in floor is caused due to the
(a) Inadequate drainage
(b) Rise in soil level
(c) Good material filling
(d) Set in water table

47. It is bulging of the tile floors
(a) Dampness
(b) Dusting
(c) Crowning
(d) All the above

48. The role of the maintenance engineer
(a) To serve the users for whom the assets have been created
(b) To adapt a negative approach to the needs and problems
(c) Without attention on rectification
(d) All the above

49. Preventive maintenance is essential to avoid
(a) Breakdowns
(b) Breakdowns and costly repairs
(c) Breakdowns and normal repairs
(d) None of the above

50. An economical and effective demolition technique is
(a) Hydro-demolition
(b) Explosive demolition
(c) Blasting demolition
(d) Chemical demolition

ANSWERS

1. d	2. a	3. b	4. c	5. b	6. b	7. d	8. a	9. b	10. b
11. d	12. c	13. a	14. a	15. d	16. a	17. a	18. a	19. c	20. b
21. a	22. c	23. d	24. a	25. b	26. d	27. b	28. d	29. d	30. a
31. d	32. a	33. d	34. a	35. d	36. a	37.b	38. d	39. a	40. d
41. c	42. b	43. d	44. b	45. d	46. a	47. c	48. a	49. b	50. c

SHORT ANSWER QUESTIONS

1. Define Maintenance Engineering.
2. What is the necessity of Maintenance?
3. Classify the maintenance work.
4. What is preventive maintenance?
5. What are the basic steps on remedial maintenance?
6. What is Routine maintenance?
7. What is the purpose of special maintenance?
8. Name the causes, which necessitate the maintenance of civil engineering works?
9. What are the stages involved in the repair of cracks?
10. How to prepare the surface for repairs?
11. Name the repair techniques as per maintenance categories.
12. Explain repair in ordinary procedure.
13. Write about curing of conventional repair work.
14. Explain repair in special procedure.
15. How to maintain steel work?
16. How to remove the old paint on iron steel work?
17. What are the defects occurs due to painting?
18. Write about cracking and scaling
19. Explain about chalking or rubbing off the clothes by paint.
20. What is meant by exudation?
21. Define the defect discolouration
22. What is alligatoring?
23. Write about the defect wrinkling
24. What is the defect due to low-grade paint?
25. Explain about blistering and peeling
26. What is the defect checking?
27. Define the defect mildew
28. Discuss the defects running and sagging
29. List the steps in sequence in painting old timber surface
30. What are the anti-termite measures to protect woodwork?
31. How to maintain road works?
32. What are the defects occurs in concrete road?
33. Prepare the checklist to inspect building for maintenance purpose.
34. What are points to be considered to inspect a building?
35. How to inspect the flooring condition of a building?
36. List out the reasons for causing dampness in a building.
37. What are the points considered in routine building maintenance?
38. What is the necessity of routine building maintenance?
39. How to maintain woodwork?
40. State some special repair cases in a building.

41. What are the stages in departmental procedure for repair of buildings?
42. How do estimate ordinary repair works?
43. How do estimate for special repairs?
44. Write about repair orders.
45. Explain the maintenance allowance for building.
46. How to control the expenditure on ordinary repairs?
47. What is the role of maintenance engineer?
48. What is the scope of maintenance?
49. What is the scale of maintenance funds?
50. What are the tools for effective maintenance?
51. What are the causes of accidents?
52. What are the remedies to avoid accidents?
53. How to prevent accidents?
54. Classify construction accidents.
55. State the safety programmes for construction.
56. Write short notes on safety in building maintenance.
57. List the sequence of dismantling.
58. What is the need of demolition work?
59. Which is an effective demolition technique?
60. How to conduct demolition technique in construction industry?

THEORY QUESTIONS

1. State the causes, which necessitate maintenance of civil engineering works.
2. Write about periodical inspection of structures.
3. Explain about steel work maintenance.
4. Discuss about woodwork maintenance.
5. Explain the points should be made the inspection of a building.
6. Discuss about routine building maintenance.
7. Write about distempers and distempering.
8. Explain any two special repair cases in a building.
9. Discuss about the departmental procedure for repair of buildings.
10. Explain about the integral maintenance of building.
11. State the role of maintenance engineer in maintenance of building.
12. Explain about normal break-up in maintenance.
13. Discuss the management tools for effective management.
14. Explain the causes of accidents and its remedies to avoid accidents.
15. Discuss about the safety programme for construction accidents.
16. Explain the safety precautions prior to dismantling and during dismantling.
17. Explain the dismantling programme on wall and floors.
18. Discuss about explosive demolition and chemical demolition agent techniques.
19. Discuss about blasting demolition and hydro-demolition techniques.
20. Review recent advanced demolition techniques.

Bibliography

1. M.S.Shetty, Concrete Technology – Theory and Practice, S.Chand and Company Ltd, New Delhi, 2005.
2. B.L.Gupta and Amit Gupta, Concrete Technology, Standard Publishers Distributors, Delhi, 2004.
3. B.L. Gupta and Amit Gupta, Maintenance and Repair of Civil Structures, Standard Publishers Distributors, Dehli.
4. M.L.Gambhir, Concrete Technology, Tata McGraw–Hill Publishing Company Limited, New Delhi, 2004.
5. Johnson, S.M., Deterioration, maintenance, and repair of structures, Mc-Graw-Hill Book Co., New York, 1965.
6. R.T.L.Allen and S.C.Edwards, The Repair of Concrete Structures, Blackie & Son Ltd, London, 1987.
7. N.P.Mailvaganam, Repair and Protection of Concrete Structures, CRC Press, 1991.
8. Peter H.Emmons, Concrete Repair and Maintenance Illustrated, Galgotia Publications, New Delhi, 2001.
9. B.S.Nayak, A Manual on Maintenance Engineering, Khanna Publishers, Delhi.
10. W.L. Sheppard: Corrosion and Chemical Resistant Masonry Materials Handbook, Standard Publishers Distributors, Delhi.
11. Proceedings of Advanced course on High Performance Materials and Methodologies for Construction and Rehabilitation of Concrete Structures, Structural Engineering Research Centre, Chennai, January 2000.
12. Proceedings of National Conference on Civil Engineering Materials and Structures, January 19-21, 1995, Osmania University, India.
13. BIS 456-2000; Code of Practice for Plain and Reinforced Concrete.
14. BIS 13311 Part 1 & 2, 1992 Recommendation for Non-Destructive method test for Concrete.
15. SP 24-1984; Handbook on causes and prevention of cracks in buildings, Bureau of Indian Standards.
16. Ferrocement Model Code, Reported by International Ferrocement Society, Committee 10, January 2001.
17. ACI Committee 201, Guide to Durable Concrete, American Concrete Institute, Detroit, 1982.
18. ACI Committee 548, Polymer concrete, Polymer modified concrete, ACI Manual of Concrete Practice, Part 5, 1998.
19. L Hollaway, University of Surrey and M Leeming, Mouchel Consulting Ltd, Strengthening of reinforced concrete structures: Using externally-bonded FRP composites in structural and civil engineering, Woodhead Publishing Limited, UK, 2005.
20. H.S. Lew, Handbook for Seismic Rehabilitation of Existing Buildings, by National Institute of Standards and Technology, Gaithersburg, MD 20899, USA.
21. General Awareness Information, Rapid Structural Safety Assessment & Monitoring for Reinforced Concrete Structure, Engineer Professional Advisory Committee, U.S. Public Health Service, June 2003.
22. Fritz, C., Naaman, A.E., and Reinhardt, H.W., "SIFCON Matrix in RC beams in High Performance Fiber Reinforced Cement Composites, Proceedings of the International Workshop held by RILEM,

ACI and others, June 23-26, 1991, Mainz, Germany; Ed. by H.W. Reinhardt and A.E. Naaman; 1st ed.; E & FN Spon, London, 1992, pp 518-528. (RILEM proceedings, vol. 15)

23. Lin, Mark W.; Abatan, Ayo O.; Zhang, Wei-Min., Crack damage detection of concrete structures using distributed electrical time domain reflectometry (ETDR) sensors, Proc. SPIE Vol. 3671, p. 297-304, Smart Structures and Materials 1999: Smart Systems for Bridges, Structures, and Highways, Steve C. Liu; Ed.
24. Smoak, W. Glenn Guide to Concrete Repair, published by the US Department of the Interior, Bureau of Reclamation, Technical Service Center, April 1997.
25. Tarek Alkhrdaji and Jay Thomas, Design, application techniques key to successful structural repair, strengthening of aging concrete facilities, By Concrete Monthly, February 2004
26. Naaman, A.E. SIFCON: Tailored properties for structural performance in High Performance Fiber Reinforced Cement Composites, Proceedings of the International Workshop held by RILEM, ACI and others, June 23-26, 1991, Mainz, Germany, 1992, pp. 18-38. (RILEM proceedings, vol. 15)
27. Structural Applications of Glass-fibre Reinforced Concrete Components (STRUCTA-GRC), EC Proposal N° CRAFT-BES2-5315 / Contract N° BRST-CT98-5232.
28. Hackman, L.E., Farrell, M.B., and Dunham, O.O., Slurry infiltrated mat concrete (SIMCON), Concrete International, Dec 1992, Vol. 14, No. 12, pp 53-56.
29. Nanni, A. Properties of aramid-fiber reinforced concrete and SIFCON, Journal of Materials in Civil Engineering, Feb 1992, Vol. 4, No. 1, pp 1-15.
30. Balaguru, P.N., Narahari, R., and Patel, M., Flexural toughness of steel fiber reinforced concrete, ACI Materials Journal, Nov-Dec 1992, Vol. 89, No. 6, pp 541-545.
31. David R. Lan Kar, Properties, applications: Slurry infiltrated Fiber Concrete (SIFCON), Concrete International, No 6, volume 12, December 1, 1984.
32. Taerwe, L.R. Influence of steel fibers on strain-softening of high-strength concrete, ACI Materials Journal, Jan-Feb 1992, Vol. 89, No. 1, pp 54-60.

Index